A TEXT BOOK OF

DESIGN OF CONCRETE STRUCTURES - I

FOR
SEMESTER – VI

THIRD YEAR (T.Y) B. TECH COURSE IN CIVIL ENGINEERING

Strictly According to New Revised Credit System Syllabus of Babasaheb Ambedkar Technological University (BATU), Lonere, (Dist. Raigad) Maharashtra,
(w.e.f. June 2019-20)

Dr. SACHIN M. PORE
M.E. (Structures), Ph. D. (Structural Dynamics, IIT Roorkee)
Associate Professor, Deptt. of Civil Engg.,
Associate Dean (R & D)
Dr. Babasaheb Ambedkar Technological University, (BATU)
LONERE, Dist. RAIGAD.

PRAKASH. A. VEDPATHAK
M. E. (Civil)
Assistant Professor, Deptt. of Civil Engg.,
Mahatma Basveshwar Education Society's
College of Engineering
AMBAJOGAI, Dist. BEED.

NIRALI PRAKASHAN
ADVANCEMENT OF KNOWLEDGE

N1101

First Edition	:	**January 2020**
©	:	**Authors**

Published By :

NIRALI PRAKASHAN

Abhyudaya Pragati, 1312, Shivaji Nagar,

Off J.M. Road, Pune – 411005

Tel - (020) 25512336/37/39, Fax - (020) 25511379

Email : niralipune@pragationline.com

➢ **DISTRIBUTION CENTRES**

Pune

Nirali Prakashan (Local)	:	119 Budhwar Peth, Jogeshwari Mandir Lane, Pune 411002, Maharashtra
		Tel : (020) 2445 2044, Mobile : 9657703145, Email : niralilocal@pragationline.com
Nirali Prakashan (Outstation)	:	S. No. 28/27 Dhayari, Near Asian College, Dhayari, Pune 411041, Maharashtra
		Tel : (020) 2469 0204, Fax : (020) 2469 0316, Mobile : 9657703143
		Email : bookorder@pragationline.com

MUMBAI

Nirali Prakashan	:	385 S.V.P. Road, Rasadhara Co-op. Hsg. Society Ltd., Girgaum, Mumbai 400004, Maharashtra
		Tel : (022) 2385 6339 / 2386 9976, Fax : (022) 2386 9976, Mobile : 9320129587
		Email : niralimumbai@pragationline.com

➢ **DISTRIBUTION BRANCHES**

JALGAON

Nirali Prakashan	:	34 V. V. Golani Market, Navi Peth, Jalgaon 425001, Maharashtra
		Tel : (0257) 222 0395, Mob : 94234 91860, **Email** : niralijalgaon@pragationline.com

KOLHAPUR

Nirali Prakashan	:	New Mahadvar Road, Kedar Plaza 1st Floor, Opp. IDBI Bank
		Kolhapur 416012, Maharashtra. Mobile : 9850046155
		Email : niralikolhapur@pragationline.com

NAGPUR

Nirali Prakashan	:	Above Maratha Mandir, Shop No 3, Second Floor,
		Rani Jhanshi Square, Sitabuldi, Nagpur 440012, Maharashtra
		Tel : (0712) 254 7129, Email : niralinagpur@pragationline.com
Nirali Prakashan		**DELHI**
		4593/15 Basement, Agarwal Lane, Ansari Road, Daryaganj
		Near Times of India Building, New DelhiV 110002 Mobile : 8505972553
		Email : niralidelhi@pragationline.com

BENGALURU

Nirali Prakashan	:	Maitri Ground Floor, Jaya Apartments, No. 99, 6th Cross, 6th Main,
		Malleswaram, Bengaluru 560003, Karnataka
		Mobile : 9449043034, **Email** : niralibangalore@pragationline.com

niralipune@pragationline.com | www.pragationline.com
Also find us on   www.facebook.com/niralibooks

Dedicated to ....

Our Beloved Students

...Authors

PREFACE

It gives us great pleasure to present the book **" Design of Concrete Structures – I "** for the students of **Semester VI Third Year (T.Y.) B. Tech. Course Civil Engineering of Dr. Babasaheb Ambedkar Technological University (BATU), Lonere, Dist. Raigad (Maharashtra)**. This book is strictly as per the new revised syllabus 2019-20 Pattern, effective from the Academic Year July 2019-20.

In New Revised Syllabus, there will be Internal Assessment (20 Marks), Mid Sem. Exam. (20 Marks) and End Sem. Exam. (60 Marks). End Sem. Exam. will be based on all six units and each unit will carry 12 Marks.

The Theory Course will have 3 Credits.

The book aims at explaining basic concepts in a simplified manner. For a successful structural design, one need to know physics of the problem what we mean by structural behavior. Then a formal mathematical process falls in a more conceptual manner rather than just computational procedures, as required by the new examination system. It is our objective to keep the presentation systematic, consistent, intensive and clear through explanatory notes and figures. Main feature of this book is, **complete coverage** of New Credit System Syllabus with large number of solved examples and exercise. **Model Question Papers** for practice are included at the end of book.

We are also thankful to **Authorities of DBATU** Lonere and **Dr. B. I. Khadakbhavi** (Principal, MBES COE, Ambajogai), **Dr. V. S. Rajamanya** (HOD, MBES COE, Ambajogai), for their motivational support. Support and wishes of our **Family Members**, **Friends** have gone a long way in making this book possible.

We take this opportunity to express our thanks to Shri. Dineshbhai Furia, Shri. Jignesh Furia, Mrs. Nirali Verma, Shri. M. P. Munde and entire team of Nirali Prakashan, namely Mrs. Deepali Lachake (Co-ordinator), and her colleagues who really have taken keen interest and untiring efforts in publishing this text.

Suggestions of our esteemed readers are most welcome and will be highly appreciated.

Pune   **Authors**

SYLLABUS

Module 1: **(5 Lectures)**

Basic Aspects of Structural Design, Introduction to Design Philosophies, Stress Strain behaviour of Materials, Permissible stresses, Comparison of Different Philosophies, Estimation of Loads

Working Stress Method

Module 2: **(5 Lectures)**

Stress block parameters, Balanced, under reinforced and over reinforced section: Modes of failure, properties of singly and doubly reinforced rectangular section beams, Analysis and Design of Singly and Doubly Reinforced Beams One Way and Two Way Slab: Behavior of slabs, types, support conditions, analysis and design with various conditions

Module 3: **(4 Lectures)**

Analysis and Design of Axially and Eccentrically Loaded Columns, Isolated Column Footings, Staircases, Design of doglegged and open well stair case, effective span and load distribution

Limit State Method

Module 4: Introduction to Limit State Approach **(5 Lectures)**

Introduction to Limit State Approach, Types and Classification of Limit States, Characteristics Strength and Characteristics Load, Load Factor, Partial Safety Factors

Module 5: Limit State of Collapse (Flexure) **(7 Lectures)**

Limit State of Collapse (Flexure): Analysis and Design of Singly and Doubly Reinforced Rectangular Beam Sections, properties of Flanged (L and T) sections, Analysis and Design of Flanged Beams

Module 6: Limit States of Collapse (Shear and Bond) **(7 Lectures)**

Limit States of Collapse (Shear and Bond): Shear Failure, Types of Shear Reinforcement, Design of Shear Reinforcement, Bond – Types, Factors Affecting, Resistance, Check for Development Length, detailing of reinforcement

CONTENTS

MODULE III

(A) COLUMN

(B) FOOTING

MODULE IV

MODULE V

(A) Singly Reinforced Rectangular Section (SRRS)

(B) Doubly Reinforced Rectangular Section (DRRS)

MODULE VI

CHAPTER - 1
INTRODUCTION OF STRUCTURAL DESIGN

1.1 INTRODUCTION

- The beginning of reinforced concrete construction was made by Lambot, a French man, who in 1850 constructed a small boat of Reinforced Cement Concrete (R.C.C.) for Paris exhibition. The credit of its invention, however, is generally given to Monjer who constructed a number of structures such as beams, pipes, tanks etc., in reinforced concrete. Later some structural formulae were furnished in Germany and Austria, which brought R.C.C. into scientific use.

- The British and American engineers, however, made use of concrete, to start with, only as a protective covering to steel against fire. Ransome, an American, who built the California Academy of Science and some other buildings, was one of the pioneers in the scientific use of R.C.C. Edwin Thacher, another American, built the first R.C.C. bridge. Reinforced concrete has been developed into a standard form of construction only in the twentieth century, and is now based on rational theory.

- Before the advent of reinforced concrete, materials in use for constructing buildings and bridges were bricks, stone, plain concrete and steel. The first three suffered from the lack of strength in tension, although they were very suitable for taking compressive stresses. Steel, on the other hand, is more suitable for taking tensile stresses, and thus a combination of concrete and steel has proved to be ideal as the two materials are utilised to take up stresses which they are most suitable for.

- The R.C. construction has the advantage over steel in being fire proof and incorrodible, but in structures subjected to excessive vibrations and alternating stress, its use will be limited.

1.2 STRUCTURAL DESIGN

- **Design :** It is a science and art of determining dimensions of element.

- **Structure :** It is an assembly of structural member transferring the load and serving desired function.

- **Structural Design :** It is an art of science of determining dimensions of different structural members with safe, serviceable durable, economical and aesthetics etc.

1.3 OBJECTIVE OF STRUCTURAL DESIGN

Basic requirements objective of structural design:

- Safety (adequate strength).

- Serviceability (structure shall not fail during its expected life span. The performance is rated by the fitness of the structure to maintain deflections, deformations, and cracking and vibration effects within acceptable limits. It is achieved by providing adequate stiffness and cracking resistance).

- Durability (the structure shall resist effectively environmental action).

- Economy.

- Aesthetics (pleasing appearance).

- Feasibility, practicability and acceptability.

1.4 REINFORCED CEMENT CONCRETE

- It is obtained by reinforcing steel in the mixture of cement, sand, aggregate with water.

- It has high compressive as well as tensile strength.

- Used for buildings component such as beams, slabs, columns, retaining walls, dams, roads, machine foundation, building foundations, staircases etc.

- It can be cast in-situ or precast.

1.5 NECESSITY OF USING STEEL BARS IN CONCRETE

- R.C.C. having high strength in compression but very weak in tension and for reinforcement vice-versa.

- The tensile strength of concrete is about 10-15% of its compressive strength, to overcome this difficulty it becomes necessary to reinforce the plain cement concrete by placing steel bars in tensile zone of the concrete. (i.e. to increase tensile strength of tensile zone of the concrete section, concrete is to be reinforced.

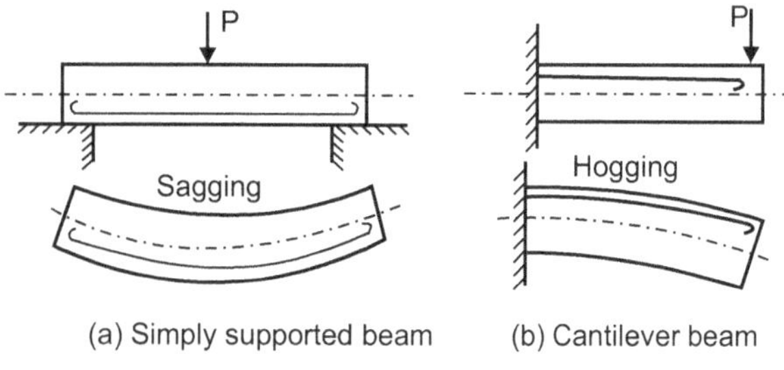

Fig. 1.1

- In both the cases steel reinforcement is provided in tensile zone only, such beam is known as singly reinforced beams. However, steel bars are also provided in compression zone is termed as anchor bars, to hold the stirrups in position.
- If the steel bars are provided on compression side which will assist the concrete in taking compression is known as doubly reinforced section.

1.6 GRADES OF CONCRETE

- According to IS 456 : 2000, the concrete mixes are designated as M10, M15, M20 –Ordinary concrete.
- M25, M30, M35, M40, M45, M50, M55 - standard concrete.
- M60, M65, M70, M75, M80-High strength concrete.

  where,

  M - mix

  Number - Ultimate compressive strength of 15cm cube at 28 days expressed in N/mm^2.

Note : When the compressive strength of any mix lies in between the strength of two grades specified above. The concrete mix is classified as concrete belonging to lower grade of two.

1.7 FUNCTIONS OF REINFORCEMENT

- To resist direct or bending tension and compression.
- To strengthen the concrete in compression also.
- To resist diagonal tension due to shear.
- To prevent buckling of main bars in column.
- To resist spiral cracking due to torsion.

1.8 ADVANTAGES AND DISADVANTAGES OF R.C.C.

Advantages :

Reinforced cement concrete has the following advantages over other construction materials.

- **Strength :** R.C.C. has very good strength in tension as well as compression.
- **Durability :** R.C.C. structures are durable if designed and laid properly. They can last up to 100 years.

- **Mouldability :** R.C.C. sections can be given any shape easily by properly designing the formwork. Thus, it is more suitable for architectural requirements.
- **Ductility :** The steel reinforcement imparts ductility to the R.C.C. structures.
- **Economy :** R.C.C. is cheaper as compared to steel and prestressed concrete. There is an overall economy by using R.C.C. because its maintenance cost is low.
- **Transportation :** The raw materials which are required for R.C.C. i.e. cement, sand aggregate, water and steel are easily available and can be transported easily. Nowadays Ready Mix Concrete, (RMQ) is used for faster and better construction. (RMC is the concrete which is manufactured in the factory and transported to the site in green or plastic state).
- **Fire Resistance :** R.C.C. structures are more fire resistant than other commonly used construction materials like steel and wood.
- **Permeability :** R.C.C. is almost impermeable to moisture.
- **Seismic Resistance :** Properly designed R.C.C. structures are extremely resistant to earthquakes.

Disadvantages :

Despite the above mentioned advantages, R.C.C. has the following disadvantages :

- R.C.C. structures are heavier than structures of other materials like steel, wood and glass etc.
- R.C.C. needs lot of formwork, centering and shuttering to be fixed, thus require lot of space and skilled labour.
- Concrete takes time to attain its full strength. Thus, R.C.C structures can be used immediately after construction unlike steel structures.

1.9 ASSUMPTIONS FOR DESIGN OF MEMBERS (WSM)

(Refer Cl. No. B-1.3, Pg. No. 80, IS 456:2000)

1. At any cross section, plane sections before bending remain plain after bending.

2. All tensile stresses are taken by reinforcement and none by concrete.

3. The stress-strain relationship of steel and concrete under working loads, is a straight line.

4. There is perfect bond between steel and concrete and no slip takes place between steel and concrete.

5. The modular ratio 'm' has the value $\dfrac{280}{3\,\sigma_{cbc}}$

   where, σ_{cbc} = Permissible compressive stress due to bending in concrete in N/mm^2

1.10 MODULAR RATIO

- From assumption (1) it is clear that the strain in steel and the concrete surrounding it will be equal, so that if E_C and E_S are the modulus of elasticity of concrete and steel respectively, the stresses in the materials, having the same strain say e, will be

  Stress in concrete,

$$\sigma_C = E_C \cdot e$$

  and

  stress in steel,

$$\sigma_S = E_S \cdot e$$

$$\therefore \quad \frac{\sigma_S}{\sigma_C} = \frac{E_S}{E_C} = m \text{ (modular ratio)} \qquad \ldots (1.1)$$

or, $\sigma_S = m \, \sigma_C$, i.e. the stress in steel is 'm' times the stress in concrete surrounding it.

- The modulus of elasticity of all grades of steel is 2×10^5 N/mm^2. The short term modulus of elasticity of concrete is given by IS 456.

$$E_C = 5000 \sqrt{f_{ck}}$$

  Also, the long term modulus of elasticity of concrete is given by

$$E_{ce} = \frac{E_C}{(1 + \theta)}$$

  where, θ is the creep coefficient given in IS 456.

- The long term modulus of elasticity increases with the time. Thus, modular ratio, as defined above, decreases with time. However, for the purpose of design, the IS 456 gives an empirical value **$m = \dfrac{280}{3\,\sigma_{cbc}}$** where σ_{cbc} is the permissible stress in concrete in bending compression. This empirical formula takes into account, the long term modulus of elasticity of concrete. Therefore, this 'm' is not the same as the modular ratio derived based on the value of **$E_C = 5000 \sqrt{f_{ck}}$** .

1.11 STRUCTURAL CONCRETE

The main ingredients required for making concrete are coarse aggregate, fine aggregate, cement and water. Coarse aggregate provide the main bulk while fine aggregate fill the voids in them and hydrated cement acts as the binder. In general, strength, behaviour under the load and durability are considered as the most important properties of concrete. The designer is concerned with the following structural properties :

1. Compressive strength.
2. Tensile strength.
3. Modulus of elasticity and Poisson's ratio.
4. Stress-strain relationship.

Besides the above mentioned structural properties, other important properties concerning the performance are

5. Shrinkage of concrete.

6. Creep of concrete

These properties are discussed in brief as follows :

1. Compressive Strength of Concrete :

- The compressive strength of hardened concrete is found by testing to failure 150 mm cubical (according to B.I.S.) specimens after 28 days of standard curing. At least three specimens are made for testing at a specified age.

- The compressive strength of concrete can also be found by testing to failure cylindrical specimens of diameter = 150 mm and height = 300 mm. Such test is carried out in United States. In India, the concrete grade is based on cube strength and if the cylinder is tested, the strength should be modified into equivalent cube strength. The ratio of the cylinder strength to the cube strength may be taken to be 0.8.

- It is observed that the strength of concrete increases with age because the process of hydration continues over a longer period.

- Also, the most important factor which affects the strength of concrete is the water-cement ratio.

2. Tensile Strength of Concrete :

- The tensile strength of concrete is comparatively low and hence it is usually not taken into account. But the knowledge of the tensile strength of concrete is important because it affects the extent and size of cracking in concrete.

- Measurement of tensile strength by subjecting the specimen to direct tension is extremely difficult. Therefore, indirect measurements for tensile strength are made. Tensile strength of concrete can be measured by two methods

  (i) Split-cylinder test (IS : 5816-1999)

  (ii) Standard beam-test (modulus of rupture test) (IS : 516).

- Flexural test on plain concrete beam specimens for determining the tensile strength of concrete is most popular laboratory method.

- The flexural tensile stress in concrete at failure is termed as modulus of rupture of concrete, can be obtained as :

$$f_{cr} = \frac{M}{Z}$$

where, M = bending moment at failure

   Z = elastic section modulus of beam

- An estimate of the same can be made from the known compressive strength of concrete by means of the empirical relationship.

$$f_{cr} = 0.7\sqrt{f_{ck}} \ N/mm^2$$

where, f_{ck} = characteristic compressive strength of concrete

3. Elastic Modulus of Concrete :

- The stress-strain curve for concrete is shown in Fig. 1.2. It is obtained from a compression test on cylindrical specimens. Since, concrete is not an elastic material, therefore the obtained stress-strain curve for concrete is non-linear.

- The slope of the tangent drawn to the stress-strain curve is maximum at the origin and reduces to zero at the peak. To adopt this slope as the elastic modulus of concrete would be both erroneous and inconvenient. However, the slope of the secant which is the line joining a point on the curve to the origin, does not vary too widely between the origin and the peak. As the lower portion of the stress-strain curve is relatively straight, an elastic modulus may be conveniently defined in that region.

- Thus, the elastic modulus of concrete is taken to be the slope of the secant drawn to the stress-strain curve at a point corresponding to 40% of the maximum stress. However, in the absence of test results, the modulus of elasticity is normally related to the compressive strength of concrete as

$$E_c = 5000\sqrt{f_{ck}}$$

where, E_c = short term static modulus of elasticity in N/mm^2

- The elastic modulus of concrete varies with the strength of concrete, its age, properties of aggregates, type of specimen and the speed of loading. Hence, it is almost impossible to predict the value of the elastic modulus of concrete.

- The Poisson's ratio for concrete ranges from 0.15 to 0.20. It is normally taken as 0.15 for strength design and many times it is neglected.

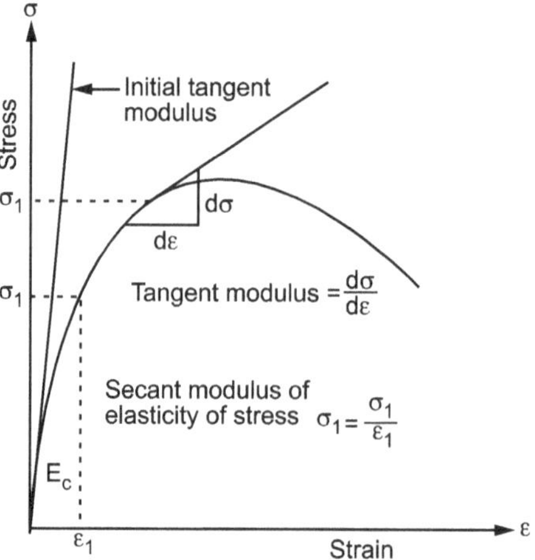

Fig. 1.2 : Stress-strain behaviour of concrete

4. Stress-Strain Curve for Concrete :

- For the experimental determination of the stress-strain curve for concrete, the specimens are tested in compression after 28 days of curing. The stress-strain curve for concrete is non-linear (Fig. 1.3). However, the lower portion of the curve is relatively straight for a stress upto 40% of the ultimate stress.

- The curve has an ascending part and a descending part separated by a peak which represents the ultimate strength of concrete in compression. For higher strength concretes, the stress-strain curve has a sharp peak whereas for lower strength concretes, the curve is comparatively flat. The maximum stress in different strength concretes has been found to occur at nearly the same strain ε_O ($\approx$ 0.002). The descending part of the curve terminates at a point which corresponds to the crushing of concrete at an ultimate strain ε_u (= 0.003 to 0.005).

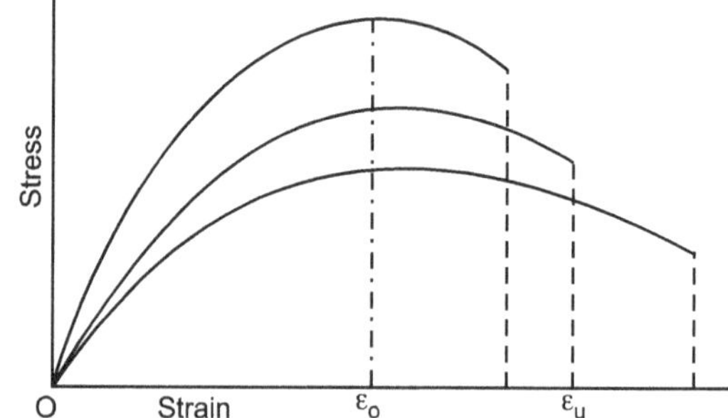

Fig. 1.3 : Stress-strain curves for concrete

- From above curves the following points should be noted :

  ➢ Maximum compressive stress occurs approximately at a strain value of 0.002. The value of stress at 0.002 strain is known as the strength of concrete.

  ➢ Lower strength of concrete has greater ductility than higher strength of concrete.

➢ Failure strain varies from 0.003 to 0.005. IS : 456 permits the maximum strain in concrete in axial compression equal to 0.002, but for bending compression the stress-strain curve is 0.0035.

Idealised Stress-Strain Curve

- The idealised stress-strain curve for concrete as adopted by IS : 456 (See Fig. 1.4). In design, the stress-strain curve is assumed parabolic upto strain value of 0.002 and then a straight line upto a strain value of 0.0035.

- The equation of the idealised stress-strain curve is given by

$$\sigma = \left[\frac{2\varepsilon}{\varepsilon_O} - \left(\frac{\varepsilon}{\varepsilon_O}\right)^2\right]\sigma_O \text{ for } 0 < \varepsilon < \varepsilon_O$$

$$\sigma = \sigma_O \text{ for } \varepsilon_O < \varepsilon < \varepsilon_u$$

where, σ_O = Idealised maximum compressive stress

ε_O = Strain corresponding to maximum stress

σ = Stress at any point

ε = Strain at any point

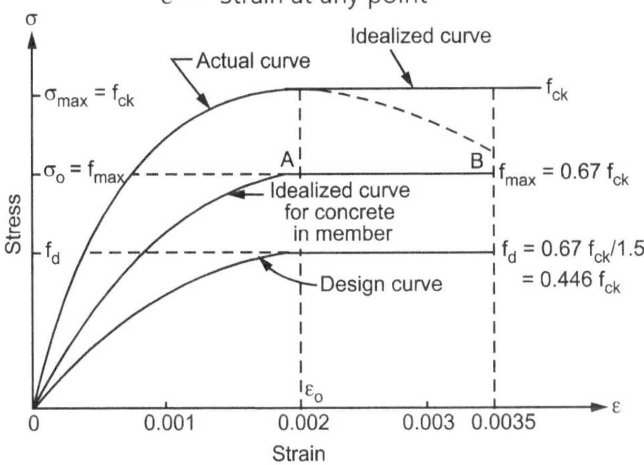

Fig. 1.4 : Idealised stress-strain curves for concrete

- The stress-strain behaviour of concrete is dependent on its strength, age at loading, rate of loading, properties of ingredients and the type and size of specimen.

5. Shrinkage of Concrete :

- The contraction of concrete per unit length during the process of hardening is known as **Shrinkage**. The total shrinkage of concrete depends upon the constituents of concrete, size of the members, environmental conditions and percentage of steel.

- The shrinkage is a long process and continues for many years particularly for mass concrete. The designer must provide the shrinkage steel to prevent shrinkage cracks. Greater the percentage of steel, lesser is the shrinkage because the reinforcement

restrains the shrinkage. A curve showing shrinkage strain against time after commencement of drying is shown in Fig. 1.5

- From Fig. 1.5, it is clear that the rate of shrinkage decreases with time. The total shrinkage strain may be in the range of 0.0002 to 0.0007.

- The IS : 456 recommends a value for shrinkage to be 0.0003 for the purpose of design.

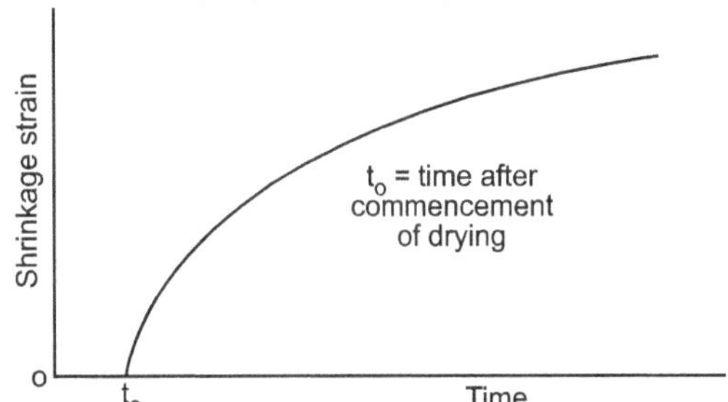

Fig. 1.5: Shrinkage curve for concrete

6. Creep of Concrete :

- Creep is that property of concrete by which it continues to deform with time under sustained stress. The creep deformation is a long process. About 75% of the ultimate creep deformation occurs during first year and continues at a slower rate for many years. The variation of creep deformations with time is schematically shown in Fig. 1.6.

- From graph in Fig. 1.6, it is clear that creep strain increases with time, though, the rate of increase becomes smaller and smaller. If the specimen, which is maintained under a constant stress for a long time, is finally unloaded, there is an instantaneous partial recovery of the elastic strain followed by the recovery of a small proportion of the total creep strain. A very large amount of creep is irreversible.

- Creep of concrete depends on constituents of concrete, size of member, environmental conditions, the stress in concrete, the age at loading and the duration of loading. As long as the stress in concrete does not exceed one-third of its characteristic compressive strength, creep may be assumed to be proportional to the stress.

- In the absence of experimental data and detailed information on the effect of the variables, the ultimate creep strain may be estimated from the following values of creep coefficient, where,

$$\text{Creep coefficient} = \frac{\text{Ultimate creep strain}}{\text{Elastic strain at the age of loading}}$$

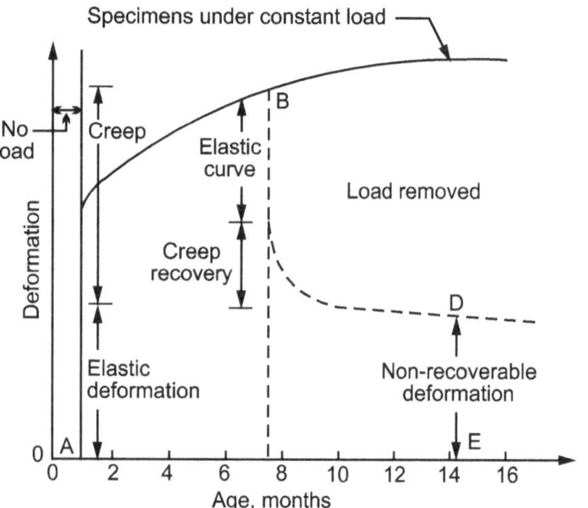

Fig. 1.6 : Elastic and creep deformations of concrete

The value of creep coefficient corresponding to age are given in Table 1.1.

Table 1.1

Age at Loading	Creep Coefficient
7 days	2.2
28 days	1.6
1 year	1.1

Notes :

- The ultimate creep strain, estimated as described above, does not include the elastic strain.

- The long term modulus of elasticity for concrete including creep value is given by,

$$E_{ce} = \frac{E_C}{1 + \theta}$$

where, E_{ce} = Long term modulus of elasticity of concrete

E_C = Short term static modulus of elasticity of concrete

θ = Creep coefficient

1.12　STEEL REINFORCEMENT

- In reinforced concrete, concrete being weak in tension, steel bars are used to carry the tension. Steel bar possess high tensile strength, ability to develop a good bond with concrete, high modulus of elasticity and nearly the same temperature coefficient of expansion and contraction as concrete.

- That is why the steel is used as a reinforcing material. The steel reinforcement for concrete is generally round in cross-section, to avoid any possible stress concentration.

- The following three types of steel reinforcement are used in practice :

　1.　Mild steel plain round bars (Fe 250).

　2.　High yield strength deformed bars (Fe 415, Fe 500).

　3.　Welded wire fabric.

- Fe refers to ferrous metal and the number indicates yield stress in N/mm^2.

1.　Mild Steel Plain Round Bars

- Mild steel bars are manufactured by hot rolling process. The stress-strain curve is obtained by testing the bars in uniaxial tension. The behaviour in compression is assumed to be the same as that in tension. A typical stress-strain curve for mild steel is shown in Fig. 1.7.

- The curve exhibits a linearly elastic behaviour upto the yield point stress f_y. Beyond this point the strains become plastic and increase considerably without increase of stress. This region of the curve is known as the **Yield Plateau**.

- The plastic strain within the yield plateau may be as much as 10 times the elastic strain occurring before the yield stress. On further loading, the stress again increases with strain on account of strain hardening. It then reaches the peak value known as the **Ultimate Stress** and then decreases until the bar finally fractures.

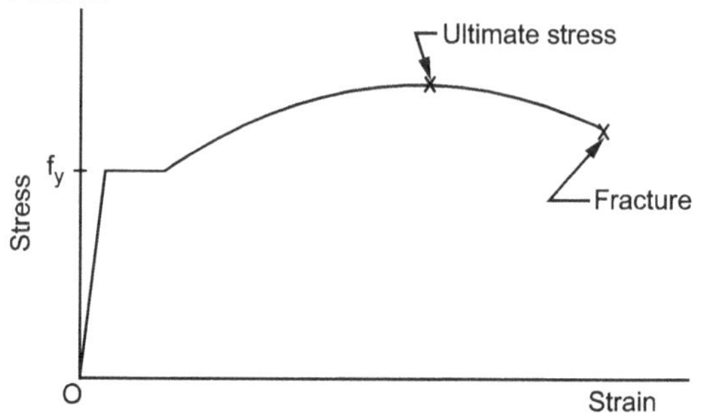

Fig. 1.7 : Stress-strain curve for mild steel

- Once the mild steel bars in a loaded reinforced concrete specimen are stressed to the yield point, they considerably stretch, causing excessive cracking of concrete and permanent deformation in the member. The yield point is indicative of the limit of usefulness of mild steel reinforcement, and as such, it is an important limit.

- An idealised stress-strain diagram for mild steel is adopted by the IS code : 456. (See Fig. 1.8).

- Plain mild steel bars have smooth surface. Load transference from concrete to steel takes place on account of bond strength between the two materials at the interface.

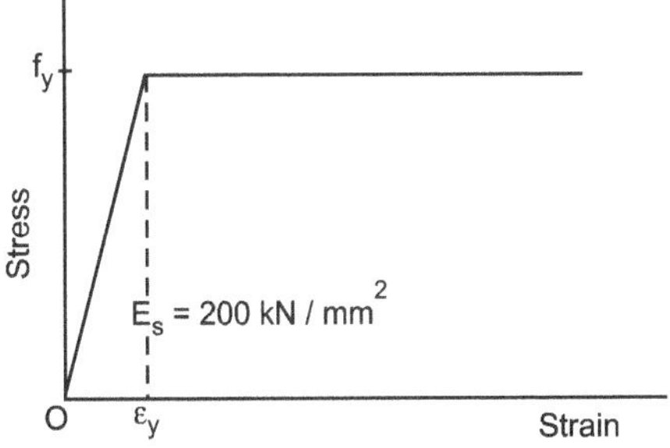

Fig. 1.8 : Idealised stress-strain curve for mild steel

2. High Yield Strength Deformed Bars

- If a mild steel bar is strained beyond the yield plateau and then unloaded, it is found on reloading that the yield plateau has completely vanished. This process is known as **Cold Working**.

- A typical stress-strain curve for a cold worked bar is shown in Fig. 1.9. Cold working may be done either by stretching or twisting of the bar. By this process, greater stresses can be attained in the reinforcement bar for smaller strains, than the yield point stress of a mild steel bar.

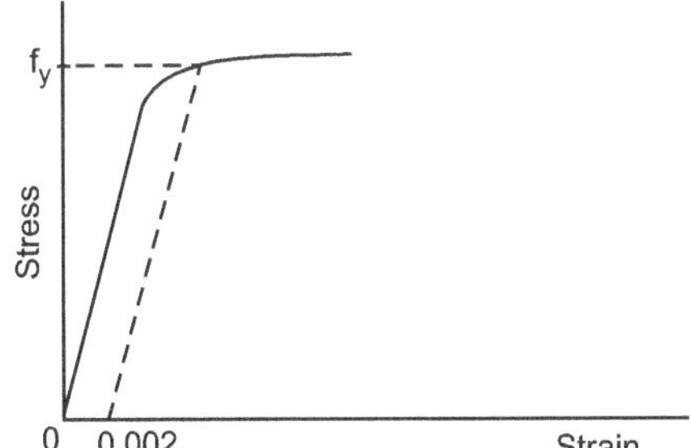

Fig. 1.9 : Stress-strain curve for cold worked steel

- Cold worked bars do not show a well defined yield point. As such, their strength is fixed up in terms of proof stress which is the stress level causing a definite residual strain upon unloading. IS code : 456 has accepted 0.2% proof stress as the standard value for the cold worked bars.

- It is the stress level corresponding to a residual strain of 0.002. An idealised stress-strain curve for HYSD bars as adopted by IS code : 456 is shown in Fig. 1.10.

- High Yield Strength Deformed (HYSD) bars have protrusions and indentations rolled on the surface, which produce mechanical interlocking with concrete in addition to plain bonding. This enhanced gripping of the bar in concrete makes it possible to develop high magnitude of axial stress in the reinforcement without the risk of slipping.

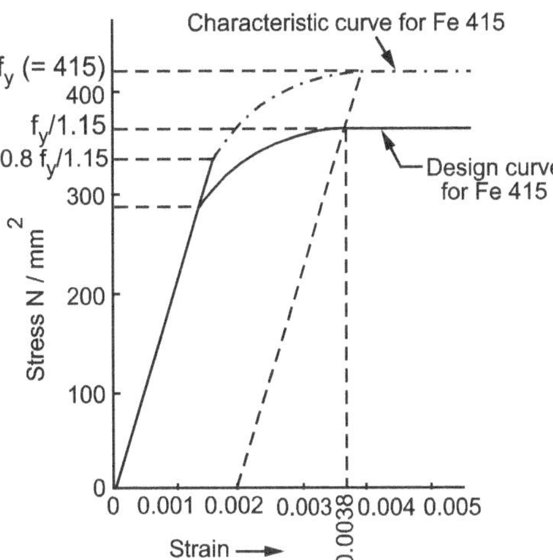

Fig. 1.10 : Idealised stress-strain curve for HYSD bars

3. Welded Wire Fabric

- Welded wire fabric is made from steel wires running in two orthogonal directions and welded at their intersections. The wire for the fabric is obtained by cold drawing mild steel bars through a series of dies of successively reduced sizes. The wire fabric may be of square mesh type. In square mesh, the wire diameter ranges from 3 mm to 10 mm, whereas the pitch ranges from 50 mm to 200 mm. Welded wire fabric is used for reinforcing slabs, pavements and shell structures.

1.12.1 Properties of Round Bars

The properties of round bars required in design are diameter, cross-sectional area, perimeter and weight per unit length, compiled in Table 1.2.

Table 1.2 : Area, Perimeter and Weights of Round Bars

Diameter of Bar (mm)	Area (mm²)	Perimeter (mm)	Weight (kg/m)
6	28.3	18.8	0.222
8	50.3	25.1	0.395
10	78.5	31.4	0.616
12	113.1	37.7	0.888
16	201.1	50.3	1.578
20	314.2	62.8	2.466
22	380.1	69.1	2.984
25	490.9	78.5	3.853
28	615.7	88.0	4.834
32	804.2	100.5	6.313
36	1017.9	113.1	7.990
40	1257	125.7	9.860

- Weight of bar in kg per metre length = $\dfrac{\phi^2}{162.2}$.

  where, ϕ is diameter of bar in mm.

- For deformed bar, the nominal diameter of the bar is obtained by measuring the weight of bar of unit length and dividing it by unit weight steel.

- The basic weight of steel is taken equal to 7850 kg/m^3.

1.13 PERMISSIBLE STRESSES

1.13.1 Permissible Stresses in Concrete

(Refer Cl. No. B-2.1, Pg. No. 80, IS 456 : 2000)

- In working stress method, the stresses in materials are not exceeded beyond their permissible values. The permissible stress in a material is given by,

$\therefore$ Permissible stress = $\dfrac{\text{Limiting strength}}{\text{Factor of safety}}$

- For concrete, the limiting strength is the crushing strength in compression the factor of safety is higher than for steel. This is so because concrete suffers from higher degree of variability regarding its strength and properties than steel which is produced under well controlled conditions. The factor of safety for flexural compressive in concrete is 3. Thus, the permissible compressive stress in concrete in flexural compression

is $\sigma_{cbc} = \dfrac{f_{ck}}{3} = 0.333\ f_{ck}$ following tables gives

permissible stresses in concrete.

Table 1.3 : Permissible stresses in concrete in compression and average bond stress (Refer Table 21, Pg. No. 81, IS 456 : 2000)

Grade of Concrete	Permissible Stress in Compression		Average Bond Stress for Plan Tension Bars (Mild) N/mm²
	Bending 'σ_{cbc}' N/mm²	Direct Direct 'σ_c' N/mm²	
M-10	3.0	2.5	-
M-15	5.0	4.0	0.6
M-20	7.0	5.0	0.8
M-25	8.5	6.0	0.9
M-30	10.0	8.0	1.0
M-35	11.5	9.0	1.1
M-40	13.0	10.0	1.2
M-45	14.5	11.0	1.3
M-50	16.0	12.0	1.4

Note :

1. For deformed bars (HYSD) the bond stress shall be increased by 40 %.

2. For bars in compression, the bond stress shall be increased by 25%.

Table 1.4 : Permissible direct stress in tension in concrete

Grade of Concrete	M-10	M-15	M-20	M-25	M-30	M-35	M-40	M-45	M-50
Tensile Stress (N/mm²)	1.2	2.0	2.8	3.2	3.6	4.0	4.4	4.8	5.2

Tensile stress = $\dfrac{\text{Tension}}{[(\text{Actual concrete Area}) + m \times (\text{Steel Area})]}$

1. Permissible Shear Stress in Concrete

(Refer Cl. No. B-5, Pg. No. 83, IS 456 : 2000)

The I. S. code has introduce the concept of nominal shear stress defined by the following relation :

Nominal shear, stress (τ_v) in beam or slabs of uniform depth shall be calculated as

$\therefore$ $\tau_v = \dfrac{V}{bd}$

where, V = shear force due to design load

        b = breadth of member (breadth of web)

        d = effective depth

In case of beams of varying depth,

$\tau_v = \dfrac{V \pm \dfrac{M}{d} \tan \beta}{bd}$

where M = Bending moment at the section

        β = Angle between the top and bottom edges of the beam

The negative sign in the formula applies when the bending moment M increases numerically in the same direction as the effective depth d increases, and the positive sign when the moment decreases numberically in this direction.

For soild slabs the permissible shear stress in concrete shall be $k.\tau_c$, where 'k' has the value given below :

Overall depth of slab 'mm'	300 or more	275	250	225	200	175	150 or less
k	1.00	1.05	1.10	1.15	1.20	1.25	1.30

Note : This does not apply for flat slab.

Table 1.5 : Permissible shear stress in concrete (τ_c)
(Refer table 23, Pg. No. 84, IS 456 : 2000)

$100 \times \dfrac{A_s}{bd}$	Permissible Shear Stress in Concrete, τ_c, N/mm²					
	Grade of Concrete					
	M15	M20	M25	M30	M35	M40 and above
≤ 0.15	0.18	0.18	0.19	0.20	0.20	0.20
0.25	0.22	0.22	0.23	0.23	0.23	0.23
0.50	0.29	0.30	0.31	0.31	0.31	0.32
0.75	0.34	0.35	0.36	0.37	0.37	0.38
1.00	0.37	0.39	0.40	0.41	0.42	0.42
1.25	0.40	0.42	0.44	0.45	0.45	0.46
1.50	0.42	0.45	0.46	0.48	0.49	0.49
1.75	0.44	0.47	0.49	0.50	0.52	0.52
2.00	0.44	0.49	0.51	0.53	0.54	0.55
2.25	0.44	0.51	0.53	0.55	0.56	0.57
2.50	0.44	0.51	0.55	0.57	0.58	0.60
2.75	0.44	0.51	0.56	0.58	0.60	0.62
3.00 and above	0.44	0.51	0.57	0.60	0.62	0.63

Note : 'As' is that area of longitudinal tension reinforcement which continues at least one effective depth beyond the section being considered except at supports where the full area of tension reinforcement may be used.

1.13.2 Permissible Stresses in Steel

(Refer Cl. B – 2.2, Pg. No. 80, IS 456 : 2000)

In case of steel reinforcement, the limiting strength is either yield stress (for mild steel) or 0.2 % of proof stress (for deformed bars). The factor of safety in case of tensile steel reinforcement is approximately 1.82. Hence, the permissible tensile stress in steel is $\sigma_{st} = \dfrac{f_y}{1.8} = 0.55\, f_y$. The permissible stress in steel are discussed in below table.

Table 1.6 : Permissible stresses in steel reinforcement
(Refer Table 22, Pg. No. 22, IS 456 : 2000)

Sr. No.	Types of Stress in Steel Reinforcement	Permissible Stresses in N/mm²		
		Mild Steel Bars Conforming to Grade 1 of IS 432 (Part I)	Medium Tensile Steel Conforming to IS 432 (Part I)	High Yield strength Deformed bars conforming to IS 1786 (Grade Fe 415)
1.	Tension (σ_{at} or σ_{sv}) (a) Up to and including 20 mm (b) Over 20 mm	140 } 130	Half the guaranteed yield stress subject to a maximum of 190.	230 230
2.	Compression in column bars (σ_{ac})	130	130	190
3.	Compression in bars in a beam or slab when the compressive resistance of the concrete is taken into account.	The calculated compressive stress in the surrounding concrete multiplied by 1.5 times the modular ratio or σ_{ac} whichever is lower		
4.	Compression in bars in a beam or slab where the compressive resistance of the concrete is not taken into account. (a) Up to and including 20 mm (b) Over 20 mm	140 } 130	Half the guaranteed yield stress subject to a maximum of 190	190 190

Notes :

- For high yield strength deformed bars of Grade Fe 500 the permissible stress in direct tension and flexural tension shall be 0.55 f_x. The permissible stresses for shear and compression reinforcement shall be as for Grade Fe 415.
- For welded wire fabric conforming to IS 1566, the permissible value in tension σ_{at} is 230 N/mm².
- For the purpose of this standard, the yield stress of steel for which there is no clearly defined yield point should be taken to be 0.2 percent proof stress.
- When mild steel conforming to Grade II of IS 432 (Part I) is used, the permissible stresses shall be 90 percent of the permissible stresses in col 3, or if the design details have already been worked out on the basis of mild steel conforming to Grade I of IS 432 (Part I); the area of reinforcement shall be increased by 10 percent of that required for Grade I steel.

1.14 DESIGN PHILOSOPHIES

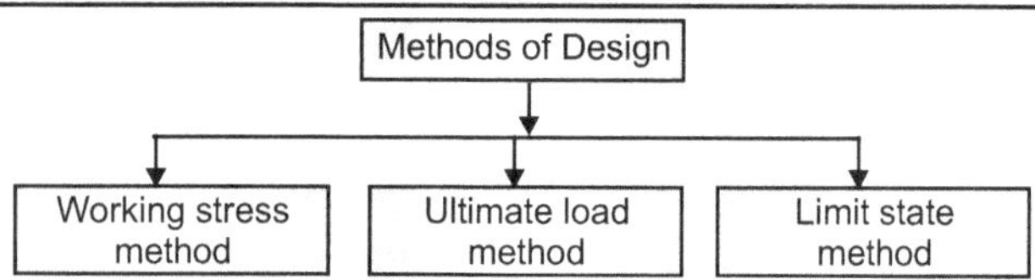

1. Working Stress Method :

- Over the years, various design philosophies have evolved in different parts of the world, with regard to reinforced concrete design.

- A 'design philosophies' is built up on a few fundamental assumptions, and is reflective of a way of thinking. The earliest codified design philosophy is the Working Stress Method (WSM) of design. Close to a hundred years old, this traditional method of design, based on linear elastic theory, is still surviving in some countries including India.

- In India, before 1964, most of the structures were designed by working stress method. In working stress method, it is assumed that concrete and steel are elastic. At worst combination of working loads, the stresses in the materials are not exceeded beyond permissible stresses. So the reserve strength of materials beyond yield point is not considered in this design.

- Working stress method is however, of great importance. For calculations of serviceability requirement like deflection and crack width under service load conditions and for designing a few structures like liquid retaining structures, highway and bridges working stress method is used. It is therefore necessary to study this method.

Design Criteria :

1. Load carrying capacity > Working load

2. Permissible stress = $\dfrac{\text{Yield stress}}{\text{Factor of safety}}$

3. Working stress due to working load < permissible stress < yield stress.

Advantages :

- It is simple method.

- Due to simplicity, it is still used for design of some complex structures such as water tank, bunkers, silos etc.

- There is no need to check serviceability requirement. The same is automatically satisfied as the working stresses are low and sections are bigger in this method.

Disadvantages :

- It assumes that stress strain relationship for steel and concrete is constant, which is not correct.

- Sections are bigger, thus it is uneconomical.

Limitations :

- The main assumption of linear elastic behaviour is not true.

- The assumption that stresses can be kept within permissible stresses is not realistic (True).

- The long term effect of creep and shrinkage are not considered.

- In WSM there is uncertainty of many factors that influence safety and serviceability.

- It proves safe but uneconomical.

2. Ultimate Load Method :

- The limitations of the working stress method to determine actual load carrying made researches to develop ultimate load method, which is also known as plastic design method.

- The second revision of IS : 456 (IS 456 : 1964), introduced Ultimate Load Method (ULM) of design and was at once adopted to many design offices for it gave more economical results than working stress method.

- In this method, the inelastic behaviour of concrete is taken into account and therefore, reserve strength of concrete can be used leading to the economical designs.

- In design, the loads on the structure are increased by suitable load factors and the structure is loaded with these increase loads called ultimate load.

- For materials (steel and concrete), the ultimate behaviours are taken into account. The ultimate load method leads the designer to economy but at the same time to very slender sections, larger deflection and larger crack width in concrete. This created a necessity of refining the method to achieve rational approach to the design of R.C.C. structures.

Design Criteria :

1. Non linear stress strain behaviour of concrete is taken into account.

2. Load factor = $\dfrac{\text{Collapse load}}{\text{Working load}}$

3. Σ (Working load × load factor) ≤ Ultimate load

Advantages :

- Due to plastic hinge concept, redistribution of internal forces is accounted.

- It allows varied selection of load factor.

- The ULM makes it possible to consider the effects of different loads acting simultaneously thus solving the shortcoming of WSM.

- This method gives exact margin of safety in terms of load.

Disadvantages / Limitations :
- This method gives very thin sections which leads to excessive deformations and cracking, thus making the structure unserviceable.
- No factor of safety are used for material stresses.

3. Limit State Method :
- The acceptable limit for the safety and serviceability requirements before occurs is called as Limit State Method.

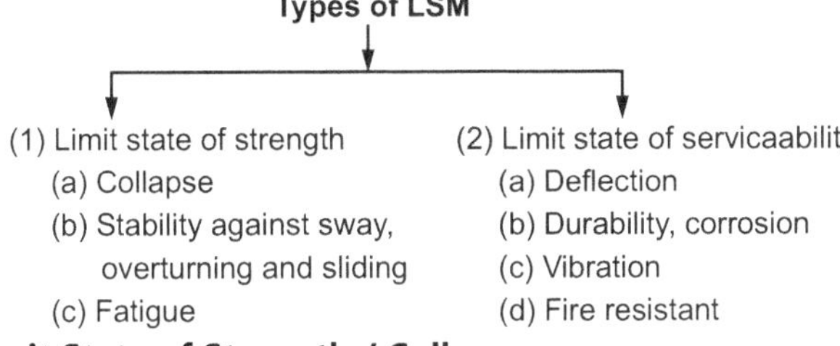

Limit State of Strength / Collapse :
- The resistance to axial load, bending, shear and torsion at every section shall not be less than appropriate value at that section produced by probable and most unfavorable combination of loads on the structure using appropriate partial safety factor.

(i) Limit State of Strength Includes :
- Loss of equilibrium of the structure as a whole or its components.
- Loss of stability of the structure including the effect of sway, overturning and sliding.
- Failure of structure due to excessive deformation, rupture of the structure or its components.
- Cracks due to fatigue.
- Brittle fractures.

(ii) Limit State or Serviceability :
- Limit state of serviceability is related to the satisfactory performance of the structure at working load.

Limit State of Serviceability Includes :
- Deformation within limit is acceptable. If it exceeds certain limit it will cause discomfort to occupants. Excessive deformation may also cause damage to finishes and non structural members.
- Durability and corrosion.
- Vibrations in the structure or any of its components causing discomfort to people or damages to the structure.
- Repairable damages or cracks due to fatigue.
- Fire resistant.

Design Criteria :
1. Load carrying capacity > Ultimate load
2. Permissible stress = $\dfrac{\text{Ultimate stress}}{\text{Partial safety factor}}$

Advantages :
- Limit state method helps to design structures based on both safety and serviceability.

- Structures do not undergo massive deflection and cracks.
- It proves to be economical as compared to other methods.
- Exact margin of safety is known.

Disadvantages :
- Design procedure is quite difficult.

1.15 COMPARISON BETWEEN WORKING STRESS METHOD AND LIMIT STATE METHOD

Table 1.7

Sr. No.	Working Stress Method	Limit State Method
1.	In WSM, the structures are designed for working loads i.e. stresses below ultimate.	In LSM, the acceptable limit for safety and serviceability requirement before failure occurs is called as LSM.
2.	In WSM, true behaviour of material is not considered, consider up to elastic limit so 100 % strength of material is not utilized. Hence the section become uneconomical.	In LSM, behaviour of material up to failure is taken into account, so that 100 % strength of material is utilized. Hence the section become economical.
3.	WSM is stress based method.	LSM is strain based method.
4.	In WSM, factor of Safety : Concrete = 3.0, Steel – 1.8	In LSM, Partial Safety factor : Concrete = 1.5, Steel = 1.15
5.	The method follows linear stress-strain behaviour of both the materials.	In this method, it follows linear strain relationship but not linear stress relationship.
6.	Stress vs Strain	Stress vs Strain
7.	It is based on elastic theory	It is based on plastic theory
8.	Non-cracking zone is generally considered	Cracking zone is considered
9.	WSM is deterministic method	LSM is non-deterministic method
10.	Design procedure is simple.	Design procedure is quite difficult.
11.	Modular ratio is used in the design of members.	Actual test strength of two materials are uses in the design of members.
12.	Design procedure is based on law of mechanics.	Design procedure is based on actual experimental observations.
13.	Exact margin of safety is not known.	Exact margin of safety is known.

1.16 TYPES OF LOADS ACTING ON STEEL STRUCTURES

- A structure is subjected to various types of loads like dead load, live load and wind load etc.
- All structure are designed for the worst combination of dead load. Live load and wind load as per IS 875. Seismic loads are also to be considered in design wherever applicable.

1. **Dead Load (DL) :** (IS 875 : 1987 Part – I)
- It is the self weight of structural members constant in magnitude and fixed in position.
- It is also called fixed / permanent load.
- It includes weight of slab, beam column, footing, staircase, permanent partitions etc.

For Example :

$$\text{Weight of concrete} = 24 \text{ kN/m}^3$$
$$\text{Weight of R.C.C.} = 25 \text{ kN/m}^3$$
$$\text{Wt. of steel} = 78.5 \text{ kN/m}^3$$
$$\text{Wt. of brick masonry} = 16 - 19 \text{ kN/m}^3$$
$$\text{Wt. of stone masonry} = 18 - 22 \text{ kN/m}^3$$
$$\text{Wt. of flooring} = 0.2 - 0.5 \text{ kN/m}^3$$

2. **Live Load / Imposed Load (L.L/I.L) :** (IS 875 : 1987 Part II)
- Live loads are variable as they depend on usage and capacity.
- It is also called not fixed / variable load.
- It includes weight of stored material, movable equipments, furniture and occupants etc.

For Example :

L.L. on floor (Residential building) = 2 kN/m^2
L.L. on floor (office, shop, assembly ball) = 2.5 to 10 kN/m^2
L.L on balcony / staircase (residential) = 3 kN/m^2

3. **Wind Load (W.L) :** (IS 875 : 1987 Part III)
- Load placed by the wind speed and its air density onto a building. (i.e. wind pressure exerted on structure is called as wind load.)
- Wind load depends on / factors affecting wind load :
  - ➢ Shape of the structure
  - ➢ Height of the structure
  - ➢ Topography / Surrounding area
  - ➢ Wind velocity
  - ➢ Orientation
  - ➢ Permeability of the structure

4. **Snow Load :** (IS 875 : 1987 Part IV)
- In the area of mountaneous region (i.e. northern part of India), the structure are subjected to snow which depends on slope of the roof.

5. **Impact Load :**
- An impact load is one whose time of application on material is less than one-third of the natural period of vibration of that material.

- Cyclic loads on a structure can lead to fatigue damage, cumulative damage or failure.

6. **Earthquake / Seismic Load :** (IS 1893 : 2002)
- The seismic loads on the structure during an earthquake results from inertia forces which were created by ground acceleration.
- The magnitude of these loads is a function of the following factors :
  - ➢ Mass of the building
  - ➢ Dynamic properties of the building.
  - ➢ Intensity-duration and frequency of ground motion.
  - ➢ Sail structure interation.

1.17 LOAD COMBINATIONS

- A judicious combination of the loads is necessary to ensure the required safety and economy in the design keeping in view the probability of :
  - ➢ Their acting together.
  - ➢ Their deposition in relation to other loads and severity of stresses or deformation caused by the combination of various loads.
- The recommended load combinations by IS 875 are given below :

1.	DL	2.	DL + IL
3.	DL + WL	4.	DL + EL
5.	DL + TL	6.	DL + IL + WL
7.	DL + IL + EL	8.	DL + IL + TL
9.	DL + WL + TL	10.	DL + EL + TL
11.	DL + IL + WL + IL	12.	DL + IL + EL + TL

Where, DL – Dead load
        IL – Imposed load
     WL – Wind load
     EL – Earthquake load
     TL – Temperature load

EXERCISE

1. Define modular ratio with reference to working stress method.
2. What is reinforced concrete ? Why concrete is to be reinforced ?
3. What are the grades of concrete?
4. State advantages and disadvantages of R.C.C.
5. State the assumptions in WSM.
6. Explain design philosophies.
7. Explain permissible stresses in steel and concrete.
8. Differentiate between WSM, ULM and LSM.
9. Enlist types of loads acting on structure with there IS code provision.
10. State different load combinations.
11. State advantages of LSM over WSM.
12. State limitations of WSM.
13. Explain stress-strain behaviour of steel and concrete.
14. Explain types of limit state method.

CHAPTER - 2
DESIGN OF BEAM (WSM)

2.1 INTRODUCTION

- This chapter explains the basic governing equations and the computation of parameters required for the design of beams and one-way slabs employing limit state of collapse in flexure. After learning the basic assumptions, the three equations of equilibrium and the computations of the total compressive and tensile forces, it is now required to determine the depth of Neutral Axis (N.A.) and then to estimate the moment of resistance of the beams using the two equations of equilibrium.

- It has been explained that the depth of neutral axis has an important role to estimate the moment of resistance. Finally, numerical problems on singly reinforced rectangular beams are taken upto illustrate the applications of the theories.

- They are, therefore, designed on the basis of limit state of collapse in flexure. The beams are also to be checked for other limit states of shear and torsion. Slabs under normal design loadings (except in bridge decks etc.) need not be provided with shear reinforcement. However, adequate torsional reinforcement must be provided wherever necessary.

Fig. 2.1

RCC Beams are of the following Three Types :

1. **Singly Reinforced Beams :** The beams in which steel reinforcement is placed in the tensile zone only are singly reinforced beams.

2. **Doubly Reinforced Beams :** The beams in which reinforcement is placed in the tensile as well as compression zone are called as doubly reinforced beams.

3. **Flanged Beams (T-Beams and L-Beams) :** The beams in which a portion of the slab acts together with the beam for resisting compressive stresses are called as flanged beams.

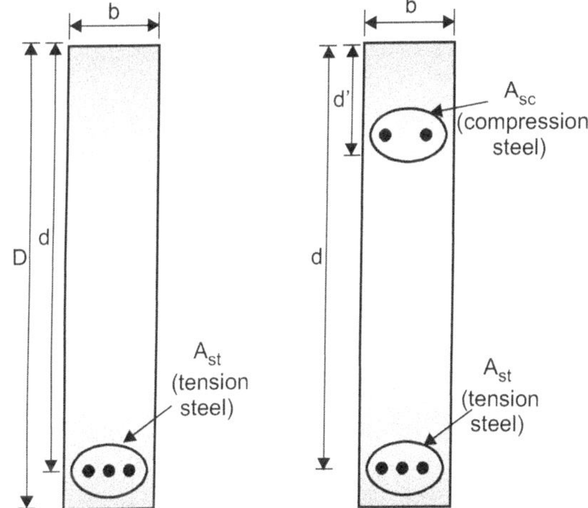

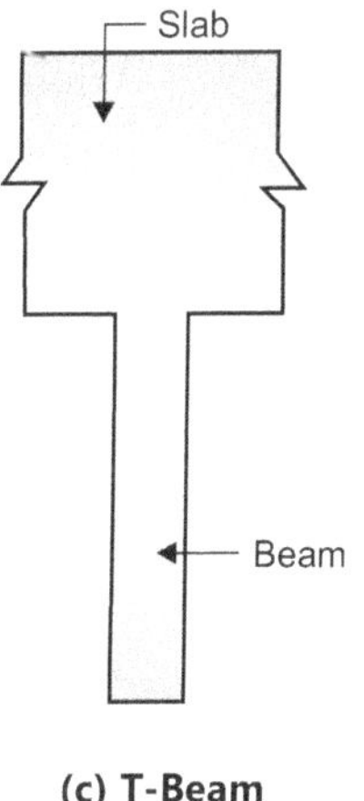

(c) T-Beam

Fig. 2.2

2.2 NECESSITY OF RECTANGULAR SECTION

- At intermediate supports of a continuous beam, the beam is subjected to negative or hogging moment causing tension in the slab region at top with the result that the slab portion become ineffective in resisting compression and the effective section of the beam is simply a rectangular section.

- Thus, at supports of a continuous beam, the beam no more acts like a flanged beam be just a rectangular beam with total depth from top of the flange to the bottom of the beam (and No the depth below the flange of the beam), even though the slab and beam are cast monolithically.

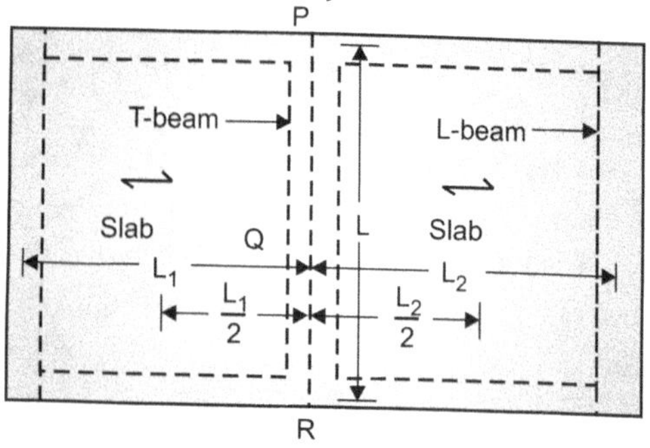

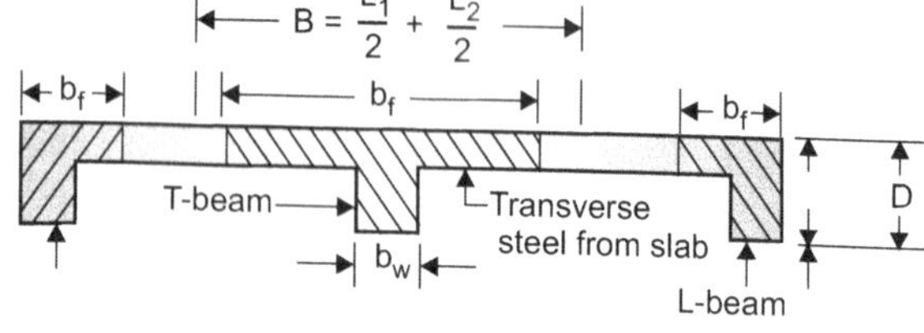

(a)

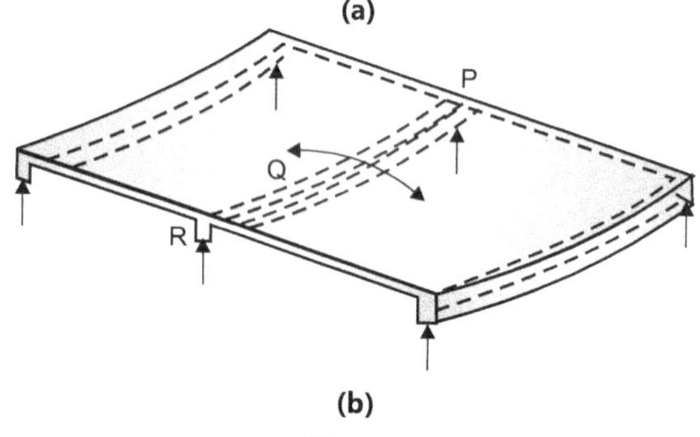

(b)

Fig. 2.3

2.3 NECESSITY OF DOUBLY REINFORCED SECTION
(Feb. 16)

Various situations under which doubly reinforced beams are used :

- When dimensions of beam are restricted by architectural or structural conditions and the section will have insufficient area of concrete to be provided for compression.
- The beams subjected to alternating loads.
- The beams subjected to impact and shocks.
- The beams subjected to eccentric moving loads such as cranes.
- When the bending moment due to external load is too much.

2.4 ROLE OF COMPRESSION STEEL

- Due to compression reinforcement, the moment of resistance can be increased without altering the dimensions of section, to meet the subjected bending moment.

2.5 NECESSITY OF FLANGED SECTION

- In R.C.C. construction, when slabs and beams are cost together the construction becomes monolithic some portion of the slab above the beam becomes the part of the beam and helps in resisting compressive stresses and it also deflects with the deflection of beam under loads.

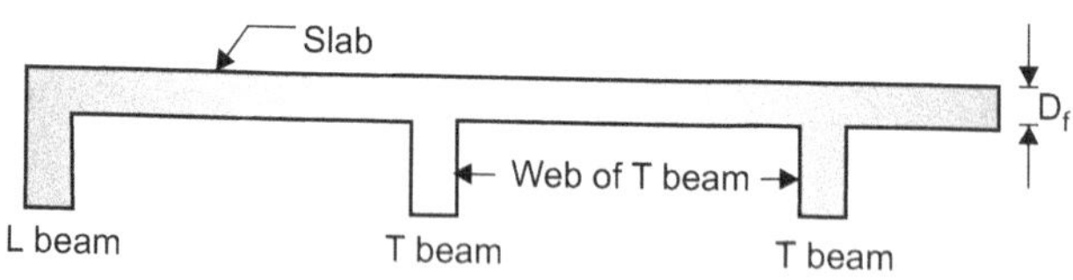

Fig. 2.4 : T beams or L beams

- Therefore, the effective section of beam becomes a flanged section having T and L shape. The slab portion of the section is called 'flange' and the portion below the slab is called 'web'.

- The intermediate beams having slab on both sides are known as T beams and the end beams having slab on one side only are known as L beams. The L-beams received loads from one side only and they are subjected to torsion also. If the torsion is less it can resisted by nominal shear reinforcement provided in the beam. If torsion is more the beam shall be designed for it.

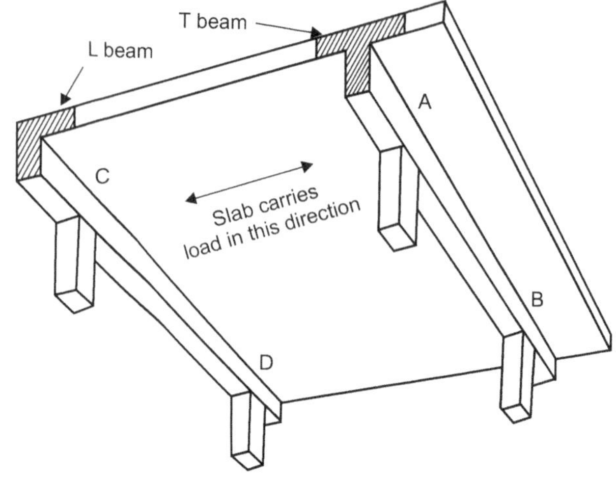

Fig. 2.5

2.6 NEUTRAL AXIS

- Neutral axis lies at the centre of gravity of the It is defined as that axis at which the

It divides the section into tension and compression zone. The position of the neutral axis depends upon the shape (dimensions) of the section and the amount of steel provided.

2.7 COMPARISON OF BALANCED, UNDER REINFORCED AND OVER REINFORCED SECTION

Sr. No.	Balanced Section	Under Reinforced Section	Over Reinforced Section
1.	$X_a = X_c$	$X_a < X_c$	$X_a > X_c$
2.	Same % of steel is provided than required for balanced section.	Less % of steel is provided than required for balanced section.	More % of steel is provided than required for balanced section.
3.	Both steel or concrete reaches the maximum permissible stress at the same time.	Tensile steel reaches maximum permissible stress in bending first. So failure of beam will be due to overstress in steel.	Here concrete reaches maximum permissible stress in bending first. So failure of beam will be due to overstress in concrete.
Mode of Failure			
4.	If the ratio of steel to concrete in a beam is such that maximum strain in two materials reach simultaneously. A sudden failure would occur with less alarming deflection.	When the amount of steel is kept less than required for balance section. The position of N.A. shifts upward to satisfy the equilibrium condition that force of compression is equal to the force of tension. In this process, C.G. of compressive force also shift upward. If the beam is further loaded, the strain in section increases. Thus result in an process of shift in N.A. continue until maximum strain in steel reaches is called as tension failure.	When the amount of steel is kept more than required for balance section. The position of N.A. shifts downward to satisfy the equilibrium condition that force of compression is equal to the force of tension. In this process C.G. of compressive force also shift upward. If the beam is further loaded, the stress and strain in section increases. Thus result in an process of shift in N.A. continue until maximum strain in concrete reaches is called as compression failure
Remark			
5.		(i) The under reinforced sections are preferred because their failure take place by yielding of steel which is a gradual type of failure and is proceeded by widening of cracks and significant increase in deflection. Hence necessary precautions can be taken before collapse take place. (ii) The term under reinforced sections refers to singly reinforced concrete sections having steel percentage less than limiting percentage.	The failure of over reinforced sections take place by crushing of concrete alone and therefore failure is sudden and explosive nature and occurs without signs. Therefore no over reinforcing of section is allowed in L.S.M. Hence when $X_a > X_c$, the section should be redesigned

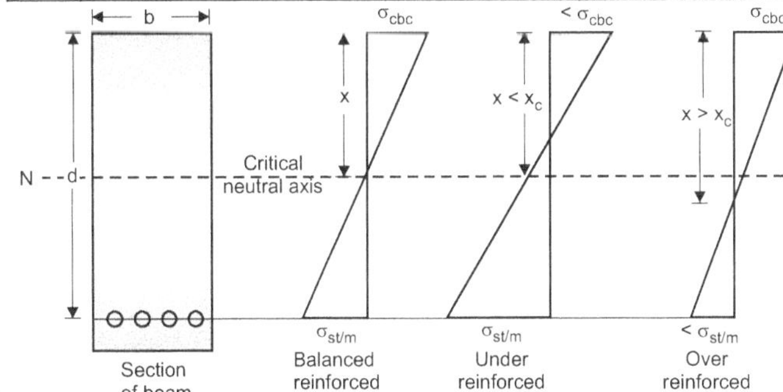

Fig. 2.6

2.8 TRANSFORMED OR EQUIVALENT SECTION

The bond between steel and concrete is assumed to be perfect so the strains in steel and the surrounding concrete will be equal

$$\varepsilon_s = \varepsilon_c \left[Strain = \frac{Stress}{Young's\ modulus} \right]$$

$$\frac{\sigma_c}{E_s} = \frac{\sigma_c}{E_c}$$

$$\sigma_s = \frac{E_s}{E_c} \cdot \sigma_c \qquad \left[\frac{E_s}{E_c} = m = Modular\ ratio \right]$$

$$\sigma_s = m \cdot \sigma_c \ or \ \sigma_c = \frac{\sigma_s}{m} \qquad \ldots(i)$$

$$P = P_s + P_c$$

$$P = \sigma_s A_{st} + \sigma_c A_c \qquad \sigma_s = m \cdot \sigma_c$$

$$P = m \cdot \sigma_c \cdot A_{st} + \sigma_c \cdot A_c$$

$$= \sigma_c (m \cdot A_{st} + A_c)$$

$$\sigma_c = \frac{P}{(A_c + m \cdot A_{st})} = \frac{P}{A_{eqc}}$$

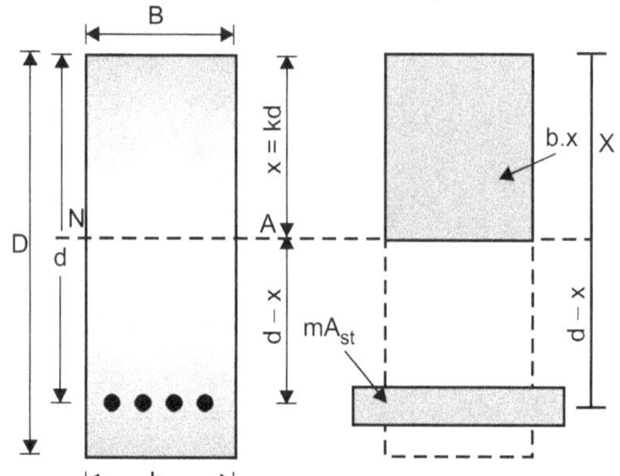

Fig. 2.7

where,

Equivalent area of section in term of concrete

$$= (A_c + m\ A_{st})$$

Equivalent concrete area of steel = $m\ A_{st}$

Equivalent concrete area = $A_c = b\ x$

Let, A = Area of cross-section

      A_c = Area of concrete

      A_{st} = Area of steel

      m = Modular ratio

      σ_s = Stress in steel

      σ_c = Stress in concrete

      ε_s = Strain in steel

      ε_c = Strain in concrete

      P_s = Load carried by steel

      P_c = Load carried by concrete

      A_{eqc} = Equivalent area of section in terms of concrete

      E_s = Young's modulus of elasticity of steel

      E_c = Young's modulus of elasticity of concrete

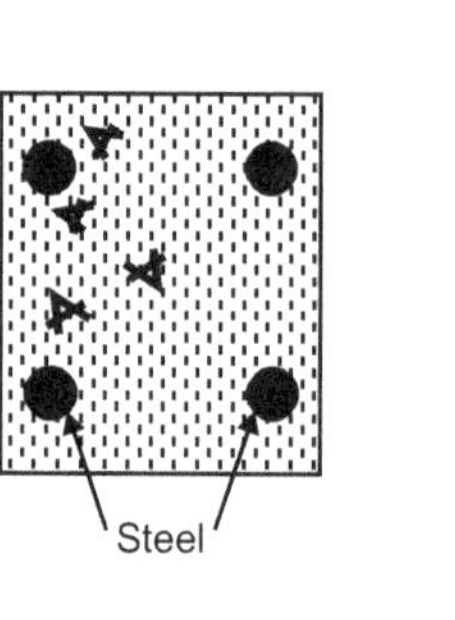
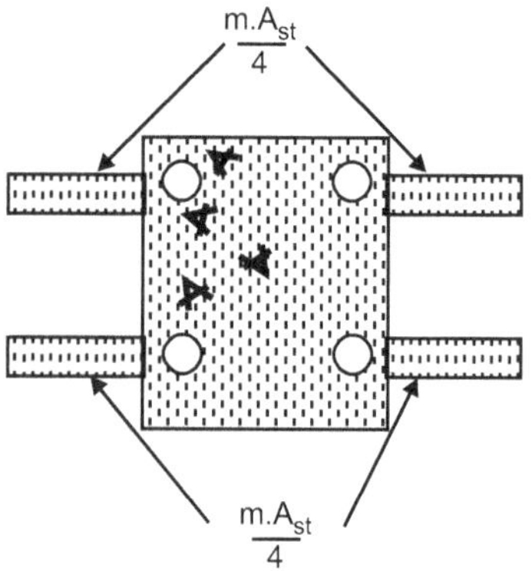

(a) R.C.C. section   **(b) Transformed section in terms of concrete**

Fig. 2.8 : Transformed section

2.9 SIZE OF BEAM

Depth of Beam :

- To get sufficient warning before the collapse of the beam under unexpected worst load case, it is desirable to keep the depth such that the percentage of steel required is around 75 per cent of that required for the balanced section. The another consideration in selecting the depth of the beam is to ensure that limit state of deflection is not reached.

- If we select depth of beam as $\frac{1}{12}^{th}$ to $\frac{1}{15}^{th}$ the of span, for simply supported beams and $\frac{1}{15}^{th}$ to $\frac{1}{20}^{th}$ of span for continuous beams, normally the above requirements are satisfied. Form the consideration of standardizing form works, overall depth is kept in multiples of 50 mm. Commonly used depths are 250 mm to 700 mm in multiples of 50 mm.

Width of Beam :

- The width of beam should accommodate the required number of bars with sufficient spacing between them and ensuring minimum side cover requirement. However, the bars may be accommodated in more than one layer also. The width is normally taken as 200 mm, 250 mm, 300 mm and 350 mm.

- Use of 230 mm is also common, since wells of width 230 mm (g" brick walls) are commonly constructed. The beam of width 230 mm will be flush with wall. The larger widths are preferred if the section is subjected to torsional moment. Clauses 23.3 in IS 456 specifies the following slenderness ratio limits to ensure lateral stability.

2.10 MOMENT OF RESISTANT OF SINGLY REINFORCED SECTION

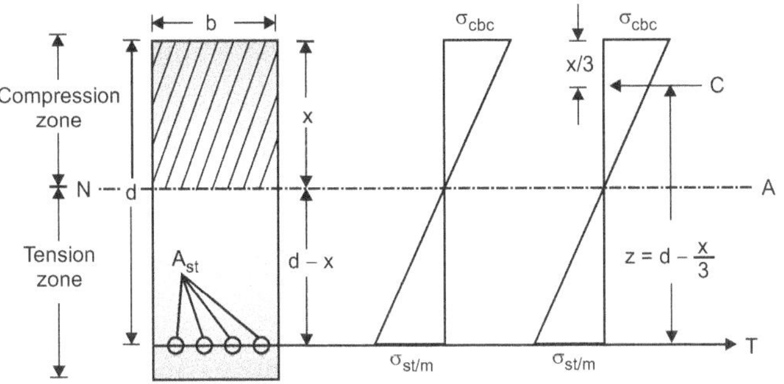

Fig. 2.9

Let, b = Breadth of a rectangular beam.

       d = Effective depth of section (measured from compression face to the C. G of reinforcement).

       x = Depth of neutral axis below compression edge.

       A_{st} = Cross-sectional area of steel in tension.

       σ_{cbc} = Permissible compressive stress in concrete in bending.

       σ_{st} = Permissible tensile stress in steel.

       D = Overall depth of beam.

       m = Modular ratio.m

    N.A = Neutral axis.

- The moment of resistance of an R.C.C. section is the moment of couple formed by total tensile force (T) acting at the C.G. of the reinforcement and the total compressive force (C) in the concrete acting at the C.G. of compressive stress diagram as shown in Fig. 2.9.

- The distance between total compressive force (C) and tensile force (T) is called the lever arm and is denoted by z.

- The C.G. of compressive stress triangle is at distance x/3 from compression edge.

$\therefore$ Lever arm $z = d - \dfrac{x}{3}$

Compressive force C

$$= \text{Average stress} \times \text{Compression area}$$

$$= \frac{0 + \sigma_{cbc}}{2} \times b \cdot x$$

$\therefore$ $C = \dfrac{bx \cdot \sigma_{cbc}}{2}$

Tensile force T

$$= \text{Stress in steel} \times \text{Area of steel}$$

$$T = \sigma_{st} \cdot A_{st}$$

$\therefore$ Moment of resistance,

$$M_r = C \times z \text{ or } T \times z$$

$\therefore$ $M_r = \dfrac{b \cdot x \cdot \sigma_{cbc}}{2} \cdot \left(d - \dfrac{x}{3}\right)$ ...(2.1)

or $M_r = \sigma_{st} \cdot A_{st}\left(d - \dfrac{x}{3}\right)$ ...(2.2)

2.11 MOMENT OF RESISTANT OF DOUBLY REINFORCED SECTION

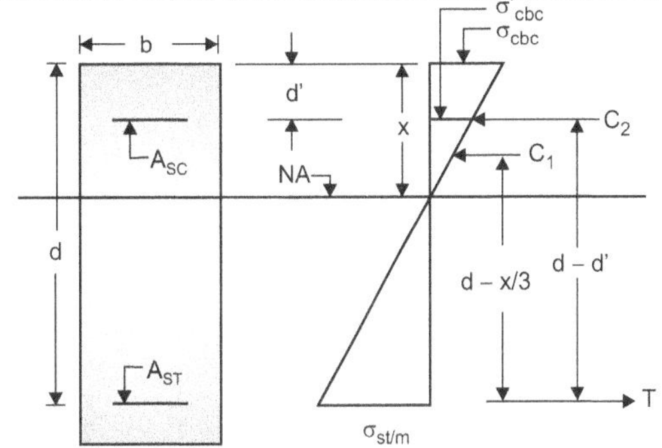

Fig. 2.10

1. Method Using Elastic Theory : In this method, the moment of resistance is calculated by taking the moments of forces about the C.G. of the tensile steel. From Fig. 2.10 we get,

M_r = Moment of compressive force of concrete about tensile steel

      + Moment of equivalent concrete force of compression steel about tensile steel

$$- \ bx \ \frac{\sigma_{cbc}}{2} \ (d - x/3) \ [1.5 \ m - 11 \ A_{sc} \ \sigma_{abc}$$

$$(d - d')]$$

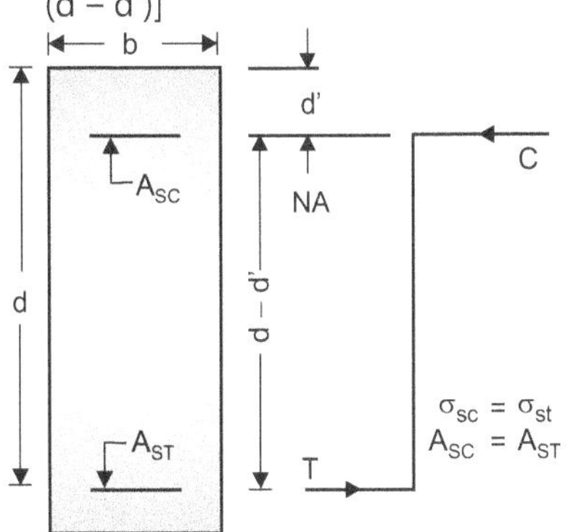

Fig. 2.11

2. Steel Beam Theory Method : The moment of resistance can be calculated by this method approximately. The whole concrete is neglected and only A_{sc} and A_{st} are considered. The following are the assumptions made.

- All the concrete area is neglected.
- Permissible stresses in tensile and compressive steel are the same.
- Permissible stresses reach their maximum stress at the same time (i.e. $A_{sc} = A_{st}$).

Taking the moment about compression steel,

$$M_r = A_{st}\, \sigma_{st}\, (d - d')$$

2.12 POSITION OF NEUTRAL AXIS OF SINGLY REINFORCED SECTION

Case I : Given

- Dimensions of section (b and d).
- Permissible stresses in concrete and steel (σ_{cbc} and σ_{st}).
- Modular ratio m.
- Referring to equivalent stress diagram, (Fig. 2.12) and considering similar triangles.

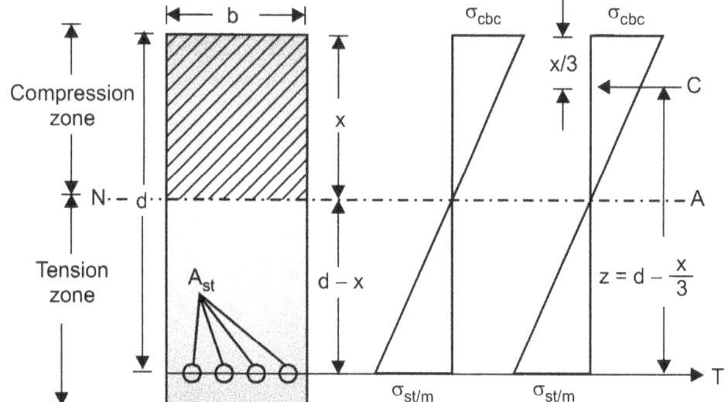

Fig. 2.12

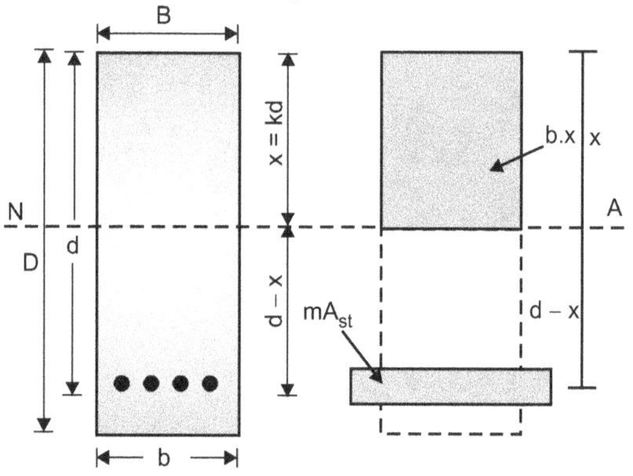

Fig. 2.13

$$\frac{\sigma_{cbc}}{\sigma_{st}/m} = \frac{x}{d - x} \qquad \ldots(2.3)$$

Case II : Given

- Dimensions of section (b and d).
- Area of tension steel (A_{st}).
- Modular ratio (m).

Moment of compression area about N.A.

$$= b \cdot x \cdot \frac{x}{2} = \frac{bx^2}{2}$$

Moment of area of tension side

$$= \text{Equivalent concrete area of tension steel} \times (d - x)$$

$$= m \cdot A_{st}\, (d - x)$$

As N.A. is situated at the C.G. of a given section, the moment of area on both sides are equal :

For determination of actual NA,

$$\frac{bx^2}{2} = m \cdot A_{st}\, (d - x) \qquad \ldots(2.4)$$

2.13 POSITION OF NEUTRAL AXIS OF DOUBLY REINFORCED SECTION (WSM)

T - Equivalent concrete area of steel in tension $= m \times A_{st}$

C_1 - Concrete area above N.A $= b.x$

C_2 - Equivalent concrete area of steel in compression

$$= (1.5\, m - 1)\, A_{sc} \qquad \text{(IS 456 states)}$$

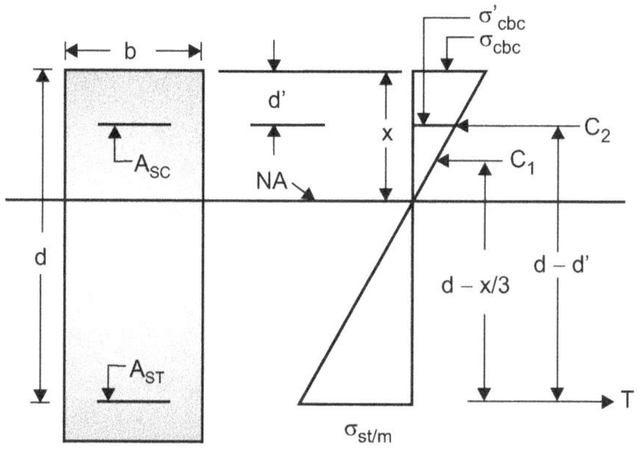

Fig. 2.14

- **To find Xc (Critical depth of N.A.)**

From similar triangles in the equivalent concrete stress diagram.

$$\frac{\sigma_{cbc}}{\sigma_{st}/m} = \frac{x_c}{d - x_c}$$

- **To find x_{a1} (Actual depth of N.A.):**

Moment of area on compression side = Moment of area on tension side.

$$bx\,\frac{x}{2} + (1.5\, m - 1)\, A_{sc}\, (x - d') = m.\, A_{st}\, (d - x)$$

$$X_a = ?$$

2.14 DESIGN CONSTANTS OF BALANCED SECTION

1. Neutral Axis Coefficient (k)

$$\frac{\sigma_{cbc}}{\sigma_{st}/m} = \frac{x_c}{d - x_c}$$

$$\therefore \quad \frac{d - x_c}{x_c} = \frac{\sigma_{st}}{m \cdot \sigma_{cbc}}$$

$$\frac{d}{x_c} - 1 = \frac{\sigma_{st}}{m \cdot \sigma_{cbc}}$$

$$\frac{d}{x_c} - 1 = \frac{\sigma_{st}}{m \cdot \sigma_{cbc}}$$

$$\frac{d}{x_c} = 1 + \frac{\sigma_{st}}{m \cdot \sigma_{cbc}} = \frac{m \cdot \sigma_{cbc} + \sigma_{st}}{m\sigma_{cbc}}$$

$$x_c = k \cdot d$$

$$\boxed{k = \frac{m \cdot \sigma_{cbc}}{m\,\sigma_{cbc} + \sigma_{st}}}$$

2. Lever Arm Constant (j)

Lever arm, $z = d - \dfrac{x}{3}$

For balanced section,

$$x = x_c = k \cdot d$$

$$z = d - \frac{k \cdot d}{3} = \left(1 - \frac{k}{3}\right) d$$

$$\therefore \quad z = jd$$

where, $\quad \boxed{j = 1 - \dfrac{k}{3}}$

3. Moment of Resistance Factor (Q) : (R)

$$M_r = \frac{b \cdot x \cdot \sigma_{cbc}}{2}\left(d - \frac{x}{3}\right)$$

For balanced section

$$x = x_c = kd$$

and $\quad z = d - \dfrac{x_c}{3} = jd$

$$M_r = \frac{b \cdot (kd) \cdot \sigma_{cbc}}{2} \times jd$$

$$= \left(\frac{\sigma_{cbc} k \cdot j}{2}\right) bd^2$$

$$\boxed{M_r = Q \cdot bd^2}$$

where, $\quad Q = \dfrac{\sigma_{cbc} \cdot k \cdot j}{2} = R$

4. Percentage of Steel (P_t)

Equating compressive force C to tensile T

$$b \cdot x \frac{\sigma_{cbc}}{2} = A_{st} \cdot \sigma_{st}$$

For balanced section,

$$x = k \cdot d$$

$$\therefore \quad \frac{bk \cdot d\,\sigma_{cbc}}{2} = A_{st} \cdot \sigma_{st}$$

$$\frac{A_{st}}{bd} = \frac{\sigma_{cbc} \cdot k}{2\sigma_{st}}$$

Percentage steel,

$$P_t = \frac{A_{st}}{bd} \times 100$$

$$\boxed{P_t = \frac{\sigma_{cbc} \cdot k}{2\,\sigma_{st}} \times 100}$$

2.15 STRESS STRAIN DIAGRAM

2.15.1 Singly Reinforced Section

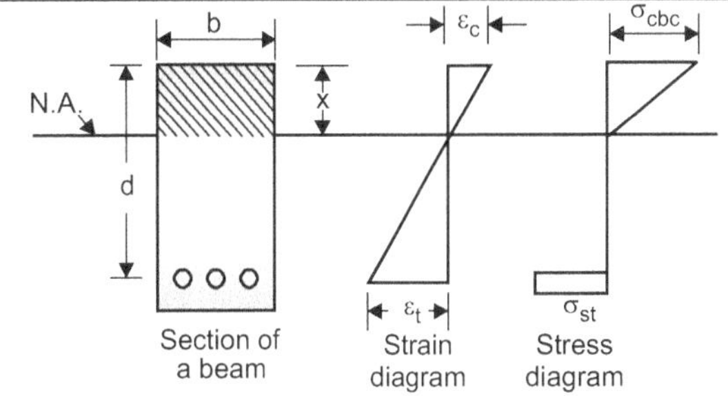

Fig. 2.15

2.15.2 Doubly Reinforced Section

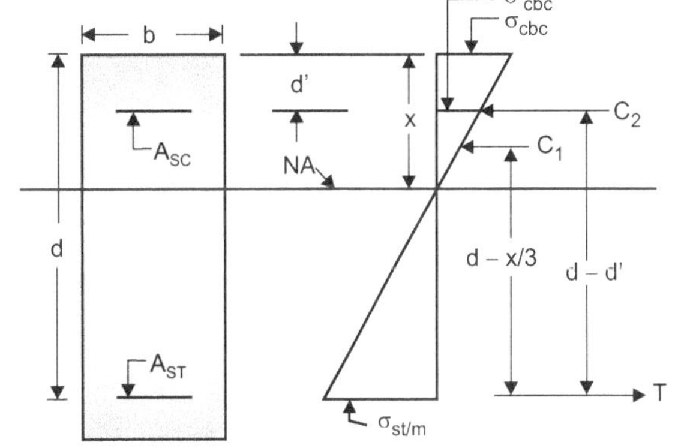

Fig. 2.16

2.15.3 Flanged Section

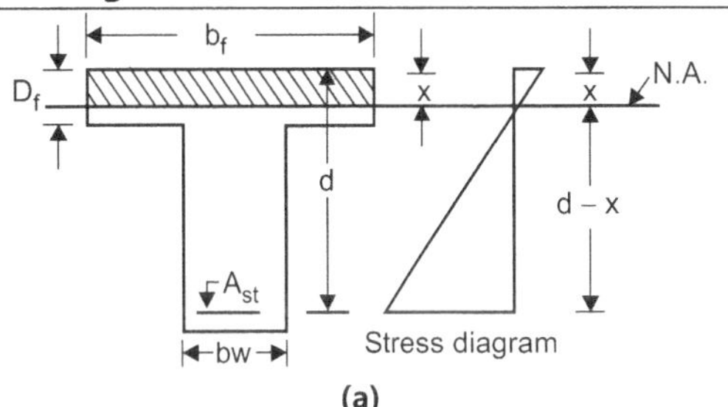

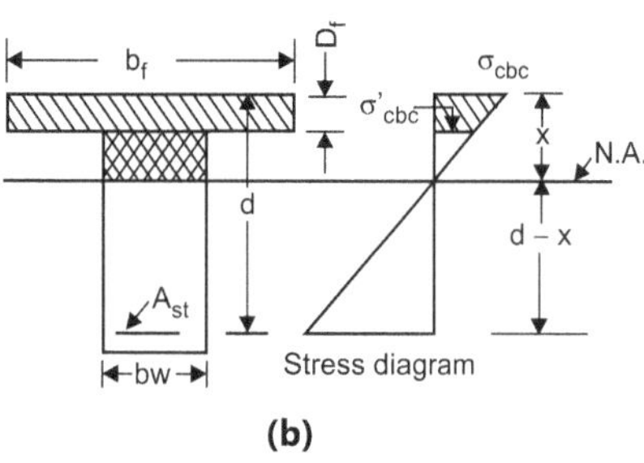

(b)

Fig. 2.17

SOLVED EXAMPLES

Type I : Singly Reinforced Section

***Example : 2.1 :** An RC beam 300 mm wide and 700 mm deep overall, is reinforced with 20 mm φ 3 bars. The centres of the bars are 50 mm from the underside of the beam. The maximum stresses are not to exceed 7 N/mm² for concrete and 190 N/mm² for steel (m = 13.33). Find the moment of resistance and the safely distributed load, the beam can carry. The span is 6 m and the weight of concrete is 25 kN/m³.*

Given :

$$b = 300 \text{ mm}$$
$$d = 700 - 50 = 650 \text{ mm}$$
$$\sigma_{cbc} = 7 \text{ N/mm}^2$$
$$\sigma_{st} = 190 \text{ N/mm}^2$$
$$m = 13.33$$
$$span = 6 \text{ m}$$
$$A_{st} = 3 \times \frac{\pi}{4} \times (20)^2 = 942 \text{ mm}^2$$

Solution :

1. To find N.A. (critical)

$$\frac{\sigma_{cbc}}{\sigma_{st}/m} = \frac{x_c}{d - x_c}$$

$$\frac{7}{190/13.33} = \frac{x_c}{650 - x_c}$$

$$\therefore \quad x_c = 214.08, \text{ or } 214 \text{ mm}$$

2. To find N.A. (actual) :

$$bx\frac{x}{2} = mA_{st}(d - x)$$

$$\frac{300 \, x^2}{2} = 13.33 \times 942 \times (650 - x)$$

$$\therefore \quad x^2 + 83.71x - 54.413 = 0$$

Solving for x, we get,

$$x = 195.115 \text{ or } 195 \text{ mm}$$

3. Compare x and x_c :

$$\therefore \quad x < x_c$$

The beam is under-reinforced and hence steel will reach its maximum permissible value first

4. Find M_r :

$$M_r = T \times z$$
$$= A_{st} \, \sigma_{st} \times z$$
$$= 942 \times 190 \times \left(650 - \frac{195}{3}\right)$$
$$= 104703300 \text{ N-mm}$$
$$\mathbf{= 104.7 \text{ kN-m}}$$

5. Let, W kN be the load on the beam per metre. Then bending moment

$$M = \frac{wl^2}{8} = \frac{w \times 6^2}{8} = 4.5 \text{ W kN-m}$$

Equating M_r to M, we get

$$104.7 = 4.5 \text{ W}$$

$$\therefore \quad \mathbf{w = 23.266 \text{ kN/m}}$$

Self-weight of the beam is

$$(0.3 \times 0.7 \times 1) \times 25 = 5.25 \text{ kN/m}$$

Net weight the beam can carry will be

$$\mathbf{23.266 - 5.25 = 18.016 \text{ kN/m or 18 kN/m}}$$

***Example 2.2 :** A section of RC beam 300 × 700 mm is reinforced four 25 mm φ bars, placed 30 mm from the bottom of the beam is subjected to a bending moment of 130 kN/m. Find the set up in concrete and steel, if m = 18.66.*

Given :

$$b = 300 \text{ mm}$$
$$d = 700 - 30 = 670 \text{ mm}$$
$$M = 130 \text{ kN-m} = 130 \times 10^6 \text{ N-mm}$$
$$m = 18.66$$
$$A_{st} = 4 \times \frac{\pi}{\Delta} \times (25)^2 = 1964 \text{ mm}^2$$

Solution :

1. To find x :

$$bx\frac{x}{2} = m A_{st}(d - x)$$

$$\frac{300x^2}{2} = 18.66 \times 1964 \times (670 - x)$$

$$\therefore \quad x^2 + 244.32 \, x - 163695 = 0$$

Solving for x **x = 300.47 or 300 mm**

2. To find z : $z = d - \dfrac{x}{3} = 670 - \dfrac{300}{3} = \mathbf{570 \text{ mm}}$

3. To find M_r :

$$M_r = T \times z$$
$$= A_{st} \, \sigma_{st} \, z$$
$$= 1964 \, \sigma_{st} \, 570 \text{ N-mm}$$

4. To find σ_{st} : Equate M_r to M.

$$1964 \times \sigma_{st} \times 570 = 130 \times 10^6$$

$$\therefore \quad \mathbf{\sigma_{st} = 116.25, \text{ or } 116 \text{ N/mm}^2}$$

5. To find σ_{cbc} :

$$\frac{\sigma_{cbc}}{\sigma_{st}/m} = \frac{x}{d-x}$$

$$\frac{\sigma_{cbc}}{116/18.66} = \frac{300}{670-300}$$

$$\therefore \quad \sigma_{cbc} = \textbf{5.04 or 5 N/mm}^2$$

Example 2.3 : *A beam 350 × 650 mm overall is provided with tensile reinforcement, consisting of four 22 mm ϕ bars. Calculate the stresses developed in steel and concrete, if the beam carries a uniformly distributed load of 20 kN/m (inclusive of self-weight). Take l_{ef} = 7m, and m = 13.33 and cover = 25 mm.*

Given :   b = 350 mm

         d = 650 − 25 = 625 mm

         m = 13.33

         W = 20 kN/m

         l_{ef} = 7 m

         A_{st} = $4 \times \frac{\pi}{4} \times (22)^2$ = 1520 mm^2

Solution :

1. To find x :

$$bx\frac{x}{2} = mA_{st}(d-x)$$

$$\frac{350\,x^2}{2} = 13.33 \times 1520 \times (625 - x)$$

$$\therefore \quad x^2 + 115.78\,x - 72.362 = 0$$

Solving for x, we get

$$x = \textbf{217.27 , or 217 mm}$$

2. To find z : $z = d - \frac{x}{3} = 625 - \frac{217}{3}$

$$= \textbf{552.66 mm}$$

3. To find M$_r$:

$$M_r = T \times z = A_{st}\,\sigma_{st}\left(d - \frac{x}{3}\right)$$

$$= 1520\,\sigma_{st} = \textbf{552.66 N-mm}$$

4. To find σ_{st} :

Bending moment, M will be

$$M = \frac{Wl^2}{8} = \frac{20 \times 7^2}{8}$$

$$= 122.5 \text{ kN-m} = \textbf{122.5} \times \textbf{10}^6 \textbf{ N-mm}$$

Equating M$_r$ to M, we get

$$1520 \times \sigma_{st} + 552.66 = 122.5 \times 10^6$$

$$\therefore \quad \sigma_{st} = \textbf{145.82 N/mm}^2$$

5. To find σ_{cbc} :

$$\frac{\sigma_{cbc}}{\sigma_{st}/m} = \frac{x}{d-x}$$

$$\frac{\sigma_{cbc}}{145.82/13.33} = \frac{217}{625-217}$$

$$\therefore \quad \sigma_{cbc} = \textbf{5.818 N/mm}^2$$

Example 2.4 : *The cross-section of a rectangular beam has to resist a bending moment of 75 kN-m. If the beam is 250 mm wide and the permissible stresses in steel and concrete are 190 N/mm^2 and 5 N/mm^2 respectively, find the effective depth and tensile reinforcement required.*

Given :   b = 250 mm

         σ_{cbc} = 5 N/mm^2

         σ_{st} = 190 N/mm^2

         M = 75 kN/m = 75×10^6 N-mm

         m = $\frac{280}{3 \times 5}$ = 18.66

Solution :

1. To find x$_c$

$$\frac{\sigma_{cbc}}{\sigma_{st/m}} = \frac{x_c}{d-x_c}$$

$$\frac{5}{190/18.66} = \frac{x_c}{d-x_c}$$

$$\therefore \quad x_c = \textbf{0.329 d, or 0.33 d mm}$$

2. Lever arm $Z = d - \frac{x_c}{3} = d - \frac{0.33\,d}{3}$

$$= 0.89\text{ d mm}$$

$$M_r = C \times Z$$

$$= bx_c\frac{\sigma_{cbc}}{2} \times z$$

$$= 250 \times 0.33\,d \times \frac{5}{2} \times 0.89\,d$$

$$= 183.56\,d^2 \text{ N/mm}$$

Equating M$_r$ to M we get

$$183.56\,d^2 = 75 \times 10^6$$

$$\therefore \quad d = \textbf{639.22 or 640 mm}$$

3. To find A$_{st}$:

$$T = C$$

$$A_{st}\,\sigma_{st} = \frac{bx_c\,\sigma_{abc}}{2}$$

$$A_{st} \times 190 = 250 \times (0.33 \times 640) \times \frac{5}{2}$$

$$\therefore \quad A_{st} = \textbf{694.7 mm}^2$$

Or $M_r = T \times z$

$$= A_{st}\,\sigma_{st} \times z$$

Equating M_r to M, we get

$$A_{st}\,\sigma_{st}\,z = M$$

$$A_{st} \times 190 \times (0.89 \times 640) = 75 \times 10^6$$

$$\therefore \quad \boxed{A_{st} = 693\ mm^2}$$

Example 2.5 : *The cross-section of an RC beam is required to resist a bending moment of 65 kN-m. Find the dimensions of the cross section and area of steel required. Assume that σ_{cbc} = 5 N/mm², σ_{st} = 140 N/mm², m = 18.66 and b = $\frac{1}{2}$ d.*

Given :

$$\sigma_{cbc} = 5\ N/mm^2$$
$$\sigma_{st} = 140\ N/mm^2$$
$$m = 18.66$$
$$M = 65\ kN\text{-}m = 65 \times 10^6\ N\text{-}mm$$
$$b = \frac{1}{2}\ d$$

Solution :

1. To find x_c

$$\frac{\sigma_{cbc}}{\sigma_{st}/m} = \frac{x_c}{d - x_c}$$

$$\frac{5}{140/18.66} = \frac{x_c}{d - x_c}$$

$$\therefore \quad \boxed{x_c = 0.399d\ or\ 0.4\ d\ mm}$$

2. Lever arm $z = d - \dfrac{x_c}{3} = d - \dfrac{0.4d}{3}$

$$= 0.866\ d,\ or\ 0.87\ d\ mm$$

$$M_r = C \times z$$

$$= bx_c\,\frac{\sigma_{cbc}}{2} \times z$$

$$= b \times 0.4d \times \frac{5}{2} \times 0.87\ d$$

Substituting $b = \dfrac{1}{2}\,d = 0.5\ d$, we get

$$M_r = 0.5\ d \times 0.4\ d \times \frac{5}{2} \times 0.87\ d$$

$$= \mathbf{0.435\ d^3\ N\text{-}mm}$$

3. Equating M_r to M, we get

$$0.435\ d^3 = 65 \times 10^6$$

$$\therefore \quad d = 530.65\ mm,\ or\ 530\ mm$$

$$\therefore \quad b = \frac{1}{2}\ d = \frac{530}{2} = \mathbf{265\ mm}$$

4. To find A_{st}

$$M_r = T \times z$$

$$= A_{st}\,\sigma_{st}\,z$$

Equating M_r to M, we get

$$A_{st}\,\sigma_{st}\,z = M$$

$$A_{st} \times 140 \times (0.87 \times 530) = 65 \times 10^6$$

$$\therefore \quad A_{st} = 1006.9\ mm^2\ or\ 1007\ mm^2$$

$$\therefore \quad Dimensions\ of\ section = 265 \times 530\ mm$$

Area of steel = $\mathbf{A_{st} = 1007\ mm^2}$

Example 2.6 : *A RC beam section of size 230 mm wide and 430 mm effective depth is reinforced with 3 no 16 mm diameter bars. Find the moment of resistance of section using WSM approach. Also find safe intensity of uniformly distributed load that can be placed on simply supported beam of effective span 4m. Use M20 grade of concrete and Fe 250 steel.*

Solution : $b = 230\ mm$

$$d = 430\ mm$$

$$A_{st} = 3\ nos.\ 16\ mm\ \phi = 3 \times 201 = 603\ mm^2$$

M20 grade of concret σ_{cbc} = MPa

Fe 250 steel for bars upto 20 mmφ

$$\sigma_{st} = 140\ MPa \quad \dots[Table\ 22,\ IS:456-2000]$$

To find M.R. = ?

$$\omega_{safe} = ?$$

1. Depth of actual netural axis,

$$b \cdot \frac{x_{act}^2}{2} = m\,A_{st}\,(d - x_{act})$$

where, $m = \dfrac{280}{3\sigma_{cbc}} = \dfrac{280}{3 \times 7} = 13.33$

$$230 \times \frac{x_{act}^2}{2} = 13.33 \times 603 \times (430 - x_{act})$$

$$\therefore \quad \mathbf{x_{act} = 141.895\ mm}$$

2. Depth of balanced netural axis,

$$x_{bal} = k_{bal} \cdot d$$

$$k_{bal} = \frac{\sigma_{cbc}}{\sigma_{cbc} + \dfrac{\sigma_{st}}{m}} = \frac{7}{7 + \dfrac{140}{13.33}} = 0.40$$

$$\therefore \quad x_{bal} = 0.40 \times 430 = \mathbf{172\ mm}$$

3. Classification of section,

As $x_{act} < x_{bal}$ the section is under-reinforced section.

4. Calculation of moment of resistance

$$M.R. = T.L.A.$$

$$= \sigma_{st} \cdot A_{st}\left(d \cdot \frac{x_{act}}{3}\right)$$

$$= 140 \times 603\ (430 - 141.895/3)$$

$$= 32.30 \times 10^6\ N.mm$$

$$\mathbf{M.R. = 32.30\ kN.m}$$

5. Safe intensity of u.d.l.

$$\text{Max. B.M.} = M = \frac{\omega \times L^2}{8} = \frac{\omega \times 4^2}{8} = 2\omega$$

Equating M.R. = Max. B.M.

$$32.30 = 2 \times \omega$$

$$\therefore \quad \omega = 16.15 \text{ kN/m}$$

Self weight of section = c/s of RC section × weight density of RCC

$$\omega_{self} = 0.230 \times 0.450 \times 25$$

(Assuming D = 430 + 20 = 450mm)

$$\omega_{self} = 2.38 \text{ kN/m}$$

$$\therefore \quad \boldsymbol{\omega_{safe} = \omega - \omega_{self} = 16.15 - 2.58 = 13.56 \text{ kN/m}}$$

Example 2.7 : *Using stress and strain diagrams for a balanced singly reinforced section as per WSM, derive the design constants (k = Neutral axis constant, j = Lever arm constant, and R = Moment of resistance constant). Use M20 grade of concrete and Fe 415 grade steel.*

Solution :

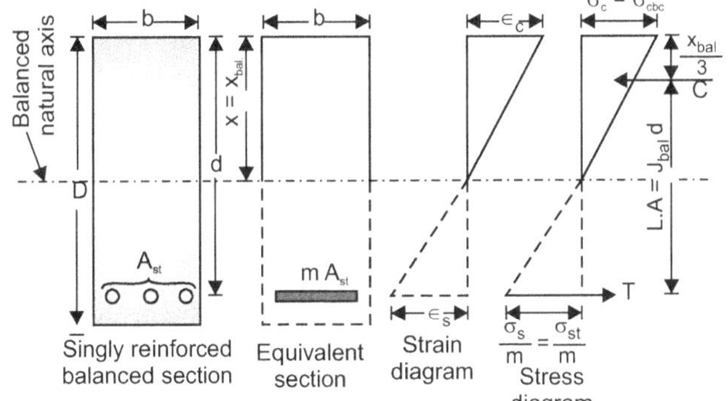

Fig. 2.18

Let,
- b = Width of RC beam
- D = Overall depth of RC beam
- d = Effective depth of RC beam
- A_{st} = Area of tension reinforcement
- m = modular ratio = $\dfrac{280}{3\sigma_{abc}}$
- x = Depth of N.A. from extreme concrete fiber in comp
- ϵ_c = Strain in extreme concrete fiber in comp
- ϵ_s = Strain in tension reinforcement
- σ_c = Stress in extreme concrete fiber in compression
- σ_s = Stress of tension reinforcement

For singly RC section as per WSM

Let, $x_{bal} = k \cdot d$ where   k = N.A. design constant

$$(\text{L.A})_{bal} = j.d \quad \text{where} \quad j = \text{L.A. design constant}$$

$$M_{bal} = R \cdot bd^2 \quad \text{where} \quad R = \text{M.R. design constant}$$

Also, $\sigma_c = \sigma_{cbc}$ and $\sigma_s = \sigma_{st}$ for balanced section.

From stress diagram, using property of similar triangle

$$\frac{\sigma_{a_{cbc}}}{x_{bal}} = \frac{\sigma_{st}/m}{d - x_{bal}}$$

$$\frac{x_{bal}}{d - x_{bal}} = \frac{\sigma_{cbc}}{\sigma_{st}/m}$$

$$\frac{x_{bal}}{d} = \frac{\sigma_{cbc}}{\dfrac{\sigma_{st}}{m} + \sigma_{cbc}}$$

$$x_{bal} = \left[\frac{\sigma_{cbc}}{\dfrac{\sigma_{st}}{m} + \sigma_{cbc}}\right] \cdot d$$

$$\therefore \quad k = \frac{\sigma_{cbc}}{\sigma_{cbc} + \dfrac{\sigma_{st}}{m}} = \text{Design constant for N.A.}$$

For M20 grade of concrete σ_{cbc} = 7 MPa and for Fe415 grade of steel σ_{st} = 230 MPa

$$m = \frac{280}{3 \times \sigma_{cbc}} = \frac{280}{3 \times 7} = 13.33$$

$$\therefore \quad \boldsymbol{k = \frac{7}{7 + \dfrac{230}{13.33}} = 0.29}$$

Now, Lever arm = $(\text{L.A})_{bal} = d - \dfrac{x_{bal}}{3}$

$$= d - \frac{k_{bal}}{3} \cdot d$$

$$= \left[1 - \frac{k_{bal}}{3}\right] d$$

$$\therefore \quad j_{bal} = \left[1 - \frac{k_{bal}}{3}\right]$$

= Design constant for lever arm

For M 20 grade of concrete and Fe 415 grade of steel,

$$k_{bal} = 0.29$$

$$\therefore \quad \boldsymbol{j_{bal} = 1 - \frac{0.29}{3} = 0.9}$$

Now,

Moment of resistance = $(\text{MR})_{balanced} = C \times \text{L.A}$

where, C = Total compression

= Area of stress diagram × Width of RC beam

$$= \frac{1}{2} \times \sigma_c \times x \times b$$

But for balanced section, $\sigma_c = \sigma_{cbc}$ and $x = x_{bal} = k_{bal} \cdot d$

$$c = \frac{1}{2} \times \sigma_{cbc} \times k_{bal} \cdot d \cdot b$$

Also $(L.A) = (L.A)_{bal} = j_{bal} \cdot d = \left[1 - \dfrac{k_{bal}}{3}\right] \cdot d$

$\therefore (MR)_{balanced} = \dfrac{1}{2} \cdot \sigma_{cbc} \times k_{bal} \times \left[1 - \dfrac{k_{bal}}{3}\right] \cdot bd^2$

$\therefore \quad R_{balanced} = \dfrac{1}{2} \sigma_{cbc} k_{bal} \left[1 - \dfrac{k_{bal}}{3}\right]$

$= $ Design constant for MR

For M_{20} grade of concrete and Fe 415 grade of steel

$$\mathbf{R_{balanced} = \dfrac{1}{2} \times 7 \times 0.29 \times 0.90 = 0.92}$$

Type II : Doubly Reinforced Section

Example 2.8 : *An RC beam 200 × 400 mm overall is reinforced with four 22 mm φ bars with centres 30 mm from the bottom edge and three 20 mm φ bars with centres 25 mm from the top edge. Find the neutral axis of the beam if m = 18.66.*

Given : $b = 200$ mm

$d = 400 - 30 = 370$ mm

$d' = 25$ mm

$m = 18.66$

$A_{sc} = 3 \times \dfrac{\pi}{4} \times (20)^2 = 942$ mm^2

$A_{st} = 4 \times \dfrac{\pi}{4} \times (22)^2 = 1520$ mm^2

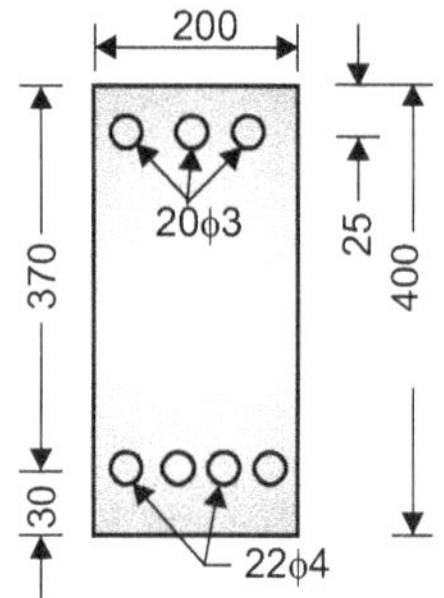

Fig. 2.19

Solution:

1. $X = ?$, $\dfrac{bxx}{2} + (1.5\,m - 1)\,A_{sc}\,(x - d') = mA_{st}\,(d - x)$

T find x : Equating moment of area on compression and tension sides about N.A.

$bx\dfrac{x}{2} + (1.5\,m - 1)\,A_{sc}\,(x - d') = m\,A_{st}\,(d - x)$

$\dfrac{200\,x^2}{2} + (1.5 \times 18.66 - 1)\,942\,(x - 25) = 18.66 \times 1520$

$(360 - x)$

$\therefore \quad x^2 + 537.87\,x - 111.299 = 0$

Solving we get $x = 159.579$ mm

From the lower edge of the beam and that on the compression side 30 mm.

Example 2.9 : *An RC beam is reinforced with five 22 mm φ bars on the tension side and four 20 mm φ bars on the compression side. The reinforcement on the tension side is placed 40 mm from the lower edge of the beam and that on the compression side 30 mm from the top edge of the beam. Find the moment of resistance of the beam. Beam section is 280 × 540 mm overall $\sigma_{cbc} = 5 N/mm^2$, $\sigma_{st} = 140\ N/mm^2$, $\sigma_{sc} = 130\ N/mm^2$ and m = 18.66.*

Given : $b = 280$ mm

$d = 540 - 40 = 500$ mm

$d' = 30$ mm

$\sigma_{cbc} = 5$ N/mm^2

$\sigma_{st} = 140$ N/mm^2

$\sigma_{sc} = 130$ N/mm^2

$m = 18.66$

$A_{st} = 5 \times \dfrac{\pi}{4} \times 22^2 = 1900$ mm^2

$A_{sc} = 4 \times \dfrac{\pi}{4} \times 20^2 = 1256$ mm^2

1. **To find x_c :**

$\dfrac{\sigma_{cbc}}{\sigma_{st}/m} = \dfrac{x_c}{d - x_c}$

$\dfrac{5}{140/18.66} = \dfrac{x_c}{500 - x_c}$

$\therefore \quad \mathbf{x_c = 199.957\ or\ 200\ mm}$

2. **To find x :**

$bx\dfrac{x}{2} + (1.5\,m - 1)\,A_{sc}\,(x - d') = mA_{st}\,(d - x)$

$\dfrac{280 x^2}{2} + (1.5 \times 18.66 - 1)\,1256\,(x - 30) = 18.66 \times 1900$

$(500 - x)$

$\therefore \quad x^2 + 495.37x - 133.885 = 0$

Solving for x, we get

$x = 194.16$ or 195 mm

$\therefore \qquad x < x_c$

Hence the beam is under-reinforced.

$\therefore \qquad \mathbf{\sigma_{st} = 140\ N/mm^2}$

3. **To find σ_{cbc}**

$\dfrac{\sigma_{cbc}}{\sigma_{st}/m} = \dfrac{x}{d - x}$

$\dfrac{\sigma_{cbc}}{140/18.66} = \dfrac{194}{500 - 194}$

$\therefore \qquad \mathbf{\sigma_{cbc} = 4.756\ N/mm^2}$

4. **To find σ'_{cbc} :**

$\sigma'_{cbc} = \sigma_{cbc} \left(\dfrac{x - d'}{x}\right)$

$$= 4.756 \left(\frac{194 - 30}{194} \right)$$

$$= \mathbf{4.02 \ or \ 4 \ N/mm^2}$$

5. Actual compressive stress in steel,

$$\sigma_{sc} = 1.5 \ m \ \sigma'_{cbc} = 1.5 \times 18.66 \times 4$$

$$= \mathbf{111.96, \ or \ 112 \ N/mm^2}$$

Hence, the actual compression stress (112 N/mm^2) < the permissible compressive stress in steel (130 N/mm^2) which is as it should be

6   To find M$_r$: Taking moments about tensile steel, we get

$$M_r = bx \frac{\sigma_{cbc}}{2} \left(d - \frac{x}{3} \right) + (1.5 \ m - 1) \ A_{sc}$$

$$\sigma_{cbc} \ (d - d')$$

$$= 280 \times 194 \times \frac{4.756}{2} \left(500 - \frac{194}{3} \right)$$

$$+ (1.5 \times 18.66 - 1) \ (1256 \times 4 \ (500 - 30)$$

$$= 119963246 \ \text{N-mm}$$

$$= \mathbf{119.96 \ kN\text{-}m}$$

Example 2.10 : *An RC beam 300 × 600 mm effective dimensions is provided with tensile and compressive reinforcement of 1256 mm^2 each. The compressive steel is placed 30 mm from the top edge of the beam. If σ_{cbc} = 7 N/mm^2, σ_{st} = 190 N/mm^2 and m = 13.33, find the moment of resistance of beam by (a) elastic theory method and (b) steel beam theory method.*

Given :

$$b = 300 \ \text{mm}$$
$$d = 600 \ \text{mm}$$
$$A_{st} = A_{sc} = 1256 \ \text{mm}^2$$
$$d' = 30 \ \text{mm}$$
$$\sigma_{cbc} = 7 \ \text{N/mm}^2$$
$$\sigma_{st} = 190 \ \text{N/mm}^2$$
$$m = 13.33$$
$$\sigma_{sc} = 130 \ \text{N/mm}^2$$

Solution :

(a) Elastic theory method :

1.   To find x$_c$:

$$\frac{\sigma_{cbc}}{\sigma_{st}/m} = \frac{x_c}{d - x_c}$$

$$\frac{7}{190/13.33} = \frac{x_c}{600 - x_c}$$

$$\therefore \qquad x_c = \mathbf{197.61 \ mm}$$

2.   To find x :

$$bx \frac{x}{2} + (1.5 \ m - 1) \ A_{sc} \ (x - d') = m \ A_{st} \ (d - x) \frac{300x^2}{2}$$

$$+ (1.5 \times 13.33 - 1) \ 1256 \ (x - 30) = 13.33 \times 1256 \ (600 - x)$$

$$\therefore \quad x^2 + 270.6x - 71.741 = 0$$

Solving for x, we get

$$x = 164.778, \ \text{or} \ 165 \ \text{mm}$$

$$\therefore \qquad x < x_c$$

Hence , the beam is under-reinforced

$$\therefore \qquad \sigma_{st} = \mathbf{190 \ N/mm^2}$$

3.   To find σ_{cbc} :

$$\frac{\sigma_{cbc}}{\sigma_{st}/m} = \frac{x}{d - x}$$

$$\frac{\sigma_{cbc}}{190/13.33} = \frac{165}{600 - 165}$$

$$\therefore \qquad \sigma_{cbc} = \mathbf{5.4 \ N/mm^2}$$

4.   To find σ'_{cbc} :

$$\sigma_{cbc} = \sigma_{cbc} \left(\frac{x - d'}{x} \right)$$

$$= 5.4 \left(\frac{165 - 30}{165} \right)$$

$$= \mathbf{4.418 \ or \ 4.42 \ N/mm^2}$$

5.   To find σ_{sc} : $\sigma_{sc} = 1.5 \ m \times \sigma_{cbc}$

$$= 1.5 \times 13.33 \times 4.42$$

$$= 88.377, \ \text{or} \ 88.38 \ \text{N/mm}^2$$

$$< 130 \ \text{N/mm}^2$$

Hence the design is all right.

6.   To find M$_r$: Taking moments about tensile steel, we get

$$M_r = b_x \frac{\sigma_{cbc}}{2} \left(\frac{d - x}{3} \right) + (1.5 \ m - 1)$$

$$A_{sc} \ \sigma'_{cbc} \ (d - d')$$

$$= 300 \times 165 \times \frac{5.4}{2} \left(600 - \frac{165}{3} \right)$$

$$+ (1.5 \times 13.33 - 1) \ 1256$$

$$\times 4.42 \ (600 - 30)$$

$$= 132946390 \ \text{N-mm}$$

$$= \mathbf{132.946 \ kN\text{-}m}$$

(b) Steel beam theory method

$$M_r = A_{st} \ \sigma_{st} \ (d - d')$$

$$= 1256 \times 190 \times (600 - 30)$$

$$= 136024000 \ \text{N-mm}$$

$$M_r = \mathbf{136.024 \ kN\text{-}m}$$

Example 2.11 : *A beam 250 × 550 mm overall is reinforced with four 25 mm ϕ bars on tension side and three 22 mm ϕ bars on compression side. The bars are at 50 mm and 30 mm centres respectively from bottom and top edge of the beam. Calculate the moment of resistance of the beam if $\sigma_{cbc} = 5$ N/mm^2, $\sigma_{st} = 140$ N/mm^2 and m = 18.66, $\sigma_{sc} = 130$ N/mm^2.*

Given :

$$b = 250 \text{ mm}$$
$$d = 550 - 50 = 500 \text{ mm}$$
$$d' = 30 \text{ mm}$$
$$\sigma_{cbc} = 5 \text{ N/mm}^2, \ \sigma_{st} = 140 \text{ N/mm}^2$$
$$m = 18.66$$
$$\sigma_{sc} = 130 \text{ N/mm}^2$$
$$A_{st} = 4 \times \frac{\pi}{4} \times 25^2 = 1963 \text{ mm}^2$$
$$A_{sc} = 3 \times \frac{\pi}{4} \times 22^2 = 1140 \text{ mm}^2$$

1. To find x_c :

$$\frac{\sigma_{cbc}}{\sigma_{st}/m} = \frac{x_c}{d - x_c}$$

$$\frac{5}{140/18.66} = \frac{x_c}{500 - x_c}$$

$$\therefore \quad \mathbf{x_c = 199.95 \text{ or } 200 \text{ mm}}$$

2. To find x :

$$bx\frac{x}{2} + (1.5m - 1) A_{sc} (x - d') = m A_{st} (d - x) \quad \frac{250\,x^2}{2} +$$

$$(1.5 \times 18.66 - 1) \ 1140 \ (x - 30) = 18.66 \times 1963 \times (500 - x)$$

$$\therefore x^2 + 539.18x - 153902 = 0$$

Solving for x, we get

$$x = 206.41 \text{ or } 206 \text{ mm}$$

$$\therefore \qquad x > x_c$$

Hence, the beam is over-reinforced and it will fail first by overstress in concrete.

$$\therefore \qquad \mathbf{\sigma_{cbc} = 5 \text{ N/mm}^2}$$

3. To find σ'_{cbc} :

$$\sigma'_{cbc} = \sigma_{cbc} \left[\frac{x - d'}{x} \right]$$

$$= 5 \left[\frac{206 - 30}{206} \right]$$

$$= \mathbf{4.27 \text{ N/mm}^2}$$

4. To find actual σ_{sc} :

$$\sigma_{sc} = 1.5 \, m\sigma_{abc}$$
$$= 1.5 \times 18.66 \times 4.27$$
$$= 119.5 \text{ N/mm}^2 < 130 \text{ N/mm}^2 \text{ O.K.}$$

5. To find M_r :

Taking moments about tensile steel, we get

$$M_r = bx \frac{\sigma_{cbc}}{2} \left(d - \frac{x}{3} \right) + (1.5 \, m - 1) A_{sc}$$
$$\sigma'_{cbc} (d - d')$$

$$= 250 \times 206 \times \frac{5}{2} \times \left(500 - \frac{206}{3} \right)$$

$$+ (1.5 \times 18.66 - 1) \ 1140 \times 4.27$$
$$\times (500 - 30)$$

$$= 117283670 \text{ N-mm}$$

$$\mathbf{M_r = 117.28 \text{ kN-m}}$$

Example 2.12 : *A beam is 250 × 450 mm effective. Tensile steel consists of four 22 mm ϕ bars and there is an equal amount of steel placed 30 mm from the top edge of the beam. Calculate the uniformly distributed load the beam can carry (including self-weight). $\sigma_{cbc} = 7$ N/mm^2, $\sigma_{st} = 140$ N/mm^2, m = 13.33, $l_{ef} = 5.7$ m, and $\sigma_{sc} = 130$ N/mm^2.*

Given :

$$b = 250 \text{ mm}$$
$$d = 450 \text{ mm}$$
$$d' = 30 \text{ mm}$$
$$\sigma_{cbc} = 7 \text{ N/mm}^2$$
$$\sigma_{st} = 140 \text{ N/mm}^2$$
$$\sigma_{sc} = 130 \text{ N/mm}^2$$
$$m = 13.33$$
$$l_{ef} = 5.7 \text{ m}$$
$$A_{st} = A_{sc} = 4 \times \frac{\pi}{4} \times 22^2 = 1520 \text{ mm}^2$$

Solution :

1. To find x_c :

$$\frac{\sigma_{cbc}}{\sigma_{st}/m} = \frac{x_c}{d - x_c}$$

$$\frac{7}{140/13.33} = \frac{x_c}{450 - x_c}$$

$$\therefore \qquad \mathbf{x_c = 179.97 \text{ or } 180 \text{ mm}}$$

2. To find x :

$$bx\frac{x}{2} + (1.5 \, m - 1) A_{sc} (x - d') = mA_{st} (d - x) \quad \frac{250x^2}{2} +$$

$$(1.5 \times 13.33 - 1) \ 520 \ (x - 30) = 13.33 \times 1520 \times (450 - x)$$

$$x^2 + 393x - 79.871 = 0$$

Solving for x, we get

$$x = 147.71 \text{ or } 148 \text{ mm}$$

$$\therefore \qquad x < x_c$$

Hence, the beam is under-reinforced. It will fail because of overstress in steel and σ_{st} = 140 N/mm^2 (limit).

3. To find σ_{cbc} :

$$\frac{\sigma_{cbc}}{\sigma_{st}/m} = \frac{x}{d-x}$$

$$\frac{\sigma_{cbc}}{140/13.33} = \frac{148}{450-148}$$

$$\therefore \quad \sigma_{cbc} = \textbf{5.146 or 5.15 N/mm}^2$$

4. To find σ'_{cbc} :

$$\sigma'_{cbc} = \sigma_{cbc}\left(\frac{x-d'}{x}\right)$$

$$= 5.15\left(\frac{148-30}{148}\right)$$

$$= \textbf{4.10 N/mm}^2$$

5. To find actual compressive stress in steel :

$$\sigma_{sc} = 1.5\, m\sigma'_{cbc}$$

$$= 1.5 \times 13.33 \times 4.10$$

$$= \textbf{81.979 or 81.98 N/mm}^2$$

This stress of 81.98 N/mm^2 < the permissible stress of 130 N/mm^2 as it should be.

6. To find M_r : Taking moments about tensile steel.

$$M_r = bx\frac{\sigma_{cbc}}{2}\left(d-\frac{x}{3}\right) + (1.5\,m - 1)$$

$$A_{sc}\,\sigma'_{cbc}\,(d-d')$$

$$= 250 \times 148 \times \frac{5.15}{2}\left(450 - \frac{148}{3}\right)$$

$$+ (1.5 \times 13.33 - 1) \times 1520 \times 4.10$$

$$\times (450 - 30)$$

$$= 87892102 \text{ N-mm}$$

$$\therefore \quad \textbf{M}_r = \textbf{87.892 kN-m}$$

7. Let w be the uniformly distributed load.
Bending moment,

$$M = \frac{wl^2}{8} = \frac{w \times (5.7)^2}{8} = 4.061\,w \text{ kN-m}$$

Equating M_r to M, we get

$$87.892 = 4.061\,w$$

$$\therefore \quad \textbf{w} = \textbf{21.64 kN/m}$$

Example 2.13 : A rectangular beam is 200 mm wide and 480 mm deep overall. It has to resist a bending moment of 100 kN-m. The reinforcement consists of four 25 mm ϕ bars on tension side and three 22 mm ϕ bars on the compression side. The centres of bars being 30 mm from the top and bottom edges of the beam. Find the stresses set up in steel and concrete, m = 18.66.

Given :

$$b = 200 \text{ mm}$$
$$d = 480 - 30 = 450 \text{ mm}$$
$$d' = 30 \text{ mm}$$
$$M = 100 \text{ kN-m}$$
$$m = 18.66$$
$$A_{st} = 4 \times \frac{\pi}{4} \times 25^2 = 1964 \text{ mm}^2$$
$$A_{sc} = 3 \times \frac{\pi}{4} \times 22^2 = 1140 \text{ mm}^2$$

Solution :

1. To find x :

$$bx\frac{x}{2} + (1.5\,m - 1)\,A_{sc}\,(x - d') = mA_{st}\,(d - x)\quad \frac{200\,x^2}{2}$$

$$+ (1.5 \times 18.66 - 1)\,1140\,(x - 30)$$

$$= 18.66 \times 1964 \times (450 - x)$$

$$\therefore \quad x^2 + 674.17x - 174147 = 0$$

Solving for x, we get

$$x = \textbf{199.36, or 199 mm}$$

2. Let σ_{cbc} be the compressive stress in concrete at the top of the beam.

$$\text{Then} \quad \sigma'_{cbc} = \sigma_{cbc}\left(\frac{x-d'}{x}\right) = \sigma_{cbc}\left(\frac{199-30}{199}\right)$$

$$= \textbf{0.849}\,\sigma_{cbc}\ \textbf{or 0.85}\,\sigma_{cbc}\ \textbf{N/mm}^2$$

3. To find M_r :

Taking moments about tensile steel,

$$M_r = bx\frac{\sigma_{cbc}}{2}\left(d-\frac{x}{3}\right) + (1.5\,m - 1)\,A_{sc}\,\sigma'_{cbc}\,(d-d')$$

$$= 200 \times 199 \times \frac{\sigma_{cbc}}{2}\left(450 - \frac{199}{3}\right)$$

$$+ (1.5 \times 18.66 - 1)\,1140 \times 0.85\,\sigma_{cbc}$$

$$\times (450 - 30)$$

$$= 18618356\,\sigma_{cbc} \text{ N-mm}$$

$$\therefore \quad \textbf{M}_r = \textbf{18618}\,\sigma_{cbc}\ \textbf{kN-m}$$

4. Equating M_r to M

$$18.618\,\sigma_{cbc} = 100$$

$$\therefore \quad \sigma_{cbc} = \textbf{5.37 N/mm}^2$$

5. To find σ_{st} :

$$\frac{\sigma_{cbc}}{\sigma_{st}/m} = \frac{x}{d-x}$$

$$\frac{5.37}{\sigma_{st}/18.66} = \frac{199}{450-199}$$

$$\therefore \quad \sigma_{st} = \textbf{126.388 or 126.39 N/mm}^2$$

6. To find σ_{sc} :

$$\sigma_{st} = 1.5\,m\,\sigma'_{cbc}$$

$$= 1.5 \times 18.66 \times (0.85 \times 5.37)$$

$$\sigma_{st} = \textbf{127.76 N/mm}^2$$

Example 2.14 : *A cross-section of an RC beam is doubly reinforced with four 20 mm ϕ bars on tension and compression side. The section is 300×500 mm effective. What will be the actual stresses in concrete and steel, if the beam is subjected to a bending moment of 120 kN-m ? m = 13.33 and d' = 25 mm.*

Given :

$$b = 300 \text{ mm}$$
$$d = 500 \text{ mm}$$
$$d' = 25 \text{ mm}$$
$$m = 13.33$$
$$M = 120 \text{ kN-m}$$
$$A_{st} = A_{sc} = 4 \times \frac{\pi}{4} \times 20^2 = 1256 \text{ mm}^2$$

Solution :

1. To find x :

$$bx \frac{x}{2} + (1.5\ m - 1)\ A_{sc}\ (x - d') = mA_{st}\ (d - x)\ \frac{300x^2}{2}$$

$$+ (1.5 \times 13.33 - 1)\ 1256\ (x - 25)$$
$$= 13.33 \times 12.56 \times (500 - x)$$
$$\therefore\ x^2 + 270.6x - 59.784 = 0$$

Solving, we get **x = 144.146 or 144 mm**

2. Let σ_{cbc} be the stress in concrete at the top of the beam.

$$\text{Then}\quad \sigma'_{cbc} = \sigma_{cbc}\left(\frac{x - d'}{x}\right)$$

$$= \sigma_{cbc}\left(\frac{144 - 25}{144}\right)$$

$$= 0.826\ \sigma_{cbc}\ \text{N/mm}^2$$

3. To find M_r : Taking moments about tensile steel,

$$M_r = bx \frac{\sigma_{cbc}}{2}\left(d - \frac{x}{3}\right) + (1.5\ m - 1)$$

$$A_{sc}\ \sigma'_{cbc}\ (d - d')$$

$$= 300 \times 144 \times \frac{\sigma_{cbc}}{2}\left(500 - \frac{144}{3}\right)$$

$$+ (1.5 \times 13.33 - 1)\ 1256 \times (0.826\ \sigma_{cbc})$$
$$(500 - 25)$$

$$= 19123776\ \sigma_{cbc}\ \text{N-mm}$$

$$= 19.124\ \sigma_{cbc}\ \text{kN-m}$$

4. Equating M_r to M,

$$19.124\ \sigma_{cbc} = 120$$

$$\therefore\ \sigma_{cbc} = 6.27\ \text{N/mm}^2$$

5. To find σ_{st}

$$\frac{\sigma_{cbc}}{\sigma_{st}/m} = \frac{x}{d - x}$$

$$\frac{6.27}{\sigma_{st}/13.33} = \frac{144}{500 - 144}$$

$$\sigma_{st} = 206.62\ \text{N/mm}^2$$

6. To find σ_{sc}

$$\sigma_{sc} = 1.5\ m\sigma'_{cbc}$$

$$= 1.5 \times 13.33 \times (0.826 \times 6.27)$$

$$\sigma_{sc} = 103.55\ \text{N/mm}^2$$

Example 2.15 : *A doubly reinforced concrete beam 250 mm wide and 600 min deep overall has to resist an external bending moment of 95 kN-m. Find the amount of tensile and compressive steel required, if cover to the centre of steel on both sides is 50 mm. $\sigma_{cbc} = 5$ N/mm^2 $\sigma_{st} = 140$ N/mm^2 and m = 18.66.*

Given :

$$b = 250 \text{ mm}$$
$$d = 600 - 50 = 550 \text{ mm}$$
$$d' = 50 \text{ mm}$$
$$\sigma_{cbc} = 5 \text{ N/mm}^2$$
$$\sigma_{st} = 140 \text{ N/mm}^2$$
$$m = 18.66$$
$$M = 95 \text{ kN-m}$$

1. To find x_c :

$$\frac{\sigma_{cbc}}{\sigma_{st}'m} = \frac{x_c}{d - x_c}$$

$$\frac{5}{140/18.66} = \frac{x_c}{550 - x_c}$$

$$\therefore\quad x_c = 219.95,\ \text{or } 220 \text{ mm}$$

2. To find A_{st1} :

$$C = T$$

$$bx_c \frac{\sigma_{cbc}}{2} = \sigma_{st}\ A_{st_1}$$

$$250 \times 220 \times \frac{5}{2} = 140 \times A_{st\,1}$$

$$\therefore\quad A_{st\,1} = 982 \text{ mm}^2$$

3. M_r of singly reinforced balance beam

$$M_r = bx_c \frac{\sigma_{cbc}}{2}\left(d - \frac{x_c}{3}\right)$$

$$= 250 \times 200 \times \frac{5}{2} \times \left(550 - \frac{220}{3}\right)$$

$$= 65541667 \text{ N-mm}$$

$$\therefore\quad M_r = 65.54 \text{ kN-m}$$

4. Remaining bending moment M_1 (to be provided)

$$M_1 = M - M_r$$

$$= 95 - 65.54 = 29.46 \text{ kN-m}$$

$$= 29.46 \times 10^6 \text{ N-mm}$$

5. To find $A_{st\,2}$:

$$A_{st\,2} = \frac{M_1}{\sigma_{st}\,(d - d')}$$

$$= \frac{29.46 \times 10^6}{140\,(550 - 50)}$$

$$= 420.85 \text{ or } 421 \text{ mm}^2$$

$\therefore$ $A_{st} = A_{st_1} + A_{st_2}$

$$= 982 + 421 = 1403 \text{ mm}^2$$

6. To find A_{sc} : Equating moments of equivalent area of tensile and compressive steel about N.A.

$$mA_{st\,2}\,(d - x_c) = (1.5\,m - 1)\,A_{sc}\,(x_c - d')$$

$$18.66 \times 421 \times (550 - 220) = (1.5 \times 18.66 - 1)\,A_{sc}\,(220 - 50)$$

$\therefore$ $\mathbf{A_{sc} = 565 \text{ mm}^2}$

Example 2.16 : *A rectangular reinforced beam is 360 × 750 mm effective. The beam has to resist a bending moment of 300 kN-m. Find the tensile and compressive steel required for the beam. Take σ_{cbc} = 7 N/mm², σ_{st} = 190 N/mm², m = 13.33 and d' = 50 mm.*

Given : b = 360 mm

         d = 750 mm

         d' = 50 mm

         σ_{cbc} = 7 N/mm²

         σ_{st} = 190 N/mm²

         m = 13.13

         M = 300 kN-m

Solution :

1. To find x_c :

$$\frac{\sigma_{cbc}}{\sigma_{st}/m} = \frac{x_c}{d - x_c}$$

$$\frac{7}{190/13.33} = \frac{x_c}{750 - x_c}$$

$\therefore$ $\mathbf{x_c = 247 \text{ mm}}$

2. M_r : For singly reinforced balanced beam

$$M_r = bx_c\,\frac{\sigma_{cbc}}{2}\left(d - \frac{x_c}{2}\right)$$

$$= 360 \times 247 \times \frac{7}{2} \times \left(750 - \frac{247}{3}\right)$$

$$= 207779122 \text{ N-mm}$$

$$\mathbf{M_r = 207.779 \text{ kN-m or } 207.8 \text{ kN-m}}$$

3. To find $A_{st\,1}$:

$$A_{st\,1} = \frac{M_r}{\sigma_{st}\,z}$$

$$= \frac{207.779 \times 10^6}{190 \times \left(750 - \dfrac{247}{3}\right)}$$

$$= \mathbf{1638 \text{ mm}^2}$$

4. Remaining bending moment M_1

$$M_1 = M - M_r$$

$$300 - 207.8 = \mathbf{92.2 \text{ kN-m}}$$

5. To find $A_{st\,2}$:

$$A_{st\,2} = \frac{M_1}{\sigma_{st}\,(d - d')} = \frac{92.2 \times 10^6}{190\,(750 - 50)}$$

$$= \mathbf{693.23 \text{ mm}^2}$$

6. To find A_{sc} : Equating moments of equivalent area of tensile and compressive steel about N.A.

$$mA_{st\,2}\,(d - x_c) = (1.5\,m - 1)\,A_{sc}\,(x_c - d')$$

$$13.33 \times 693.23 \times (750 - 247) = (1.5 \times 13.33 - 1)\,A_{sc}\,(247 - 50)$$

$\therefore$ $A_{sc} = 1242.14 \text{ mm}^2$

         $A_{st} = A_{st\,1} + A_{st\,2}$

$\therefore$ $A_{st} = 1638 + 693.23 = \mathbf{2331.23 \text{ mm}^2}$

Example 2.17 : *A rectangular beam section, 230 mm wide and effective depth 415 mm is reinforced with 3 bars of 20 mm diameter in the tensile zone and 2 bars of 16 mm in the compression zone. Determine moment of resistance of the section using WSM. Use M20 grade of concrete and Fe 415 grade of steel.*

Solution : Given

         b = 230 mm

         d = 415 mm

$$A_{st} = 3 \times \frac{\pi}{4} \times 20^2 = 942 \text{ mm}^2$$

$$A_{sc} = 2 \times \frac{\pi}{4} \times 16^2 = 401.92 \text{ mm}^2$$

         M = ? (WSM)

M20 and Fe 415

         σ_{cbc} = 7 N/mm² (Table 21, pp. 81, IS 456)

         σ_{st} = 230 N/mm² (Table 22, pp. 82, IS 456)

1. To find xu

$$\frac{\sigma_{cbc}}{\sigma_{st}/m} = \frac{x_c}{d - x_c} \quad \cdots\ m = \frac{280}{3\sigma_{cbc}} = \frac{280}{3 \times 7} = 13.33$$

$$\frac{7}{230/13.33} = \frac{x_c}{415 - x_c}$$

$\therefore$ $\mathbf{x_c = 119.79 \text{ mm or } 120 \text{ mm}}$

2. To find x

$$b.x. \frac{x}{2} + [1.5m - 1] A_{sc} (x - d') = m \cdot A_{st} (d - x)$$

Assume,   d' = Distance of compressive steel from the top edge of the beam to the centre of the steel = 30 mm

$$\frac{230 \times x^2}{2} + (1.5 \times 13.33 - 1)\, 401.92\, (x - 30) = 13.33 \times 942\, (415 - x)$$

$$115\, x^2 + 7634.47\, (x - 30) = 12556.86\, (415 - x)$$

$$115\, x^2 + 7634.47x - 229034.1 = 52110969 - 12556.86x$$

$$115x^2 + 20191.33x - 4982062.8 = 0$$

$$x^2 + 175.576x - 43322.28 = 0$$

$$x = \frac{-175.576 \pm \sqrt{175.576^2 + 173289.12}}{2}$$

$$= \frac{-175.576 \pm 451.79}{2}$$

$$\therefore \qquad x = 138.108 > x_c$$

$\therefore$ Hence the beam is over reinforced and it will fail first by overstress in concrete.

3. To find $\sigma_{cbc'}$

$$\sigma_{cbc'} - \sigma_{cbc} \left(\frac{x - d'}{x} \right) = 7 \left(\frac{138.108 - 30}{138.108} \right) = 5.47\ \text{N/mm}^2$$

4. To find actual σ_{sc}

$$\sigma_{sc} = 1.5\, m\sigma_{cbc}$$

$$= 1.5 \times 13.33 \times 5.47$$

$$\mathbf{= 109.37\ \text{N/mm}^2 < 230\ \text{N/mm}^2} \ \therefore \ \text{OK}$$

5. To find M_r

Taking moments about tensile steel, we get

$$M_r = b \cdot x \cdot \frac{\sigma_{cbc}}{2} \left(d - \frac{x}{3} \right)$$

$$+ (1.5m - 1)\, A_{sc}\, \sigma_{cbc'}\, (d - d')$$

$$= 230 \times 138.108 \times \frac{7}{2} \left(415 - \frac{138.108}{3} \right)$$

$$+ (1.5 \times 13.33 - 1) \times 401.92 \times 5.47$$

$$(415 - 30)$$

$$= 41024290.86 + 16077812.94$$

$$\mathbf{M_r = 57102103.8\ \text{N.mm}} = 57.102\ \text{kN.m}$$

EXERCISE

1. Define SRRS and DRRS, state necessity of DRRS.

2. Explain concept of under-reinforced, balanced and over-reinforced section.

3. Why over-reinforced section not allowed in RCC design.

4. Draw stress strain diagram for

   (a) Singly reinforced section

   (b) Doubly reinforced section

5. A rectangular beam of breadth b mm and effective depth d mm is reinforced only on the tension side. Express its moment of resistance in terms of b and d. Assume that σ_{cbc} = 8.5 N/mm², σ_{st} = 190 N/mm² and m = 10.98.

6. An RC beam is 150 mm wide and 300 mm deep (effective)

   Calculate

   (a) The moment of resistance

   (b) Area of tensile steel and

   (c) Uniformly distributed load (including self weight)

   Take σ_{cbc} = 5 N/mm²

              σ_{st} = 140 N/mm²

              m = 18.66

              $L_{eff.}$ = 4 m

7. A rectangular beam is 360 × 750 mm effective section. The beam has to resist a bending moment of 215 kN.m calculate the tensile and compressive steel required. Take σ_{cbc} = 5 N/mm², σ_{st} = 140 N/mm², m = 18.66 and d' = 50 mm.

8. A doubly reinforced concrete beam, 300 × 680 mm effective is reinforced on tension and compression side with four 25 mm ϕ bars. Compression steel is placed 25 mm from the top of the beam. If the beam carries a bending moment of 215 kN.m. Find the stresses induced in steel and concrete , m = 13.33.

◈ ◈ ◈

3.1 INTRODUCTION

- A flat piece of concrete, put on the walls or columns of a structure. It serve as a walking surface but may also serve as a load bearing member is called as slab. It is subjected to transverse loads which are transferred to the walls or beams.

- Slab is a plate element having depth much smaller than other two dimensions so slab is a two dimensional structural element. Slabs are designed same as beam with unit width.

3.2 CLASSIFICATION OF SLABS

- Slabs are classified according to shape, support condition and spanning direction.

(1) According to shape

(2) According to support condition

(3) According to spanning direction

(a) Square
(b) Rectangular
(c) Circular

(a) Simply supported
(b) Continuous
(c) Cantilever
(d) Flat slab directly supported by column without beam

(a) One way slab
(b) Two way slab

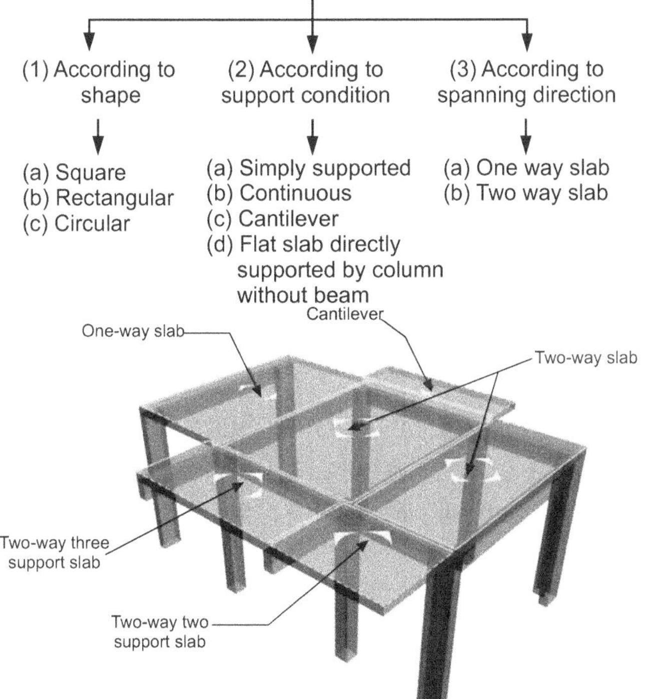

Fig. 3.1

3.3 STRUCTURAL ANALYSIS

- One way slabs subjected to mostly uniformly distributed vertical loads carry them primarily by bending in the shorter direction. Therefore, for the design, it is important to analyse the slab to find out the bending moment (both positive and negative) depending upon the supports. Moreover, the shear forces are also to be computed for such slabs. These internal bending moments and shear forces can be determined using elastic method of analysis considering the slab as beam of unit width i.e. one metre.

- However, these values may also be determined with the help of the coefficients given in Table 12 and 13 (clause 22.5.1) of IS 456 : 2000 [1]. It is worth mentioning that these coefficients are applicable if the slab is of uniform cross-section and subjected to substantially uniformly distributed loads over three or more spans and the spans do not differ by more than fifteen per cent of the longest span. It is also important to note that the average of the two values of the negative moment at the support should be considered for unequal spans or if the spans are not equally loaded. Further, the redistribution of moments shall not be permitted to the values of moments obtained by employing the coefficients of bending moments as given in IS 456 : 2000 [1].

- For slabs built into a masonry wall developing only partial resistant, the negative moment at the face of the support should be taken as $Wl/24$. Where, W is the total design loads on unit width and l is the effective span. The shear coefficient, given in Table 13 of IS 456 : 2000 [1] in such a situation may be increased by 0.05 at the end support as per clause 22.5.2 of IS 456 : 2000 [1].

3.4 DIFFERENCE BETWEEN ONE WAY SLAB AND TWO WAY SLAB

Sr. No	One Way Slab	Two Way Slab
1.	The ratio of larger span to shorter span is greater than 2, the slab is called as one way slab. $\dfrac{l_y}{l_x} > 2.0$	The ratio of larger span to shorter span is less than or equal to 2, is called as two way slab. $\dfrac{l_y}{l_x} \le 2.0$
2.	The bending takes place in one direction only i.e. shorter span.	The bending takes place in both the directions.

conti...

3.	Depth required is more.	Depth required is less.
4.	Main steel reinforcement is provided along shorter span.	Main steel reinforcement is provided along both the spans.
5.	Less economical as thickness is more and the amount of steel is also more.	More economical as thickness of slab is less and the amount of steel is also less.
6.	(a) (b) (c) **Fig. 3.2**	(a) (b) (c) **Fig. 3.3**
7.	Symbol L_y, L_x where L_x = shorter span L_y = longer span	Symbol L_y, L_x

3.5 IS SPECIFICATION REGARDING SLAB

1. Effective Span : (Refer clause 22.2, IS 456 : 2000)

The effective span of a slab depends on the boundary condition given in clause 22.2, IS 456 : 2000 to determine effective span of slab.

Table 3.1 : Effective span of slab

(Clause 22.2 of IS 456 : 2000)

Sr. No.	Support Condition	Effective Span
1.	Simply supported not built integrally with its supports.	Lesser of (i) clear span + effective depth of slab and (ii) centre to centre of supports.
2.	Continuous when the width of the support is < 1/12th of clear span.	Do.
3.	Continuous when the width of the support is > lesser of 1/12th of clear span or 600 mm. (i) For end span with one end fixed and the other end continuous or for intermediate spans. (ii) For end span with one end free and the other end continuous. (iii) Span with roller or rocker bearings.	(i) Clean span between the supports. (ii) Lesser of (a) clear span + half the effective depth of slab, and (b) clear span + half the width of the discontinuous support. (iii) The distance between the centres of bearings.
4.	Cantilever slab at the end of a continuous slab.	Length upto the centre of support.
5.	Cantilever span	Length upto the face of the support + half the effective depth.
6.	Frames	Centre to centre distance.

2. Nominal Cover to the Reinforcement :

- The minimum cover to main bars shall not be less than the following :

  (i)   15 mm ⎫

  (ii) Diameter of main bar ⎬ whichever is more

3. Maximum Diameter of Reinforcing Bar : (Clause 26.5.2.2)

The diameter of reinforcing bar shall not exceed one eight of the total thickness of the slab.

$\therefore$ Diameter of bar $< \dfrac{1}{8} \times$ Total thickness of slab

4. Reinforcement :

(i) Minimum Reinforcement : The reinforcement for a slab in each direction consist of main bars and distribution steel bars should not be less than as given below :

Types of Reinforcement	Percentage of Steel
Mild steel	0.15 % of Ag
HYSD steel	0.12 % of Ag

(ii) Maximum Spacing between Bars :

(a) The Spacing / Pitch of the Main Bars Shall not Exceed the Following :

- Three time effect depth of slab (i.e. 3d) ⎱ less
- 300 mm ⎰

(b) The Distribution / Temperature Reinforcement

These are reinforcement provided running at right angles to the main steel in order to distribute the load and the temperature and shrinkage stresses. The pitch of the distribution bars shall not exceed the following :

- Five times effective depth of slab (i.e. 5d) ⎱ less
- 450 mm ⎰

5.   Thickness of Slab

The following table gives the maximum values of the ratio of span to depth.

Sr. No.	Types of Slab	Ratio of Span to Depth
1.	Simply supported one way slab	30
2.	Continuous one way slab	35
3.	Simply supported two way slab	35
4.	Continuous two way slab	40
5.	Cantilever slab	12

Note : For normal conditions of loading the thickness of a slab may taken as the 40 mm per metre run of the span. A slab shall not ordinarily be less than 100 mm in thickness.

3.6 TWO WAY SLAB

When slabs are supported on all four sides and $\frac{l_y}{l_x} \leq 2$, the slabs are known as two way slabs. In this type, slabs are spanning in both directions. According to corner conditions, slabs are classified into two types :

1. Simply supported slabs.
2. Restrained slabs.

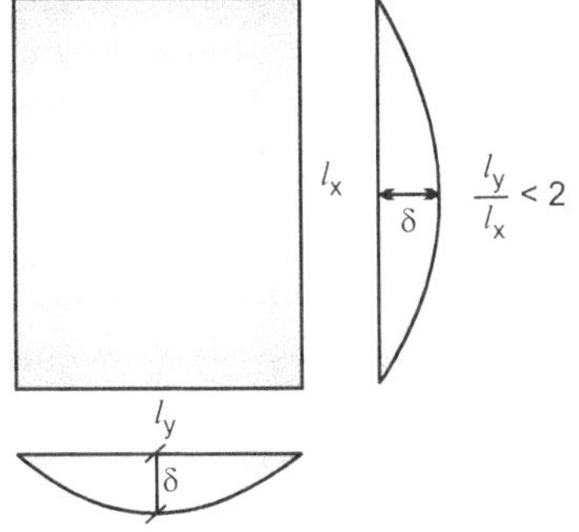

Fig. 3.4

3.7 SIMPLY SUPPORTED SLABS

When simply supported slabs do not have adequate provision to resist torsion at corners and to prevent the corners from lifting, the maximum moments per unit width are given by the following equation :

$$M_x = \alpha_x\, w\, l_x^2$$
$$M_y = \alpha_y\, w\, l_x^2$$

where, M_x, M_y, w, l_x, l_y are same as those in D–1.1 α_x and α_y are moment coefficients given in Table 6.1.

Atleast 50 percent of the tension reinforcement provided at mid–span should extend to the supports. The remaining 50% should extend to within 0.1 l_x or 0.1 l_y of the support, as appropriate.

Table 3.2 : Bending Moment Coefficients For Slabs Spanning in Two Directions at Right Angles, Simply Supported on Four Sides (Clause D–2.1)

l_y/l_x	1.0	1.1	1.2	1.3	1.4	1.5	1.75	2.0	2.5	3.0
α_x	0.062	0.074	0.084	0.093	0.099	0.104	0.113	0.118	0.122	0.124
α_y	0.062	0.061	0.059	0.055	0.051	0.046	0.037	0.029	0.020	0.014

3.8 RESTRAINED SLABS

When the corners of a slab are prevented from lifting, the slab may be designed as follows.

The maximum bending moments per unit width in a slab are given by the following equations :

$$M_x = \alpha_x\, w\, l_x^2$$
$$M_y = \alpha_y\, w\, l_x^2$$

where, α_x and α_y = Moment coefficients given in Table 6.2.

w = Total design load per unit area

M_x, M_y = Moments on strips of unit width spanning l_x and l_y respectively and

l_x and l_y = Lengths of the shorter span and longer span respectively.

- Slabs are considered as divided in each direction into middle strips and edge strips as shown in Fig. 3.5, the middle strip being three–quarters of the width and each edge strip one-eighth of the width.
- The maximum moments calculated as in D–1.1 apply only to the middle strips and no redistribution shall be made.
- Tension reinforcement provided at mid–span in the middle strip shall extend in the lower part of the slab to within 0.25 l of a continuous edge, or 0.15 l of a discontinuous edge.
- Over the continuous edges of a middle strip, the tension reinforcement shall extend in the upper part of the slab at distance of 0.15 l from the support, and at least 50% shall extend a distance of 0.3 l.

- At a discontinuous edge, negative moments may arise. They depend on the degree of fixity at the edge of the slab but, in general, tension reinforcement equal to 50% of that provided at mid–span extending 0.1 l into the span will be sufficient.
- Reinforcement in edge strip, parallel to that edge, shall comply with the minimum given in Section 3 and the requirements for torsion given are as follows.

3.9 REQUIREMENT OF TORSION REINFORCEMENT

- Torsion reinforcement shall be provided at any corner where the slab is simply supported on both edges meeting at that corner. It shall consist of top and bottom reinforcement, each with layers of bars placed parallel to the sides of the slab and extending from the edges a minimum distance of one–fifth of the shorter span. The area of reinforcement in each of these four layers shall be three–quarters of the area required for the maximum mid–span moment in the slab.
- Torsion reinforcement equal to half that described in D–1.8 shall be provided at a corner contained by edges over only one of which the slab is continuous.
- Torsion reinforcements need not be provided at any corner contained by edges over both of which the slab is continuous.

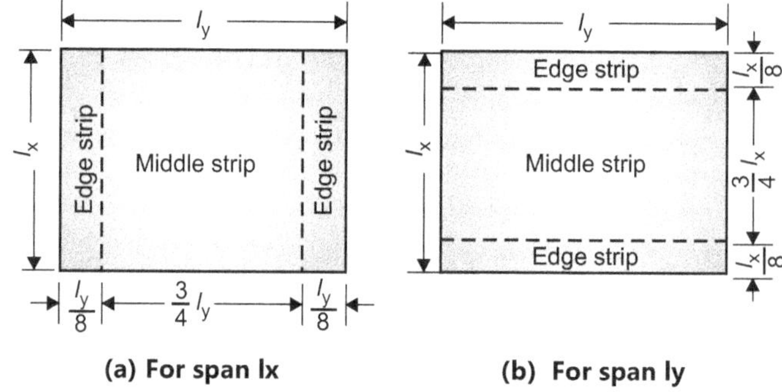

Fig. 3.5

Table 3.3 : Bending Moment Coefficients for Rectangular Panels Supported on Four Sides with Provision for Torsion at Corners

Case No.	Type of Panel and Moments Considered	Short Span Coefficients α_x (Values of l_y/l_x)								Long Span Coefficients α for all Values of l_y/l_x
		1.0	1.1	1.2	1.3	1.4	1.5	1.75	2.0	
(1)	(2)	(3)	(4)	(5)	(6)	(7)	(8)	(9)	(10)	(11)
1.	Interior Panels :									
	Negative moment at continuous edge	0.032	0.037	0.043	0.047	0.051	0.053	0.060	0.065	0.032
	Positive moment at mid-span	0.024	0.028	0.032	0.036	0.039	0.041	0.045	0.049	0.024
2.	One Short Edge Continuous :									
	Negative moment at continuous edge	0.037	0.043	0.048	0.051	0.055	0.057	0.064	0.068	0.037
	Positive moment at mid-span	0.028	0.032	0.036	0.039	0.041	0.044	0.048	0.052	0.028
3.	One Long Edge Discontinuous :									
	Negative moment at continuous edge	0.037	0.044	0.052	0.057	0.063	0.067	0.077	0.085	0.037
	Positive moment at mid-span	0.028	0.033	0.039	0.044	0.047	0.051	0.059	0.065	0.028
4.	Two Adjacent Edges Discontinuous :									
	Negative moment at continuous edge	0.047	0.053	0.060	0.065	0.071	0.075	0.084	0.091	0.047
	Positive moment at mid-span	0.035	0.040	0.045	0.049	0.053	0.056	0.063	0.069	0.035
5.	Two Short Edges Discontinuous :									
	Negative moment at continuous edge	0.045	0.049	0.052	0.056	0.059	0.060	0.065	0.069	–
	Positive moment at mid-span	0.035	0.037	0.040	0.043	0.044	0.045	0.049	0.052	0.035
6.	Two Long Edges Discontinuous :									
	Negative moment at continuous edge	–	–	–	–	–	–	–	–	0.045
	Positive moment at mid-span	0.035	0.043	0.051	0.057	0.063	0.068	0.080	0.088	0.035
7.	Three Edges Discontinuous (One Long Edge Continuous) :									
	Negative moment at continuous edge	0.057	0.064	0.071	0.076	0.080	0.084	0.091	0.097	–
	Positive moment at mid-span	0.043	0.048	0.053	0.057	0.060	0.064	0.069	0.073	0.043
8.	Three Edges Discontinuous (One Short Edge Continuous) :									
	Negative moment at continuous edge	–	–	–	–	–	–	–	–	0.057
	Positive moment at mid-span	0.043	0.051	0.059	0.065	0.071	0.076	0.087	0.096	0.043
9.	Four Edges Discontinuous :									
	Positive moment at mid-span	0.056	0.064	0.072	0.079	0.085	0.089	0.100	0.107	0.056

3.10 ARRANGEMENT OF TORSION REINFORCEMENT

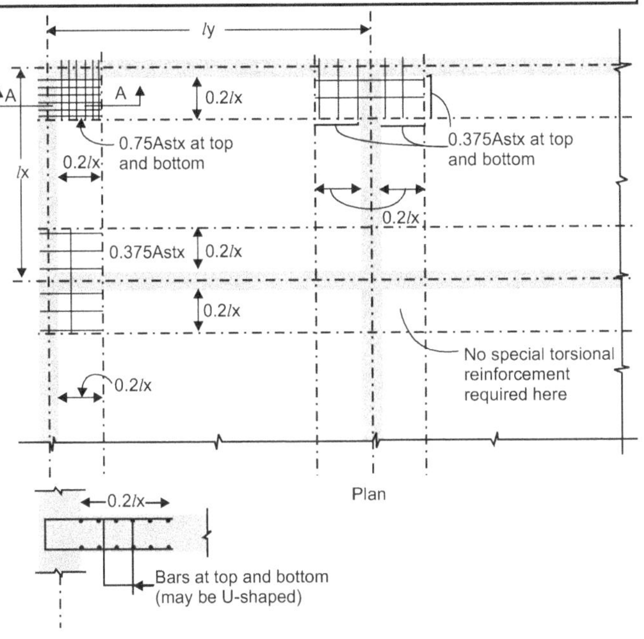

Chart

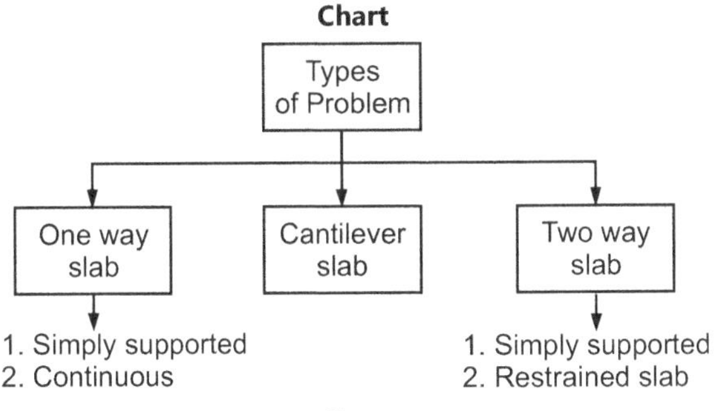

Chart

SOLVED EXAMPLES

Type I : Design of One Way Simply Supported Slab

Example 3.1 : *Design a R.C. slab for a room having size 3m × 7m. The thickness of supporting wall is 300 mm. The slab carries 100 mm thick lime concrete at its top, the unit weight of which may be taken as 19 kN/m³. The live load on the slab may be taken as 2.50 kN/m². Assume the slab to the simply supported at the ends. Use M-20 concrete and Fe-415 steel.*

Solution : Given

Room size 3m × 7m

Support width = 300 mm

$L.L. = 2.50$ kN/m²

$F.F. = 100$ mm thick, 19 kN/m³

M-20; $\sigma_{cbc} = 7$ N/mm²

Fe-415 ; $\sigma_{st} = 230$ N/mm²

(1) Aspect ratio :

$$\frac{l_y}{l_x} = \frac{7}{3} = 2.33 > 2$$

…Design slab as a one way slab.

(2) Calculation of design constant :

(i) Modular ratio (m)

$$m = \frac{280}{3 \times \sigma_{cbc}} = \frac{280}{3 \times 7} = 13.33$$

(ii) N.A. depth factor (k)

$$k = \frac{m.\sigma_{cbc}}{m \cdot \sigma_{cbc} + \sigma_{st}} = 0.29$$

(iii) Lever arm factor (j)

$$j = \left(1 - \frac{k}{3}\right) = \left(1 - \frac{0.29}{3}\right) = 0.90$$

(iv) Moment resisting factor (Q)

$$Q = \frac{1}{2} \times \sigma_{cbc} \times j \times k = 0.91$$

(3) Calculation of loading (w) :

The slab will bend over a clear span of 3m. Let the overall depth of slab should be 40 mm per metre length.

$\therefore \quad D_{assume} = 125$ mm, c = 25 mm

$\therefore \quad d_{assume} = 100$ mm

Consider '1m' width of slab.

(i) Live load = $2.50 \times 1 = 2.50$ kN/m

(ii) Floor finish = $(0.1 \times 1 \times 19) = 1.90$ kN/m

(iii) Self wt. of slab = $(0.125 \times 25) = 3.13$ kN/m

$\therefore \quad$ Total udl = w = 7.53 kN/m

(4) Effective slab (l_{eff})

(i) $l_{eff} = L + b = 3 + 0.3 = 3.30$ m ⎤ less

(ii) $l_{eff} = L + d = 3 + 0.1 = 3.10$ m ⎦

$\therefore \quad l_{eff} = 3.10$ m

(5) Calculation of bending moment and shear force :

(i) Max. B.M. = M = $\dfrac{w.l^2}{8} = \dfrac{7.53 \times 3.10^2}{8}$

$\therefore \quad M = 9.05$ kN.m

(ii) Max. S.F. = V = $\dfrac{wl}{2} = \dfrac{7.53 \times 3.10}{2}$

$\therefore \quad V = 11.67$ kN

(6) Check for depth / flexure / bending :

By equating max. B.M. and resisting moment

$\therefore \quad M = Qbd^2$

$9.05 \times 10^6 = 0.91 \times 1000 \times d^2$

$\therefore \quad d_{req} = 99.72$ mm $< d_{provided}$

… Okay safe, in bending

(7) Calculation of main steel (A_{st}) :

$$A_{st} = \frac{M}{\sigma_{st} \times j \times d} = \frac{9.05 \times 10^6}{230 \times 0.90 \times 100}$$

$$\therefore \quad A_{st} = 437.20 \text{ mm}^2$$

and $\quad A_{stmin} = 0.12\%$ of $A_g = \dfrac{0.12}{100} \times 1000 \times 125$

$$A_{stmin} = 150 \text{ mm}^2$$

$\therefore$ **Provide, A_{st} = 437.20 mm^2**

Assume, diameter of bar $= \phi = 8$ mm

- **Spacing (S) :**

(i) $\quad S = \dfrac{1000 \times \left(\dfrac{\pi}{4} \times \phi^2\right)}{A_{st}} = 114.97 \text{ mm}$

(ii) $\quad S = 3d = 3 \times 100 = 300$ mm

(iii) $\quad S = 300$ mm

$\left. \begin{array}{c} \\ \\ \\ \end{array} \right\}$ whichever is less

$$\therefore \quad S = 114.97 \approx 110 \text{ mm}$$

$\therefore$ **Provide, 8 mm ϕ @ 110 mm c/c as a main steel.**

(8) Check for shear :

(i) Nominal shear stress (τ_v) :

$$\tau_v = \frac{V}{bd} = \frac{11.67 \times 10^3}{1000 \times 100}$$

$$\boldsymbol{\tau_v = 0.12 \text{ N/mm}^2}$$

(ii) Permissible shear stress in concrete ($\tau_{c \cdot k}$) :

(refer Table - 23, Pg. No. 84 Is 456 : 2000)

$$P_{t\,max} = \frac{A_{st\,provided}}{bd} \times 100$$

$$= \frac{\left(\dfrac{1000 \times \dfrac{\pi}{4} \times 8^2}{110}\right)}{1000 \times 100} \times 100$$

$$P_{tmax} = 0.46\%$$

Ptmax	τ_c
0.25	0.22
0.46	
0.50	0.30

0.50	0.08
1.00	0.32

$$\therefore \quad \tau_c = 0.22 - (0.21 \times 0.32)$$

$$\tau_c = 0.15 \text{ N/mm}^2$$

$\therefore$ Hence,

$$\tau_{c \cdot k} = 0.15 \times 1.3$$

$$\therefore \quad \boldsymbol{\tau_{c \cdot k} = 0.20 \text{ N/mm}^2}$$

(iii) Compare τ_v and $\tau_{c \cdot k}$:

$$0.12 \;<\; 0.20$$

$$\tau_v \;<\; \tau_{c \cdot k} \qquad \text{... Okay, Safe in shear}$$

(9) Check for development length :

(Refer clause 26.2.1 and 26.2.3.3 (c), IS 456)

$$\frac{1.3\, M_1}{V} + L_o \geq L_d \qquad\qquad \text{... (a)}$$

(i) $\quad M_1 =$ moment at support

$$= \sigma_{st} \times A_{st} \times j \times d$$

$$M_1 = 230 \times \left(\frac{1000 \times \dfrac{\pi}{4} \times 8^2}{110 \times 2}\right) \times 0.90 \times 100$$

$$\therefore \quad M_1 = 4.73 \text{ kN.m}$$

(ii) $\quad L_o = \dfrac{\text{support width}}{2} - \text{cover} + 3\phi$

$$\text{... Providing } 90° \text{ bend}$$

$$\therefore \quad L_o = \frac{300}{2} - 25 + 3 \times 8$$

$$= 149 \text{ mm}$$

(iii) $\quad L_d = \dfrac{\phi \cdot \sigma_s}{4 \times \tau_{bd}}$

$$= \frac{8 \times 230}{4 \times 0.8 \times 1.6}$$

$$\therefore \quad \boldsymbol{L_d = 359.38 \text{ mm} \approx 360 \text{ mm}}$$

$\therefore$ from equation (a)

$$\frac{1.3 \times 4.73 \times 10^6}{11.67 \times 10^3} + 149 \geq 360$$

$$675.91 \;>\; 360 \qquad \text{... Okay}$$

Hence, anchorage requirement is satisfied.

(10) Calculation of distribution steel (A_{std}) :

$$\therefore \quad A_{std} = A_{stmin} = 150 \text{ mm}^2$$

Assume, diameter of bar $= \phi = 6$ mm

- **Spacing (S) :**

(i) $\quad S = \dfrac{1000 \times \left(\dfrac{\pi}{4} \times 6^2\right)}{A_{std}}$

$$= 188.50 \text{ mm}$$

(ii) $\quad S = 5 \cdot d = 5 \times 100 = 500$ mm

(iii) $\quad S = 450$ mm

$\left. \begin{array}{c} \\ \\ \\ \end{array} \right\}$ Whichever is less

$$\therefore \quad S = 188.50 \approx 180 \text{ mm}$$

∴ **Provide, 6 mm f 180 mm c/c as a distribution steel.**

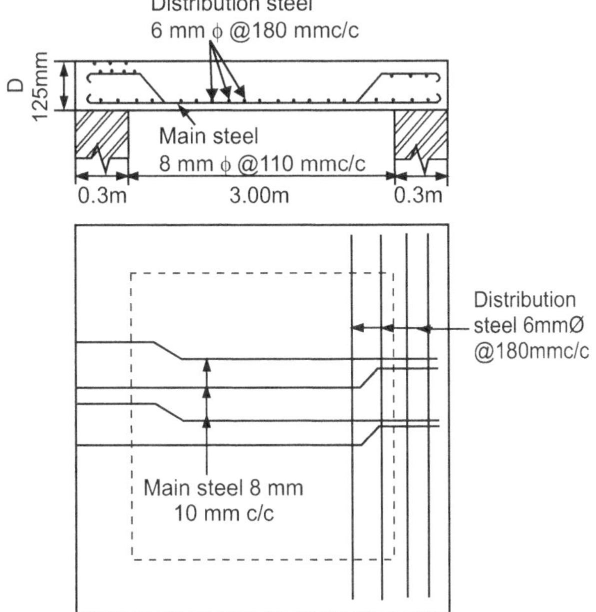

Fig. 3.6 : One way slab

Example 3.2 : *Design a simply supported slab for a passage 2.75 m wide carries a live load = 3 kN/m². The slab is supported on 230 mm thick masonry wall. Take, M-20, Fe-250.*

Solution : Given :

$$\text{Passage} = 2.75 \text{ m wide (one way slab)}$$

$$\text{support width} = 230 \text{ mm}$$

$$\text{Live load} = 3 \text{ kN/m}^2$$

M-20 ; $\sigma_{cbc} = 7 \text{ N/mm}^2$

Fe-250 ; $\sigma_{st} = 130 \text{ N/mm}^2$

(1) Calculation of design constant :

(i) Modular ratio (m) :

$$m = \frac{280}{3 \times \sigma_{cbc}} = \frac{280}{3 \times 7} = 13.33$$

(ii) N.A. depth factor (K) :

$$K = \frac{m \cdot \sigma_{cbc}}{m \cdot \sigma_{cbc} + \sigma_{st}} = 0.42$$

(iii) lever arm factor (j) :

$$j = \left(1 - \frac{K}{3}\right) = \left(1 - \frac{0.42}{3}\right) = 0.86$$

(iv) Moment resisting factor (Q) :

$$Q = \frac{1}{2} \times \sigma_{cbc} \times j \times k = 1.26$$

(2) Calculation of loading (w) :

The slab will bend over a clear span of 2.75 m. Let the overall depth of slab should be 40 mm per metre length.

∴ $D_{assume} = 110 \text{ mm}, c = 25 \text{ mm}$

∴ $d_{assume} = 85 \text{ mm}$

Consider '1m' width of slab,

(i) live load $= 3 \times 1 = 3 \text{ kN/m}$

(ii) Assume, floor finish $= 1 \times 1 = 1 \text{ kN/m}$

(iii) Self wt. of slab $= (0.11 \times 25) = 2.75 \text{ kN/m}$

....

∴ **Total udl = w = 6.75 kN/m**

(3) Effective span (l_{eff}) :

(i) $l_{eff} = L + b = 2.75 + 0.23 = 2.98 \text{ m}$

(ii) $l_{eff} = L + d = 2.75 + 0.085 = 2.84 \text{ m}$

∴ $l_{eff} = 2.84 \text{ m}$

(4) Calculation of bending moment and shear force :

(i) Max. B.M. $= M = \dfrac{w \cdot l^2}{8} = \dfrac{6.75 \times 2.84^2}{8}$

∴ **M = 6.81 kN.m**

(ii) Max. S. F. $= V = \dfrac{w \cdot l}{2} = \dfrac{6.75 \times 2.84}{2}$

∴ **V = 9.59 kN**

(5) Check for depth / flexure :

$$M = Qbd^2$$

$$6.81 \times 10^6 = 1.26 \times 1000 \times d^2$$

∴ $d_{req} = 73.52 \text{ mm} < d_{provided}$

... Okay, safe in flexure

(6) Calculation of Main reinforcement (steel) :

∴ $A_{st} = \dfrac{M}{\sigma_{st} \times j \times d} = \dfrac{6.81 \times 10^6}{130 \times 0.86 \times 85}$

∴ $A_{st} = 716.62 \text{ mm}^2$

and $A_{stmin} = 0.15 \% \text{ of } A_g = \dfrac{0.15}{100} \times 1000 \times 110$

∴ $A_{stmin} = 165.00 \text{ mm}^2$

∴ **Provide, A_{st} = 716.62 mm²**

Assume, diameter of bar $= \phi = 10 \text{ mm}$

• **Spacing (S) :**

(i) $S = \dfrac{1000 \times \left(\dfrac{\pi}{4} \times 10^2\right)}{716.62}$

$= 109.60 \text{ mm}$

(ii) $S = 3d = 3 \times 85 = 255 \text{ mm}$

(iii) $S = 300 \text{ mm}$

$\left.\right\}$ whichever is less

∴ $S = 109.60 \approx 100 \text{ mm}$

∴ **Provide, 10 mm φ @ 100 mm c/c, as a main reinforcement.**

(7) Check for shear :

(i) Nominal shear stress (τ_v) :

$$\tau_v = \frac{V}{bd} = \frac{9.59 \times 10^3}{1000 \times 85}$$

∴ τ_v = **0.11 N/mm^2**

(ii) Permissible shear stress in concrete ($\tau_{c \cdot k}$) :

$$P_{tmax} = \frac{A_{st\,provided}}{bd} \times 100$$

$$= \frac{\left(\dfrac{1000 \times \dfrac{\pi}{4} \times 10^2}{100}\right)}{1000 \times 85} \times 100$$

$$p_{tmax} = 0.92\ \%$$

Ptmax	τ_c
0.75	0.35
0.92	
1.00	0.39
0.25	0.04
1	0.16

0.17

∴ τ_c = 0.34 + (0.17 × 0.16)

∴ τ_c = 0.37 N/mm^2

∴ $\tau_{c \cdot k}$ = 0.37 × 1.3

∴ $\tau_{c \cdot k}$ = 0.48 N/mm^2

(iii) Compare τ_v and $\tau_{c \cdot k}$:

   0.11 < 0.48

   $\tau_v < \tau_{c \cdot k}$ ... Okay, safe in shear

(8) Calculation of distribution reinforcement :

∴ $A_{std} = A_{stmin}$ = 165.00 mm^2

Assume, diameter of bar = 6 mm

• **Spacing (S) :**

(i) $S = \dfrac{1000 \times \left(\dfrac{\pi}{4} \times 6^2\right)}{165.00}$

     = 171.36 mm

(ii)   S = 5d = 5 × 85 = 425 mm

(iii)   S = 450 mm

     } whichever is less

∴   S = 171.36 ≈ 170 mm

∴ **Provide, 6 mm φ @ 170 mm c/c, as a distribution steel**

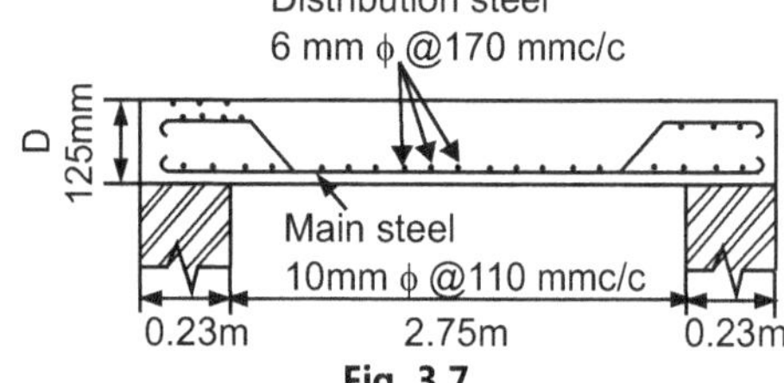

Fig. 3.7

Type II: Design of One Way Continuous Slab

Example 3.3 : *Design a continuous R.C. slab for a ball 15m × 7m. The slab is supported by R.C. beams 200 mm wide. The thickness of end wall is 300 mm. The slab carries an imposed load of 2.50 kN/m^2. The floor finish weighing 750 N/m^2. Use M-20 and Fe-415 grades material.*

Solution : Given : Hall size, 15 m × 7 m

   Support width = 300 mm

   Imposed load (L.L.) = 2.5 kN/m^2

   Floor finish = 0.75 kN/m^2

   M-20 ; σ_{cbc} = 7 N/mm^2

   Fe-415 ; σ_{st} = 230 N/mm^2

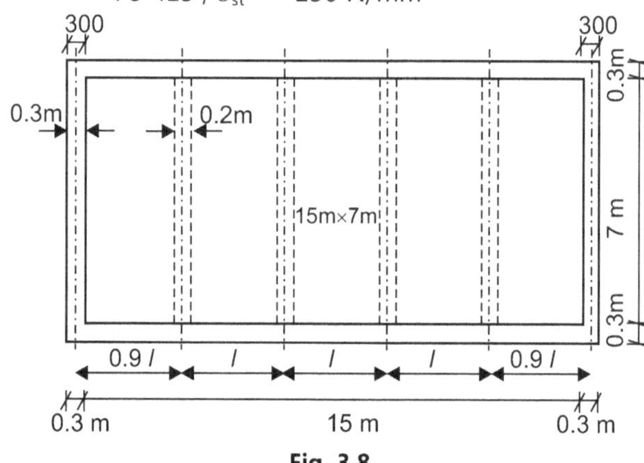

Fig. 3.8

The distance between the centres of end bearing

     = 15 + 0.30 = 15.30 m.

It is usual to make the end span length at 0.9 times intermediate span length Fig. shows the arrangement.

∴   If 'l' be the intermediate span of slab.

∴   We have,

     $3l + 2 \times 0.9 \times l$ = 15.30

         $4.8\,l$ = 15.30

∴   For intermediate slab, l = 3.19 m centre to centre distance.

∴   For end slab = $0.9l$ = 2.87 m centre to centre distance.

(1) Calculation of design constant

(a) Modular ration (m)

$$m = \frac{280}{3 \times \sigma_{cbc}} = \frac{280}{3 \times 7} = 13.33$$

(b) N.A. depth factor (k)

$$k = \frac{m \cdot \sigma_{cbc}}{m \cdot \sigma_{cbc} + \sigma_{st}} = 0.29$$

(c) Lever arm factor (j)

$$j = \left(1 - \frac{k}{3}\right) = \left(1 - \frac{0.29}{3}\right) = 0.90$$

(d) Moment resisting factor (Q)

$$Q = \frac{1}{2} \times \sigma_{cbc} \times j \times k = 0.91$$

(2) Calculation of loading (w) :

Assume, the overall depth of slab should be 40 mm per metre length.

$$\therefore \quad D_{assume} = 125 \text{ mm}$$

$$d_{assume} = 100 \text{ mm}$$

Consider '1m' width of slab.

(a)　　Dead load

(i)　　Floor finish $= 0.75 \times 1 = 0.75$ kN/m

(ii) Self wt. of slab $= 25 \times D = 25 \times 0.125$

$$= 3.13 \text{ kN/m}$$

$$\therefore \quad \text{Total D.L.} = W_D = 3.88 \text{ kN/m}$$

(b) Live load / Imposed load

$$W_L = 2.50 \text{ kN/m}$$

(3) Effective span

(a) For intermediate span

(i)　$l_{eff.}$ = c/c distance between support

$$= 3.19 \text{ m}$$

(ii) $l_{eff.}$ = L + d = (3.19 − 0.2) + 0.1 = 3.09 m

$\Big\}$ less

$$\therefore \quad l_{eff.} = 3.09 \text{ m} \dots \text{ for intermediate span}$$

(b)　For end span

(i)　$l_{eff.}$ = c/c distance between support

$$= 2.87 \text{ m}$$

(ii) $l_{eff.} = L + d = \left(2.87 - \dfrac{0.2}{2} - \dfrac{0.3}{2}\right) + 0.1$

$$= 2.72 \text{ m}$$

$\Big\}$ less

$$\therefore \quad l_{eff.} = 2.72 \text{ m} \dots \text{ for end span}$$

(4) Calculation of Max. B.M. (M) and S.F. (V)

(Refer Table 12 and 13, Pg. No. 36, TS 456 : 2000

(a) Bending moment

(A) B.M. for end span

(i)　B.M. (at middle of end span)

$$= \frac{W_D \cdot l^2}{12} + \frac{W_L \cdot l^2}{10}$$

$$= \frac{3.88 \times 2.72^2}{12} + \frac{250 \times 2.72^2}{10}$$

$$= 4.24 \text{ kN/m} \quad \dots\text{(i)}$$

(ii)　B.M. (at support next to end support)

$$= -\frac{W_D \cdot l^2}{12} - \frac{W_L \cdot l^2}{9}$$

$$= -\frac{3.88 \times 2.72^2}{10} - \frac{2.50 \times 2.72^2}{9}$$

$$= -4.93 \text{ kN.m} \quad \dots\text{(ii)}$$

(B)　B.M. for intermediate span

(i)　B.M. (at middle of interior span)

$$= +-\frac{W_D \cdot l^2}{24} + \frac{W_L \cdot l^2}{12}$$

$$= \frac{3.88 \times 3.09^2}{24} + \frac{2.50 \times 3.09^2}{12}$$

(ii) B.M. (at interior support)

$$= -\frac{W_D \cdot l^2}{12} - \frac{W_L \cdot l^2}{9}$$

$$= -\frac{3.88 \times 3.09^2}{12} - \frac{250 \times 3.09^2}{9}$$

$$= -5.74 \text{ kN/m} \quad \dots\text{(iv)}$$

(b)　　Shear force

(A) S.F. (at end support)

$$= 0.4 \, W_D \cdot l + 0.45 \times W_L \cdot l$$

$$= 0.4 \times 3.88 \times 2.72 + 0.45 \times 2.50 \times 2.72$$

$$= 7.28 \text{ kN}$$

SF (at support next to end support), outer side

$$= 0.6 \, (W_D \times l + W_L \times l)$$

$$= 0.6 \, (3.88 \times 2.72 + 2.50 \times 2.72)$$

$$= 10.41 \text{ kN}$$

(B) S.F. (at inner side of interior support)

$$= 0.55 \times W_D \times l + 0.60 \times W_L \times l$$

$$= 0.55 \times 3.88 \times 3.09 + 0.60 \times 2.50 \times 3.09$$

$$= 11.23 \text{ kN}$$

S.F. (at all other interior support)

$$= 0.5 \times W_D \times l + 0.6 \times W_L \times l$$

$$= 0.5 \times 3.88 \times 3.09 + 0.6 \times 2.50 \times 3.09$$

$$= 10.63 \text{ kN}$$

(5) Check for depth / bending / flexure

By equating Max. B.M. to resisting B.M.

$$M = Q.bd^2$$

$$5.74 \times 10^6 = 0.91 \times 1000 \times d^2$$

$$d_{req.} = 79.42 \text{ mm} < d_{provided}$$

(6) Calculation of main reinforcement (A_{st})

The maximum bending moments in the end spans (ii) and (iv) are nearly equal. Hence same reinforcement will be provided in all spans.

Consider, Max. B. M. = 5.74 kN.m for design

$$\therefore \quad A_{st} = \frac{M}{\sigma_{st} \times j \times d} = \frac{5.74 \times 10^6}{230 \times 0.90 \times 100}$$

$$\therefore \quad A_{st} = 277.29 \text{ mm}^2$$

and $A_{stmin} = 0.12$ % of $A_g = \dfrac{0.12}{100} \times 1000 \times 125$

and $\quad A_{st \, min} = 150.00 \text{ mm}^2$

$$\therefore \quad \textbf{Provide } A_{st} = \textbf{277.29 mm}^2$$

Assume, diameter of bar = ϕ = 8 mm

Spacing (s)

(i) $s = \dfrac{1000 \times \left(\dfrac{\pi}{4} \times \phi^2\right)}{A_{st}} = 181.27$ mm $\Bigg\}$ less

(ii) $s = 3d = 3 \times 100 = 300$ mm

(iii) $s = 300$ mm

∴ $s = 181.27 \approx 175$ mm

∴ **Provide, 8mm ϕ@ 175 mm c/c, as a main steel.**

(7) Check for shear

(i) Nominal shear stress (τ_v)

$$\tau_v = \frac{V_{max}}{bd} = \frac{11.23 \times 10^3}{1000 \times 100} = 0.11 \text{ N/mm}^2$$

(ii) Permissible shear strength of concrete $(\tau_c \cdot k)$

$$P_{t\,max} = \frac{A_{st\,provided}}{bd} \times 100 = \frac{\left(\dfrac{1000 \times \dfrac{\pi}{4} \times 8^2}{175}\right)}{1000 \times 100} \times 100$$

∴ $P_{t\,max} = 0.29$ %

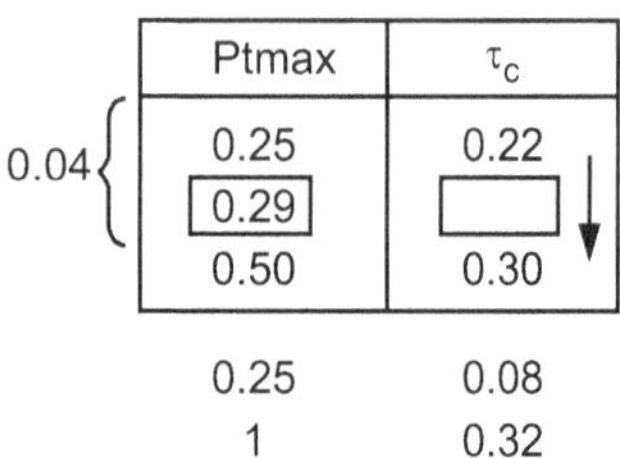

	Ptmax	τ_c
0.04 {	0.25	0.22
	0.29	
	0.50	0.30
	0.25	0.08
	1	0.32

$\tau_c = 0.22 + (0.04 \times 0.32)$

$\tau_c = 0.23$ N/mm^2

∴ $\tau_c \cdot k = 0.23 \times 1.3$

∴ $\tau_c \cdot k = 0.30$ N/mm^2

(iii) Compare τ_v and $\tau_c \cdot k$

   $0.11 < 0.30$

   $\tau_v < \tau_{c.k}$ …Okay, safe in shear

(8) Development length (L$_d$)

$$L_d = \frac{\phi \cdot \sigma_s}{4 \times \tau_{bd}} = \frac{8 \times 230}{4 \times 0.8 \times 1.6}$$

∴ $L_d = 359.38 \approx 360$ mm

(9) Calculation of distribution steel (A$_{std}$)

∴ $A_{std} = A_{st\,min} = 150$ mm^2

Assume, diameter of bar = ϕ = 6 mm

Spacing (s)

(i) $s = \dfrac{1000 \times \left(\dfrac{\pi}{4} \times \phi^2\right)}{A_{std}} = 188.50$ mm $\Bigg\}$ less

(ii) $s = 5d = 5 \times 100 = 500$ mm

(iii) $s = 450$ mm

∴ $s = 188.50 \approx 175$ mm

∴ **Provide, 6 mm ϕ @ 175 mm c/c, as a distribution steel.**

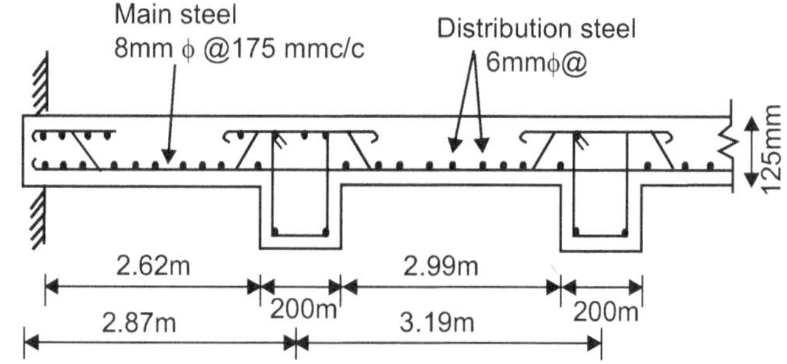

Fig. 3.9 : One way continuous slab

Type III : Design of Cantilever Slab

Example 3.4 : *Design a cantilever slab to carry a live load of 2.50 kN/m^2. The projection of slab beyond the face of the wall is 1.20 metre. Use M-20 concrete and mild steel.*

Solution : Live load = 2.50 kN/m^2

Cantilever projection = 1.20 m

M-20 ; σ_{cbc} = 7 N/mm^2

Fe-250 ; σ_{st} = 130 N/mm^2

(1) Calculation of design constant

(i)   Modular ratio (m) :

$$m = \frac{280}{3\,\sigma_{cbc}} = \frac{280}{3 \times 7} = 13.33$$

(ii)   N.A. depth factor (k) :

$$k = \frac{m \cdot \sigma_{cbc}}{m \cdot \sigma_{cbc} + \sigma_{st}} = 0.42$$

(iii)   Lever arm factor (j)

$$j = \left(1 - \frac{k}{3}\right) = \left(1 - \frac{0.42}{3}\right) = 0.86$$

(iv)   moment resisting factor (ϕ)

$$\phi = \frac{1}{2} \times \sigma_{cbc} \times j \times k = 1.26$$

(2) Calculation of loading (w)

The slab will bend over a clear span of 1.20m, let the overall depth of slab should be 60 mm per metre length.

∴ $D_{assume} = 100$ mm, C = 25 mm

∴ $d_{assume} = 75$ mm

Consider '1m' width of slab

(i)   Live load = $2.50 \times 1 = 2.50$ kN/m

(ii) Assume, F.F. = $1.00 \times 1 = 1.00$ kN/m

(iii) Self wt. of slab = $25 \times D = 2.50$ kN/m

∴   Total udl = w = 6.00 kN/m

(3) Effective span (left)

(i) $\quad l_{eff} = L + \dfrac{b}{2} = 1.20 + \dfrac{0.23}{2} = 1.32 \text{ m}$

(ii) $\quad l_{eff} = L + \dfrac{d}{2} = 1.20 + \dfrac{0.10}{2} = 1.25 \text{ m}$

$\Big\}$ less

$\therefore \quad l_{eff} = 1.25 \text{ m}$

(4) Calculation of bending moment and shear force

(i) Max. B. M. $= M = \dfrac{W \times l^2}{2} = \dfrac{6.00 \times 1.25^2}{2}$

$\therefore \qquad M = 4.69 \text{ kN.m}$

(ii) Max. S.F. $= V = wl = 6.00 \times 1.25$

$\therefore \qquad V = 7.50 \text{ kN}$

(5) Check for depth / bending

by equating max. B. M. and moment of resistance

$$M = Qbd^2$$
$$4.69 \times 10^6 = 1.26 \times 1000 \times d^2$$
$$d_{req.} = 61.01 \text{ mm} < d_{provided}$$

...okay safe in bending

(6) Calculation of main reinforcement (A_{st})

$\therefore \quad A_{st} = \dfrac{M}{\sigma_{st} \times j \times d} = \dfrac{4.69 \times 10^6}{130 \times 0.86 \times 75}$

$\therefore \quad A_{st} = 559.33 \text{ mm}^2$

and, $\quad A_{st\,min} = 0.15\% \text{ of } Ag = \dfrac{0.15}{100} \times 1000 \times 100$

$\qquad A_{st\,min} = 150 \text{ mm}^2$

Provide, $A_{st} = 559.33 \text{ mm}^2$

Assume, diameter of bar $= \phi = 8$ mm

Spacing (s) :

(i) $\quad s = \dfrac{1000 \times \left(\dfrac{\pi}{4} \times \phi^2\right)}{A_{st}} = 89.87 \text{ mm}$

(ii) $\quad s = 3d = 3 \times 75 = 225 \text{ mm}$

(iii) $\quad s = 300 \text{ mm}$

$\Big\}$ less

$\therefore \qquad s = 89.87 \approx 80 \text{ mm}$

$\therefore$ **Provide, 8mm ϕ@ 80 mm c/c, as a main steel.**

(7) Check for shear

(i) Nominal shear stress (τ_t)

$$\tau_v = \dfrac{V}{bd} = \dfrac{7.50 \times 10^3}{1000 \times 75}$$

$\therefore \qquad \tau_v = 0.10 \text{ N/mm}^2$

(ii) Permissible shear stress in concrete ($\tau_c \cdot$ k)

$$P_{t\,max} = \dfrac{A_{st\,provided}}{bd} \times 100$$

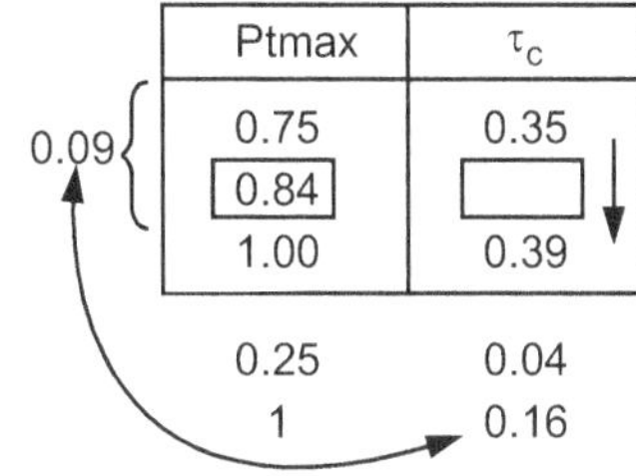

$$= \dfrac{\left(\dfrac{1000 \times \dfrac{\pi}{4} \times 8^2}{80}\right)}{1000 \times 75} \times 100$$

$\therefore \qquad P_{t\,max} = 0.84\%$

Ptmax	τ_c
0.75	0.35
0.84	
1.00	0.39

0.09 {

0.25 → 0.04
1 → 0.16

$\therefore \qquad \tau_c = 0.34 + (0.09 \times 0.16)$

$\therefore \qquad \tau_c = 0.35 \text{ N/mm}^2$

$\tau_{c \cdot k} = 0.35 \times 1.3$

$\therefore \qquad \boldsymbol{\tau_{c \cdot k} = 0.46 \text{ N/mm}^2}$

(iii) Compare τ_v and $\tau_{c \cdot k}$

$\qquad 0.10 < 0.46$

$\qquad \tau_v < \tau_{c \cdot k}$ ...okay safe in shear

(8) Development length (L_d)

$$L_d = \dfrac{\phi \cdot \sigma s}{4 \times \tau_{bd}} = \dfrac{8 \times 130}{4 \times 0.8 \times 1.6} = 203.13$$

$\therefore$ **Provide, $L_d \approx 225$ mm**

(9) Calculation of distribution steel (A_{std})

$\therefore \qquad A_{std} = A_{st\,min} = 150 \text{ mm}^2$

Assume, diameter of bar $= \phi = 6$ mm

• **Spacing (s)**

(i) $\quad s = \dfrac{1000 \times \left(\dfrac{\pi}{4} \times \phi^2\right)}{A_{std}}$

(ii) $\quad s = 5d = 5 \times 75 = 375 \text{ mm}$

(iii) $\quad s = 450 \text{ mm}$

$\Big\}$ less

$\therefore$ **Provide, 6mm ϕ @ 180 mm c/c as a distribution steel.**

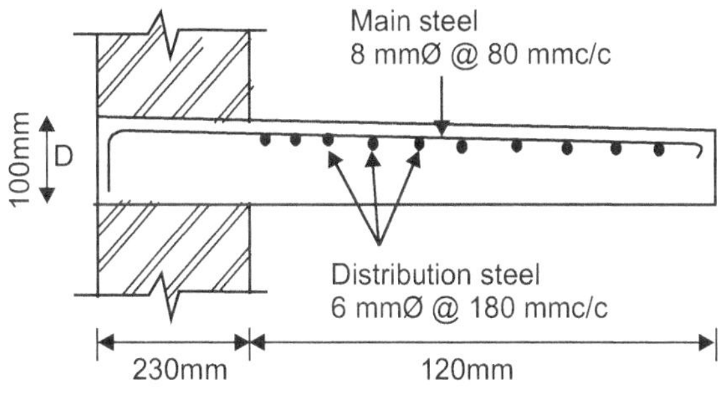

Fig. 3.10

Type IV : Design of Two Way Simply Supported Slab

Example 3.5 : *Design a R.C. slab for a room of size 4 m × 6 m. The slab is simply supported on all the four edges and carries a live load of 2 kN/m² and finishing of 1 kN/m². Take M-20 mix and Fe-415 steel.*

Solution : Given :

Room size $= 4\ m \times 6\ m$

Assume, support width $= 230$ mm

live load $= 2\ kN/m^2$

Floor finish $= 1\ kN/m^2$

M-20, $\sigma_{cbc} = 7\ N/mm^2$

Fe-415, $\sigma_{st} = 230\ N/mm^2$

(1) Aspect ratio :

$$\frac{l_y}{l_x} = \frac{6}{4} = 1.5 < 2$$

... Design slab as a "two way slab"

(2) Calculation of design constant :

(i)   Modular ratio (m) :

$$m = \frac{280}{3 \times \sigma_{cbc}} = \frac{280}{3 \times 7} = 13.33$$

(ii) N.A. depth factor (k) :

$$k = \frac{m \cdot \sigma_{cbc}}{m \cdot \sigma_{cbc} + \sigma_{st}} = 0.29$$

(iii) Lever arm factor (j) :

$$j = \left(1 - \frac{k}{3}\right) = \left(1 - \frac{0.29}{3}\right) = 0.90$$

(iv) Moment resisting factor (Q) :

$$Q = \frac{1}{2} \times \sigma_{cbc} \times j \times k = 0.91$$

(3) Calculation of load (w)

The slab will bend initially over a clear span 4m, let the overall depth of slab should be 40 mm per metre length.

$\therefore$ $D_{assume} = 175$ mm, c = 25 mm

$\therefore$ $d_{assume} = 150$ mm

Consider '1m' width of slab

(i)   live load $= 2 \times 1 = 2\ kN/m$

(ii) floor finish $= 1 \times 1 = 1\ kN/m$

(iii) self wt. of slab $= 25 \times D = 25 \times 0.175 = 4.38\ kN/m$

$\therefore$ Total udl $= W = 7.38\ kN/m$

(4) Effective span (l_{ex}) :

(i) $l_{ex} = Lx + b = 4 + 0.23 = 4.23$ m $\Big\}$

(ii) $l_{ex} = Lx + d = 4 + 0.15 = 4.15$ m $\Big\}$ Less

$\therefore$ $l_{ex} = 4.15$ m

(5) Calculation of bending moments :

For simply supported slab refer Table 27, IS 456 : 2000 for B.M. coefficient ;

$$\frac{l_{ey}}{l_{ex}} = \frac{6.15}{4.15} = 1.48$$

l_{ey}/l_{ex}	α_x	α_y
1.40	0.99	0.051
1.48		
1.50	0.104	0.046

0.08 { (for the first three rows)

0.10	0.005	0.005
1	0.05	0.05

$\therefore$ $\alpha_x = 0.099 + (0.08 \times 0.05) = 0.103$

$\therefore$ $\alpha_y = 0.051 - (0.08 \times 0.05) = 0.047$

$M_x = \alpha_x \cdot w \cdot l_{ex}^2 = 0.103 \times 7.38 \times 4.15^2$

$\therefore$ $M_x = 13.09\ kN.m$

$M_y = \alpha_y \cdot w \cdot l_{ex}^2 = 0.047 \times 7.38 \times 4.15^2$

$\therefore$ $M_y = 5.97\ kN.m$

(6) Check for depth/bending/flexure :

By equating Max. B.M. and moment resisting capacity

Max. B.M. $= Q\ bd^2$

$13.09 \times 10^6 = 0.91 \times 1000 \times d^2$

$\therefore$ $d_{req} = 119.94\ mm < d_{provided}$

... Okay, safe in flexure

(7) Calculation of main reinforcement (A_{st}) :

(A) Along x-direction (for short span) :

$$A_{stx} = \frac{M_x}{\sigma_{st} \times j \times d} = \frac{13.09 \times 10^6}{230 \times 0.90 \times 150}$$

$\therefore$ $A_{stx} = 421.58\ mm^2$

and $A_{stmin} = 0.12\%$ of $A_g = \frac{0.12}{100} \times 1000 \times 175$

$\therefore$ $A_{stmin} = 210.00\ mm^2$

$\therefore$ Provide, $A_{stx} = 421.58\ mm^2$

Assume, diameter of bar $= \phi = 8$ mm

- **Spacing (S_x) :**

(i) $S_x = \dfrac{1000 \times \left(\frac{\pi}{4} \times \phi^2\right)}{A_{stx}} = 119.23$ mm $\Big\}$

(ii) $S_x = 3d = 3 \times 150 = 450$ mm $\Big\}$ Less

(iii) $S_x = 300$ mm

$\therefore$ $S_x = 119.28 \approx 100$ mm

$\therefore$ **Provide, 8 mm ϕ @ 100 mm c/c, as a main steel along X · direction**

(B) Along Y-direction (for long span) :

$$\therefore \quad d' = d - \frac{\phi}{2} = 150 - \left(\frac{8}{2}\right) = 146 \text{ mm}$$

$$A_{sty} = \frac{M_y}{\sigma_{st} \times j \times d'} = \frac{5.97 \times 10^6}{230 \times 0.90 \times 146}$$

$$\therefore \quad A_{sty} = 197.54 \text{ mm}^2 \ngtr A_{stmin}$$

$\therefore$ **Provide, $A_{sty} = A_{stmin} = 210 \text{ mm}^2$**

Assume, diameter of bar = ϕ = 8 mm

- **Spacing (S_y) :**

$$\text{(i)} \quad S_y = \frac{1000 \times \left(\frac{\pi}{4} \times 8^2\right)}{210} = 239.36 \text{ mm}$$

$$\text{(ii)} \quad S_y = 3d' = 3 \times 146 = 438 \text{ mm}$$

$$\text{(iii)} \quad S_y = 300 \text{ mm}$$

$\left.\begin{array}{}\\ \\ \\\end{array}\right\}$ Less

$$\therefore \quad S = 239.36 \approx 225 \text{ mm}$$

$\therefore$ **Provide, 8 mm ϕ @ 225 mm c/c, as a main steel along y-direction**

(8) Check for shear

$$V = \frac{w \times l_{ex}}{2} = \frac{7.38 \times 4.15}{2} \quad \text{... (Approx.)}$$

$$\therefore \quad V = 15.31 \text{ kN}$$

(i) Nominal shear stress (τ_v) :

$$\tau_v = \frac{V}{bd} = \frac{15.31 \times 10^3}{1000 \times 150}$$

$$\therefore \quad \tau_v = 0.10 \text{ N/mm}^2$$

(ii) Permissible shear strength of concrete ($\tau_{c \cdot k}$) :

$$\therefore \quad P_{tmax} = \frac{A_{st \times provided}}{bd} \times 100$$

$$= \frac{\left(\dfrac{1000 \times \frac{\pi}{4} \times 8^2}{100}\right)}{1000 \times 150} \times 100$$

$$\therefore \quad P_{tmax} = 0.34 \text{ \%}$$

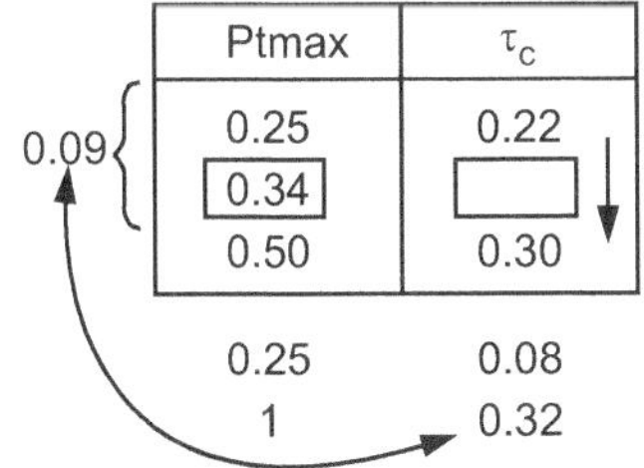

Ptmax	τ_c
0.25	0.22
0.34	
0.50	0.30

0.25	0.08
1	0.32

$$\therefore \quad \tau_c = 0.22 + (0.09 \times 0.32)$$

$$\tau_c = 0.25 \text{ N/mm}^2$$

and $\quad \tau_{c \cdot k} = 0.25 \times 1.25$

$\therefore \quad$ **$\tau_{c \cdot k} = 0.31 \text{ N/mm}^2$**

(iii) Compare τ_v And $\tau_{c \cdot k}$

$$0.10 < 0.31$$

$$\tau_v < \tau_{c \cdot k} \quad \text{... Okay, safe in shear}$$

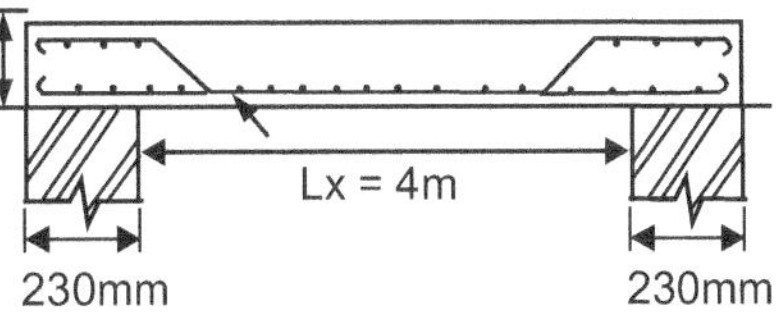

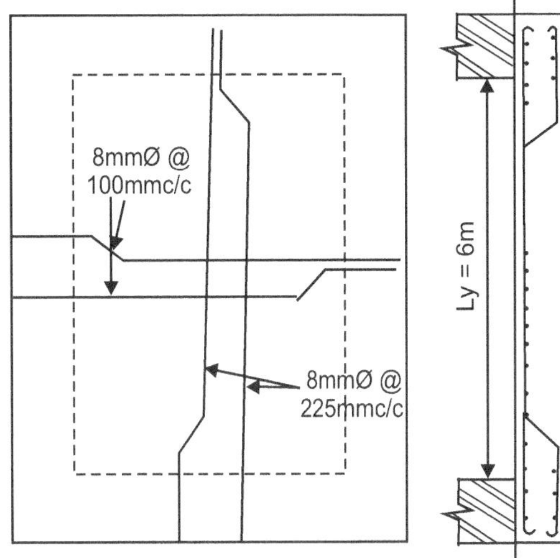

Fig. 3.11 : Two way slab

Type V : Design of Two Way Restrained Slab

Example 3.6 : *Design a R.C. slab for a room measuring 6.5 m × 5 m. The slab is to be cast monolithically over the beams with corners held down. The width of the supporting beam is 250 mm. The slab carries superimposed load of 3 kN/m². Use M-20 concrete mix and mild steel reinforcement.*

Solution : Given :

Room size 6.5 m × 5.0 m, restrained slab

i.e. (four edges discontinuous, case-9, Table 26, IS 456)

superimposed load = 3 kN/m²

　　M-20 ; σ_{cbc} = 7 N/mm²

　　Fe-250; σ_{st} = 130 N/mm²

(1) Aspect ratio :

$$\frac{l_y}{l_x} = \frac{6.50}{5.00} = 1.3 < 2$$

　　　　... Design slab as a "Two way restrained slab".

(2) Calculation of design constant :

(i) Modular ratio (m) :

$$m = \frac{280}{3 \times \sigma_{cbc}} = \frac{280}{3 \times 7} = 13.33$$

(ii) N.A. depth factor (k) :

$$K = \frac{m \cdot \sigma_{cbc}}{m \cdot \sigma_{cbc} + \sigma_{st}} = 0.42$$

(iii) Lever arm factor (j) :

$$j = \left(1 - \frac{k}{3}\right) = \left(1 - \frac{0.42}{3}\right) = 0.86$$

(iv) Moment resisting factor (Q) :

$$Q = \frac{1}{2} \times \sigma_{cbc} \times j \times k = 1.26$$

(3) Calculation of load (w) :

The slab will bend initially over a clear span 5 m, let the overall depth of slab should be 40 mm per metre length.

$\therefore \quad D_{assume} = 200$ mm, c = 25 mm

$\therefore \quad d_{assume} = 175$ mm

Consider '1m' width of slab

(i)   Superimposed load = (L.L × F.F.) = 3 kN/m

(ii) Self wt. of slab = 25D = 25 × 0.2 = 5 kN/m

....

$\therefore \quad$ Total udl = w = 8 kN/m

(4) Effective span (l_{ex}) :

(i) $l_{ex} = L_x + b = 5 + 0.23 = 5.23$ m ⎫

(ii) $l_{ex} = L_x + d = 5 + 0.175 = 5.18$ m ⎬ Less

$\therefore \quad l_{ex} = 5.18$ m

(5) Calculation of bending moment :

For restrained slab refer Table-26, Case-9, IS 456 : 2000 for B.M. coefficients :

$$\frac{l_{ey}}{l_{ex}} = \frac{6.68}{5.18} = 1.29$$

$\alpha_x = 0.079 \; ; \; \alpha_y = 0.056$

$M_x = \alpha_x \cdot w \cdot l_{ex}^2 = 0.079 \times 8 \times 5.18^2$

$\therefore \quad M_x = 16.96$ kN.m

$M_y = \alpha_y \cdot w \cdot l_{ex}^2 = 0.056 \times 8 \times 5.18^2$

$\therefore \quad M_y = 12.02$ kN.m

(6) Check for depth/bending :

By equating Max. B.M. and resisting B.M.

$M = Q\,bd^2$

$16.96 \times 10^6 = 1.26 \times 1000 \times d^2$

$\therefore \quad d_{req} = 116.02$ mm < $d_{provided}$

... Okay, Safe in bending

(7) Calculation of main reinforcement (A_{st}) :

(A) Along x-direction (for short span) :

$$A_{stx} = \frac{M_x}{\sigma_{st} \times j \times d} = \frac{16.96 \times 10^6}{130 \times 0.86 \times 175}$$

$\therefore \quad A_{stx} = 866.85$ mm^2

And, $A_{stmin} = 0.15$ % of $A_g = \frac{0.15}{100} \times 1000 \times 200$

$\therefore \quad A_{stmin} = 300.00$ mm^2

$\therefore \quad$ **Provide, A_{stx} = 866.85 mm^2**

Assume, diameter of bar ϕ = 10 mm

- **Spacing (S_x) :**

(i) $S_x = \dfrac{1000 \times \left(\frac{\pi}{4} \times \phi^2\right)}{A_{stx}} = 90.60$ mm ⎫

(ii) $S_x = 3d = 3 \times 175 = 525$ mm ⎬ Less

(iii) $S_x = 300$ mm ⎭

$\therefore \quad S_x = 90.60 \approx 80$ mm

$\therefore \quad$ **Provide, 10 mm ϕ @ 80 mm c/c, as a main steel along x-direction**

(B) Along y-direction (for long span) :

$$A_{sty} = \frac{M_y}{\sigma_{st} \times j \times d'} \quad ... (\because d' = d - \phi = 165 \text{ mm})$$

$$= \frac{12.02 \times 10^6}{130 \times 0.86 \times 165}$$

$\therefore \quad A_{sty} = 651.60$ mm^2 > A_{stmin}

Assume, diameter of bar = ϕ = 10 mm

- **Spacing (S_y) :**

(i) $S_y = \dfrac{1000 \times \left(\frac{\pi}{4} \times \phi^2\right)}{A_{sty}} = 120.53$ mm ⎫

(ii) $S_y = 3d' = 3 \times 165 = 495$ mm ⎬ Less

(iii) $S_y = 300$ mm ⎭

$\therefore \quad S_y = 120.53 \approx 110$ mm

$\therefore \quad$ **Provide, 10 mm ϕ @ 110 mm c/c, as a main steel along y-direction.**

(8) Calculation of Torsional reinforcement at corners :

All the edges are discontinuous

$\therefore \quad$ Area of steel required at the corner containing both edges are discontinuous $= \dfrac{3}{4} \times A_{stx}$

$$= \frac{3}{4} \times \left(\frac{1000 \times \left(\frac{\pi}{4} \times 10^2\right)}{80} \right)$$

$$= 736.31 \text{ mm}^2$$

Assume, diameter of bar = ϕ = 8 mm

$$\text{Spacing} = S = \frac{1000 \times \left(\frac{\pi}{4} \times 8^2\right)}{736.31} = 68.27 \text{ mm} \approx 65 \text{ mm}$$

$$\text{Size of torsional steel mesh} = \frac{l_x}{5} = \frac{5.18}{5} = 1.04 \text{ m}$$

$\therefore \quad$ **Provide, 8 mm ϕ @ 65 mm c/c, torsional steel for steel mesh of size 1.04 m × 1.04 m**

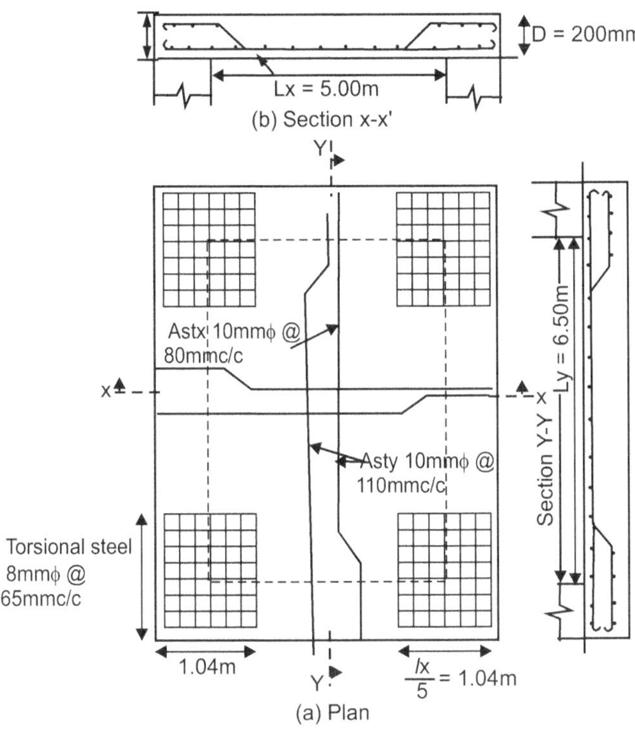

Fig. 3.12 : Two way slab (Restrained)

Example 3.7 : *Design an R.C.C. slab has to be provided over a room 4m × 5m. The slab is continuous over two adjacent supports and discontinuous at the other two supports. The live load on the slab is 3.5 kN/m². Use M-20 and Fe415.*

Solution : Given : Room size 4 m × 5 m

    (Two adjacent edges discontinuous)

$$\text{live load} = 3.50 \text{ kN/m}^2$$

$$\text{M-20}; \ \sigma_{cbc} = 7 \text{ N/mm}^2$$

$$\text{Fe-415}; \ \sigma_{st} = 230 \text{ N/mm}^2$$

(1) Aspect ratio :

$$\frac{l_y}{l_x} = \frac{5}{4} = 1.25 < 2$$

    Design slab as a "Two way restrained slab"

(2) Calculation of design constant :

(i)   Modular ratio (m) :

$$m = \frac{280}{3 \times \sigma_{cbc}}$$

$$= \frac{280}{3 \times 7} = 13.33$$

(ii)   N.A. depth factor (K) :

$$K = \frac{m \cdot \sigma_{cbc}}{m \cdot \sigma_{cbc} + \sigma_{st}}$$

$$= 0.29$$

(iii)   Lever arm factor (j) :

$$j = \left(1 - \frac{K}{3}\right) = \left(1 - \frac{0.29}{3}\right)$$

$$= 0.90$$

(iv)   Moment resisting factor (Q) :

$$Q = \frac{1}{2} \times \sigma_{cbc} \times j \times k$$

$$= 0.91$$

(3) Calculation of loading (w) :

The slab will bend initially over a clear span 4 m, let the overall depth of slab should be 40 mm per metre length.

$$\therefore \quad D_{assume} = 175 \text{ mm,}$$

$$c = 25 \text{ mm}$$

$$d_{assume} = 150 \text{ mm}$$

Consider, '1m' width of slab

(i)   Live load = 3.50 × 1 = 3.50 kN/m

(ii)   Assume, F.F. = 1.00 × 1 = 1.00 kN/m

(iii)   Self wt. of slab = 25 × 0.175 = 4.38 kN/m

....

$$\therefore \qquad \text{Total udl} = w = 8.88 \text{ kN/m}$$

(4) Effective span (l_{ex}) :

(i) $l_{ex} = L_x + b = 4 + 0.23 = 4.23$ m ⎫

(ii) $l_{ex} = L_x + d = 4 + 0.15 = 4.15$ m ⎬ Less

$$\therefore \quad l_{ex} = 4.15 \text{ m}$$

(5) Calculation of bending moment :

For restrained slab refer Table-26, case-4, IS-456 : 2000 for B. M. coefficients ;

$$\therefore \quad \frac{l_{ey}}{l_{ex}} = \frac{5.15}{4.15} = 1.24$$

l_{ey}/l_{ex}	$-\alpha_x$	$+\alpha_x$
1.20	0.060	0.045
1.24		
1.30	0.065	0.049

0.04 {

0.10	0.005	0.004
1	0.05	0.04

$$-\alpha_x = 0.060 + (0.04 \times 0.05) = 0.062$$

$$+\alpha_x = 0.045 + (0.04 \times 0.04) = 0.047$$

(A)   Negative moment at the continuous edge :

(i)   short span moment (M_x) :

$$= \alpha_x \cdot w \cdot l_{ex}^{\,2}$$

$$= 0.062 \times 8.88 \times 4.15^2$$

$$\therefore \qquad M_x = 9.48 \text{ kN.m}$$

(ii) long span moment (M_y) :

$$= \alpha_y \cdot w \cdot l_{ex}^2$$
$$= 0.047 \times 8.88 \times 4.15^2$$
$$\therefore \quad M_y = 7.19 \text{ kN.m}$$

(B) Positive moment at mid span :

(i) Short span moment (M_y) :

$$M_x = \alpha_x \cdot w \cdot l_{ex}^2$$
$$= 0.047 \times 8.88 \times 4.15^2$$
$$\therefore \quad M_x = 7.19 \text{ kN.m}$$

(ii) long span moment (M_y) :

$$= \alpha_y \cdot w \cdot l_{ex}^2$$
$$= 0.035 \times 8.88 \times 4.15^2$$
$$\therefore \quad M_y = 5.35 \text{ kN.m}$$

(6) Check for depth / bending / flexure :

By equating Max. B.M. and resisting B.M.

$$\therefore \quad M = Q \, bd^2$$
$$9.48 \times 10^6 = 0.91 \times 1000 \times d^2$$
$$d_{req} = 102.07 \text{ mm} < d_{provided}$$

$$\text{... Okay... safe in bending}$$

(7) Calculation of Main reinforcement :

(A) Negative moment at continuous edge :

(a) Short span :

$$A_{stx} = \frac{M_x}{\sigma_{st} \times j \times d}$$
$$= \frac{9.48 \times 10^6}{230 \times 0.90 \times 150}$$
$$\therefore \quad A_{stx} = 305.31 \text{ mm}^2$$

and $\quad A_{stmin} = 0.12 \%$ of A_g

$$= \frac{0.12}{100} \times 1000 \times 175$$

$$A_{stmin} = 210.00 \text{ mm}^2$$

∴ Provide, A_{stx} = 305.31 mm^2

Assume, diameter of bar = ϕ = 8 mm

- **Spacing (S) :**

(i) $\quad S_x = \dfrac{100 \times \left(\frac{\pi}{4} \times \phi^2\right)}{A_{stx}} = 164.64 \text{ mm}$

(ii) $\quad S_x = 3d = 3 \times 150 = 450 \text{ mm}$ $\quad$ ... Less

(iii) $\quad S_x = 300 \text{ mm}$

$$\therefore \quad S_x = 164.64 \approx 150 \text{ mm}$$

∴ Provide, 8 mm ϕ @ 150 mm c/c, along short span at edge

(b) Long span :

$$A_{sty} = \frac{M_y}{\sigma_{st} \times j \times d}$$
$$= \frac{7.19 \times 10^6}{230 \times 0.90 \times 150}$$
$$\therefore \quad A_{sty} = 231.56 \text{ mm}^2 > A_{stmin}$$

Assume, diameter of bar = ϕ = 8 mm

- **Spacing (S_y) :**

(i) $\quad S_y = \dfrac{1000 \times \left(\frac{\pi}{4} \times \phi^2\right)}{A_{sty}}$

$$= 217.07 \text{ mm} < 3d \text{ or } 300 \text{ mm}$$
$$\therefore \quad S_y = 217.07 \approx 200 \text{ mm}$$

∴ Provide, 8 mm ϕ @ 200 mm c/c, along long span at edge

(B) Positive moment at mid span :

(a) Short span :

$$A_{stx} = \frac{M_x}{\sigma_{st} \times j \times d}$$
$$= \frac{7.19 \times 10^6}{230 \times 0.90 \times 150}$$
$$\therefore \quad A_{stx} = 231.56 \text{ mm}^2 > A_{stmin}$$

∴ Provide, 8 mm ϕ @ 200 mm c/c, along short span at mid span

(b) long span :

$$A_{sty} = \frac{M_y}{\sigma_{st} \times j \times d} = \frac{5.35 \times 10^6}{230 \times 0.90 \times 160}$$
$$\therefore \quad A_{sty} = 172.30 \text{ mm}^2 \not> A_{stmin}$$

∴ Provide, A_{sty} = A_{stmin} = 210 mm^2

Assume, $\quad \phi = 8$ mm

- **Spacing (S_y) :**

(i) $\quad S_y = \dfrac{1000 \times \left(\frac{\pi}{4} \times \phi^2\right)}{A_{sty}}$

$$= 239.36 \text{ mm} < 3d \text{ or } 300 \text{ mm}$$
$$S = 239.36 \approx 225 \text{ mm}$$

∴ Provide, 8 mm ϕ @ 225 mm c/c, along short span at mid span

(8) Calculation of Torsional reinforcement at corners :

(A) At corner having both edges continuous :

Torsion reinforcement need not be provided at corners contained both edges continuous.

(B) At corner having both edges discontinuous :

Area of steel required $= \dfrac{3}{4} \times A_{stx}$

$$= \dfrac{3}{4} \times \left(\dfrac{1000 \times \dfrac{\pi}{4} \times 8^2}{150} \right)$$

$$= 251.33 \text{ mm}^2$$

Assume, diameter of bar = 8 mm

$$\text{Spacing} = S = \dfrac{1000 \times \left(\dfrac{\pi}{4} \times 8^2 \right)}{251.33}$$

$$= 200 \text{ mm}$$

$\therefore$ Size of torsional steel mesh $= \dfrac{l_x}{5}$

$$= \dfrac{4.15}{5}$$

$$= 0.83 \text{ m} \approx 0.85 \text{ m}$$

$\therefore$ **Provide, 8 mm ϕ @ 200 mm c/c, torsional steel mesh of size 0.85 m $\times$ 0.85 m**

(C) At corner having one edge continuous and one edge discontinuous :

Area of steel required $= \dfrac{1}{2} \times \left(\dfrac{3}{4} \times A_{stx} \right)$

$$= 125.67 \text{ mm}^2$$

$\therefore$ **Provide, 8 mm ϕ 400 mm c/c, torsional steel mesh of size 0.85 m $\times$ 0.85 m**

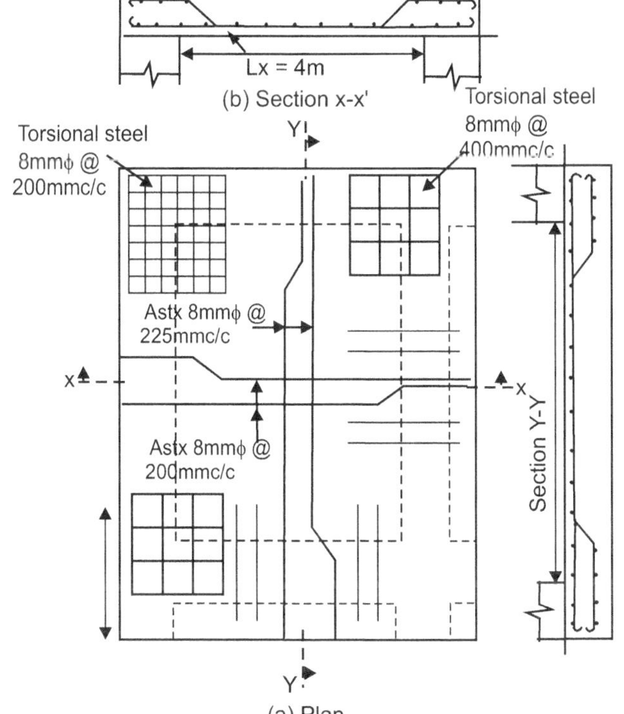

Fig. 3.13 : Two way restrained slab

EXERCISE

1. Define slab, state classification of slab.

2. Differentiate between one way slab and two way slab.

3. Why distribution steel is provided in slab and how much its amount ?

4. Explain design procedure of 'Design of one way slab'.

5. Why torsion reinforcement is provided is R.C.C. slab?

6. Design a slab for a room of size 2.8 m $\times$ 6.00 m, carries a superimposed load of 4 kN/m^2. Use M-15 and M.S. reinforcement.

7. Design a simply supported slab supported on masonry walls to the following requirements :

    clear span = 3.75 m

    live load = 2 kN/m^2

    partition load = 2.5 kN/m^2

    floor finish = 0.6 kN/m^2

    Materials : M-25, Fe-415.

8. Design a cantilevering chajja projecting for a length of 0.9 metre beyond a 230 mm $\times$ 230 mm lintel. The live load on the chajja is 800 N/m^2. Use M-20 and Fe 415.

9. Design a cantilever slab of span 1.00 m carries superimposed load 2.00 N/m^2. Use M-25, Fe-250.

10. Design a continuous R.C.C. slab for a community hall of size 21 m $\times$ 8 m, carries a superimposed load of 4000 N/m^2, the beams are spaced at 3 m c/c. Take M-25 and HYSD reinforcement.

11. Design a slab for a room of size 4 $\times$ 4 carries superimposed load of 4.50 kN/m^2. The slab is simply supported over 250 mm thick brick wall. Use M-15 and M.S. reinforcement.

12. Design a slab for a room 4.5 m $\times$ 5.5 m carries live load of 3.5 kN/m^2. Take M-25 and Fe-415. (corners are not held down).

13. Design a slab 4.5 m $\times$ 5.5 m interior panel in a building. The slab carries a L.L. of 3.5 kN/m^2, F.F. = 0.6 kN/m^2 and partition load of 2.00 kN/m^2. Take M-25 and Fe-415.

14. Design the slab for the floor of a residential building for a live load of 3000 N/m^2. Figure shows plan of a residential building. Use M-20 and Fe-415.

    Design slab A, B, C. Also design the slab torsional corner.

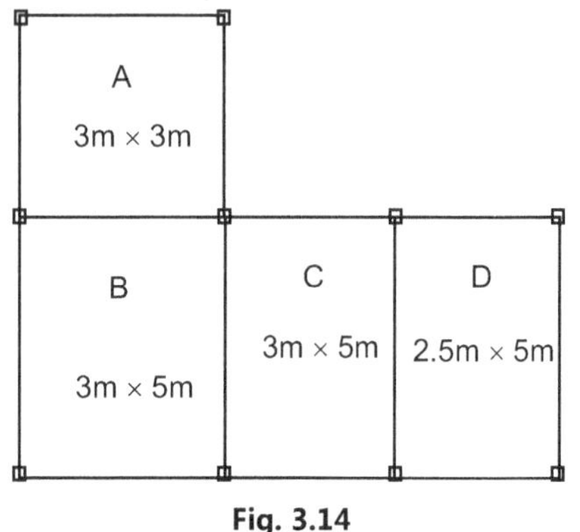

Fig. 3.14

◇ ◇ ◇

CHAPTER - 4
DESIGN OF COLUMN & FOOTING (WSM)

(A) COLUMN

4.1 INTRODUCTION

- A compression member i.e. column is an important element of every reinforced concrete structure. These are used to transfer the load of superstructure to the foundation safely. Mainly columns, struts and pedestals are used as compression members in buildings, bridges, supporting system of tanks, factories and many more such structures.

- A column is defined as a vertical compression member which is mainly subjected to axial loads and the effective length of which exceeds three times its least lateral dimension. The compression member whose effective length is less than three times its least lateral dimension is called as Pedestal. The compression member which is inclined or horizontal and is subjected to axial load is called Strut. Struts are used in trusses.

4.2 CLASSIFICATION OF COLUMNS

- The columns may be classified based on different criteria as follows :

  1. Based on type of reinforcement.
  2. Based on type of loading.
  3. Based on slenderness ratio.

1. **Based on Type of Reinforcement :** Reinforced concrete columns may be classified into the following three types based on the type of reinforcement provided : (See Fig. 4.1).

   (i) Tied Column : Where the main longitudinal bars are enclosed within closely spaced lateral ties.

   (ii) Spiral Column : Where the main longitudinal bars are enclosed within closely spaced and continuously wound spiral reinforcement.

   (iii) Composite Column : Instead of longitudinal bars, if the column is reinforced with structural steel shapes, known as composite column.

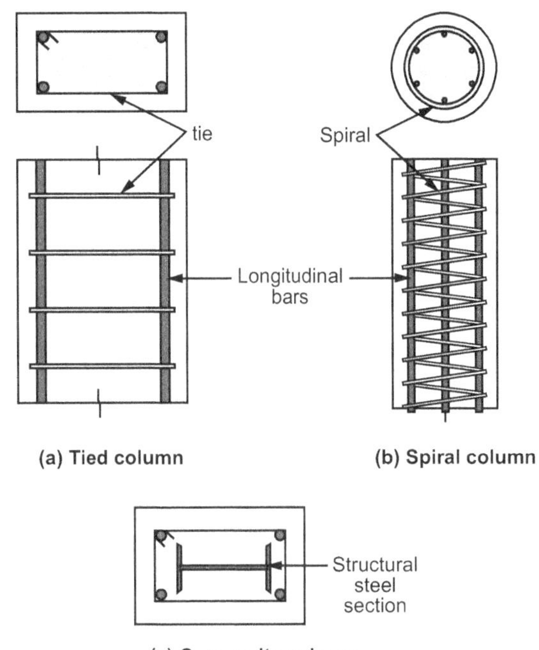

Fig. 4.1 : Types of columns

2. **Based on Type of Loading :** Columns may be classified, based on the nature of loading as : (See Fig. 4.2).

   (i) Columns with axial loading.

   (ii) Columns with uniaxial eccentric loading.

   (iii) Columns with biaxial eccentric loading.

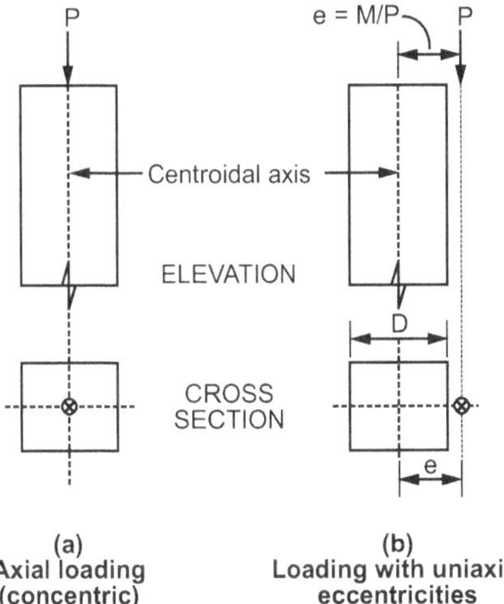

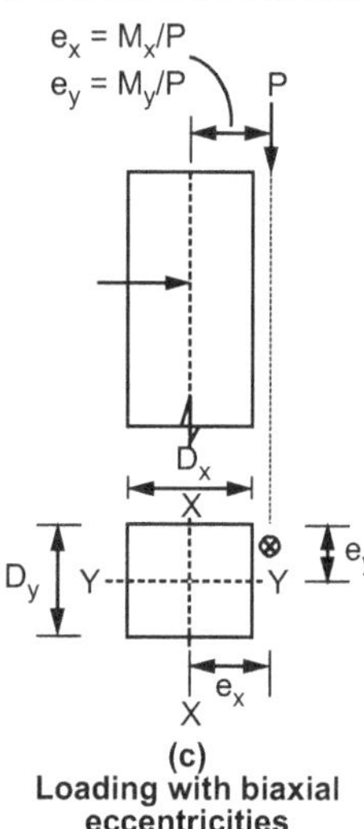

(c)
Loading with biaxial
eccentricities

Fig. 4.2 : Different loading situations in columns

3. **Based on Slenderness Ratio :** Columns may be classified into the following two types, based on slenderness ratio, are considered insignificant or significant.

   (i) Short columns.

   (ii) Slender (Long) columns.

(i) Short Columns : When the ratio of effective length of the column to its least lateral dimension does not exceed 12, it is termed as short column.

(ii) Long Columns : A long column is the one whose ratio of effective length to its lateral dimension end exceeds 12.

4.3 CATEGORIZATION OF COLUMN

Categorisation of column is extremely helpful as it decides the approach to column design at a latter stage because the procedure for design of column in each of the three categories is different. The column shall be first divided into the following three categories.

1. **Internal Columns or Axially Loaded Columns :** Internal columns carrying beam either in all four directions or only in opposite directions.

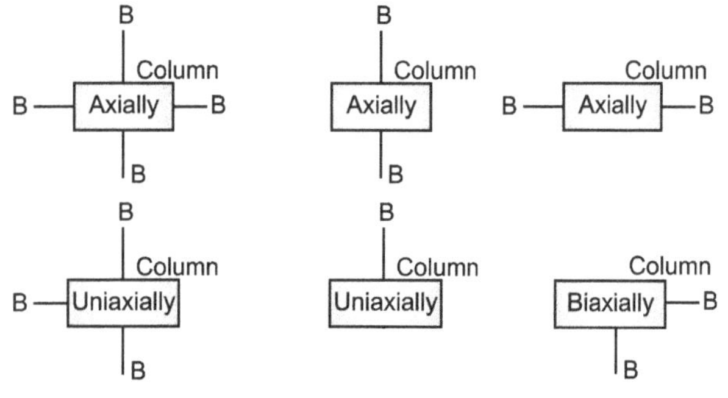

Fig. 4.3

2. **Side Columns or Columns Subjected to Axial Load and Uniaxial Bending :** Columns, along the sides of a building, which carry beams either in three orthogonal directions or a single beam in one direction.

3. **Corner Columns or Columns Subjected to Axial Load and Biaxial Bending :** Corner columns or the columns which carry beams in two perpendicular directions.

4.4 MATERIAL OF CONSTRUCTION

Columns may be classified as following, as per the material used for construction :

- **Timber Columns :** Timber columns are generally used for light loads. They are used in small trusses and wooden houses. These are called as posts.

- **Masonry Columns :** These are used for light loads.

- **R.C.C. Columns :** R.C.C. columns are used for mostly all types of buildings and other R.C.C. structures like tanks, bridges etc.

- **Steel Columns :** Steel columns are used for heavy loads.

- **Composite Columns :** Composite columns are used for very heavy loads. They consist of steel sections, like joists (I or H sections) embedded in R.C.C. section as shown in Fig. 4.4.

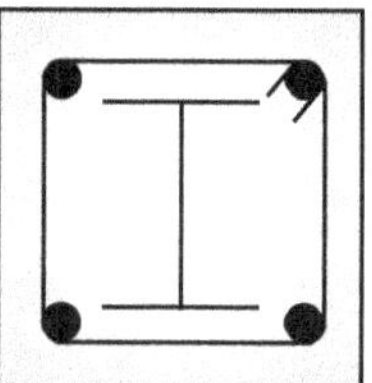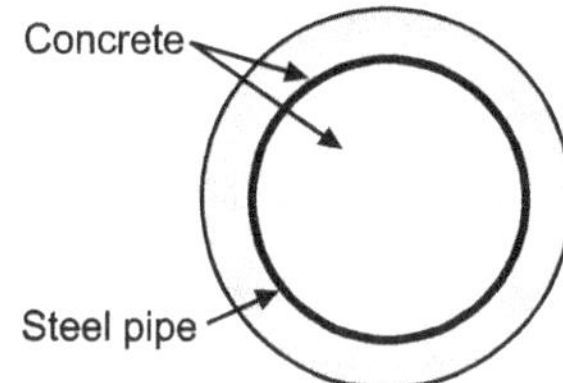

Fig. 4.4 : Composite columns

4.5 NECESSITY OF R.C.C. COLUMN

- Concrete is strong in compression. Thus a column can be made up of plain concrete but it is always advisable to use RCC columns instead of plain concrete columns because of following reasons :

  ➢ A plain concrete column requires very large area as compared to RCC columns. Thus by using R.C.C. columns a lot of space can be saved as the size of the column will be less.

  ➢ To resist tensile stresses which may be caused due to eccentricity of loads.

  ➢ To resists bending stresses which may be caused due to eccentricity of loads.

  ➢ To resist twisting moment which may be caused due to earthquakes or horizontal movement.

4.6 NECESSITY OF CONSTITUENTS MATERIALS IN COLUMN

(Function of concrete, longitudinal reinforcement and transverse reinforcement) :

1. **Concrete :**

- Concrete being very strong in compression should not theoretically, require any reinforcement but for a given load the column may require a large section, further its short-coming is that it is very weak in tension and therefore, in all circumstances leading to development of tension need to be examined. In view of this reinforcement is required to be provided to resist tension and to reduce the size of the column.

- The tension is likely to develop in column due to accidental transverse loads, eccentric loads and the end moments which cause bending in the member. Further since bending can occur in any direction, inducing tension on any face, the reinforcement is required to all the extreme faces running parallel to the axis of the member.

- This, thus, forms a longitudinal steel. For short members, there is a possibility of failure by longitudinal splitting due to development of transverse tension or failure by shear i.e. diagonal tension. This therefore, requires transverse reinforcement. The two reinforcements are obviously required to hold each other.

- The various functions served by the longitudinal and transverse reinforcement can be summarized as under.

2. **Longitudinal Reinforcement :**

   (i) To assist concrete, in resisting compression, so as to reduce the overall size, of the column.

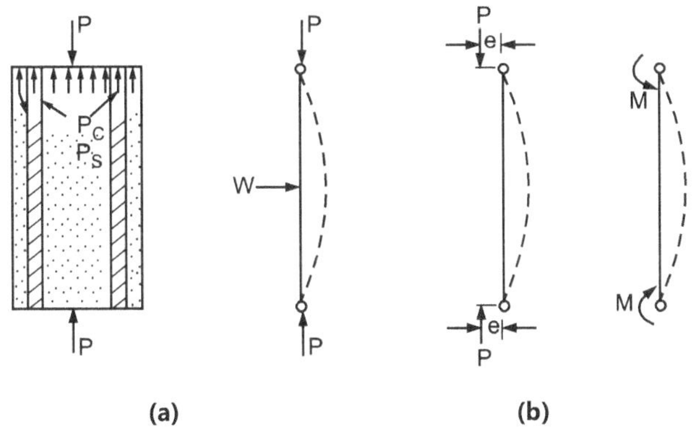

Fig. 4.5 : Functions of longitudinal reinforcement

   (ii) To resist any tension that might develop due to bending caused by transverse load, eccentric load or the moments.

(iii) To reduce the effect of creep and shrinkage due to sustained loading.

(iv) To prevent or delay sudden brittle collapse.

(v) To impart necessary ductility to the column.

(vi) To hold the transverse reinforcement.

3. **Transverse Reinforcement :**

   (i) To prevent buckling of longitudinal bars.

   (ii) To prevent longitudinal splitting of concrete.

   (iii) To resist diagonal tension due to transverse shear.

   (iv) To confine the concrete.

   (v) To hold the longitudinal reinforcement in position.

   (vi) To prevent or delay sudden collapse and impart necessary ductility to the members.

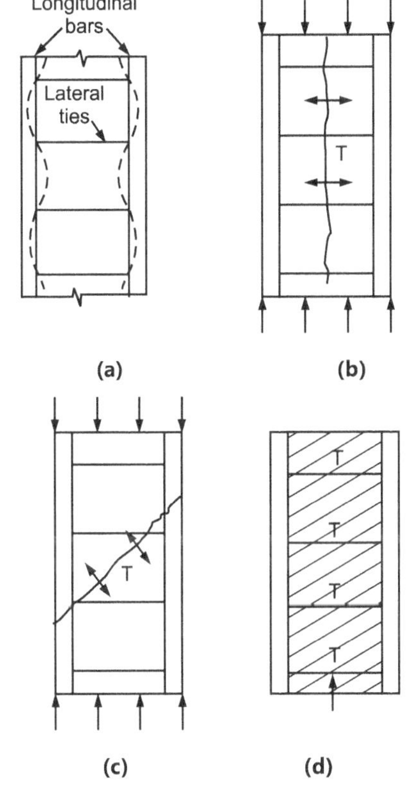

Fig. 4.6 : Functions of transverse and reinforcement

(a) Buckling of L-bar prevented (b) Longitudinal splitting arrested (c) Diagonal tension resisted

(d) Concrete confined

Design of a R.C. column, therefore, involves determination of size of concrete section and the amount and arrangement of longitudinal and transverse reinforcement.

4.7 LENGTH IN BEAM AND SLAB CONSTRUCTION

- **Beam and Slab Construction :** In this case l is the clear distance between the floor and the underside of the shallow beam framing into the column in each direction at the next higher level as shown in Fig. 4.7.

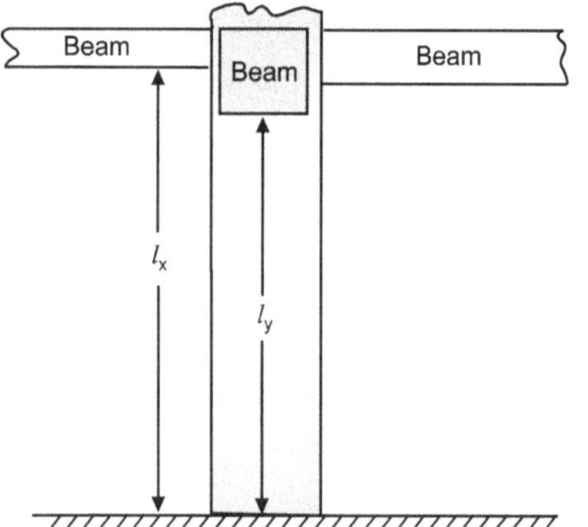

Fig. 4.7 : Length in beam and slab construction

- **Unsupported Length :** Clear distance between the end restrains (i.e. between the floor and underside of the beam).
- **Effective Length :** length between the point of contra flexures of a buckled column.

Effective Length IS : 456 : 2000 (Annexure - E) gives the values of effective length for various end conditions in terms of its unsupported length (Table 4.1).

Table 4.1 : Effective Length of Column as per IS : 456-2000

Degree of End Restraint of Compression Member	Symbol	Theoretical Value of Effective Length	Recommended Value of Effective Length
Effectively held in position and restrained against rotation at both ends (i.e. both ends are fixed).		0.5l	0.65l
Effectively held in position at both ends, restrained against rotation at one end (i.e., fixed at one end and hinged at the other end).		0.7l	0.80l
Effectively held in position at both ends, but not restrained against rotation (i.e., both ends are hinged).		1.00l	1.00l
Effectively held in position and restrained against rotation at one end, and at the other end restrained against rotation, but not held in position.		1.00l	1.20l
Effectively held in position and restrained against rotation at one end, and, at the other end partially restrained against rotation but not held in position.		–	1.50l
Effectively held in position at one end but not restrained against rotation, and at the other end restrained against rotation but not held in position.		2.00l	2.00l
Effectively held in position and restrained against rotation at one end but not held in position nor restrained against rotation at the other end (i.e., fixed at one end and free at the other end)		2.00l	2.00l

Note : l = Unsupported length of column.

4.8 SLENDERNESS RATIO

The slenderness ratio of a compression member is defined as the ratio of effective length to the least lateral dimension.

4.9 RADIUS OF GYRATION

It is the distance of area from centroidal axis at which get constant moment of inertia. The radius of gyration of a given area about any axis is that distance from the given axis at which the entire area is assumed to be concentrated without changing the moment of inertia about given axis. It is denoted by 'k'

$$I = AK^2 \quad \therefore k = \sqrt{\dfrac{I}{A}}$$

The unit of radius of gyration will be mm, cm, m etc.

4.10 POSITION OF COLUMN

- At the corners of a building.
- At the intersections of wall.
- Near the corners of a building (along a cross wall) if difficulties are encountered in providing footing.
- At such a position which can avoid larger span beam.

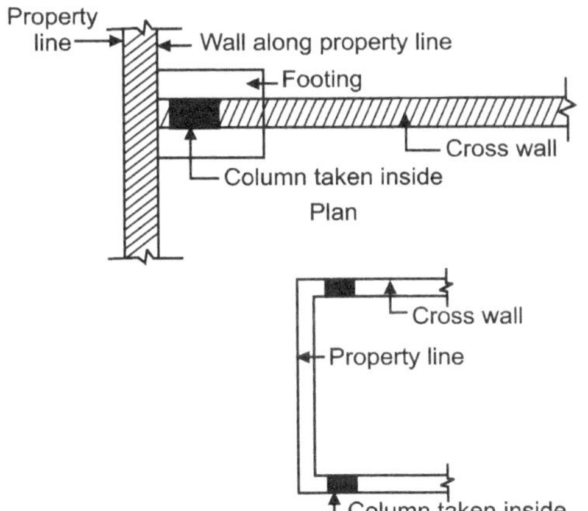

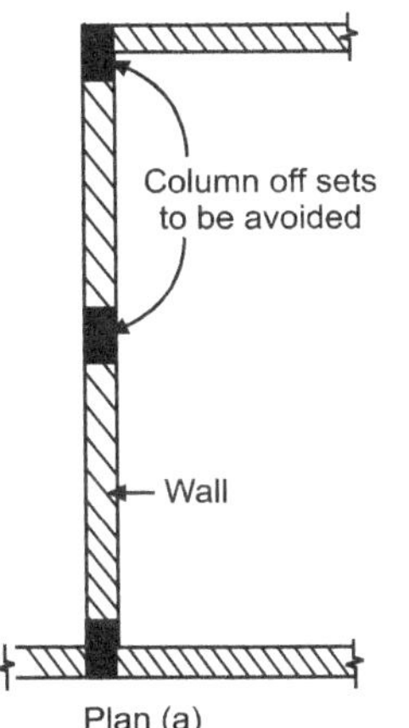

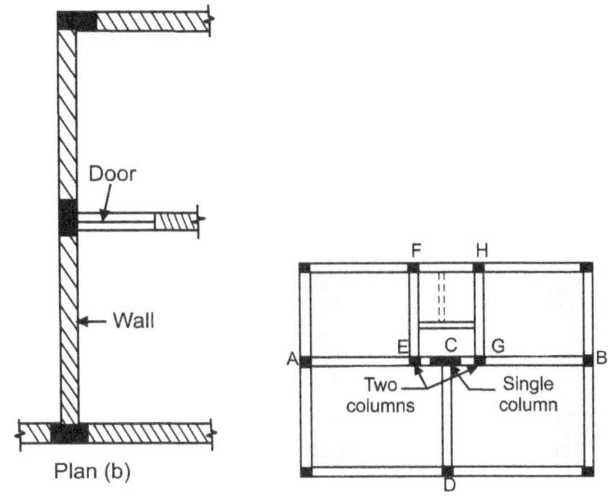

Fig. 4.8

Width (Small Side) of Column

- Width of column > Width of beam.
- Width of column < thickness of wall

Depth (Long Side) of Column (Orientation)

- Depth of column shall be in the plane of wall to avoid offset.
- Depth of column shall be in the plane of bending (along larger beam) so as to get larger moment resisting capacity.

4.11 DIFFERENCE BETWEEN SHORT COLUMN AND LONG COLUMN

Sr. No.	Short Column	Long Column
1.	A column is considered to be short if the ratio of effective length to its least lateral dimension is less than or equal to 12.	A column is considered to be long if the ratio of effective length of column to its least lateral dimension is greater than 12.
2.	The ratio of effective length of a short column to its least radius of gyration is less than or equal to 40.	The ratio of effective length of a long column to its least radius or gyration is greater than 40.
3.	Buckling tendency is very low.	Long and slender columns buckle easily.
4.	The load carrying capacity is high as compared to long column of the same cross-sectional area.	The load carrying capacity of a long column is less as compared to short column of same cross-sectional area.
5.	The failure of the short column is by crushing.	The column generally fails in buckling.

4.12 TYPES OF LATERAL REINFORCEMENT

An R.C.C. column has longitudinal and lateral reinforcement. They can also be classified according to the manner in which the longitudinal steel is laterally supported or tied.

- Column with longitudinal steel and lateral ties.
- Column with longitudinal steel and spiral ties.
- In this type of arrangement the longitudinal bars are tied laterally at suitable internals with the help of ties as shown in Fig. 4.9.

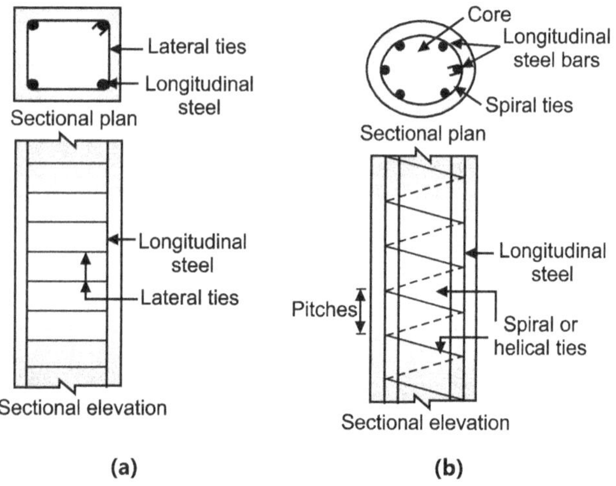

Fig. 4.9

- Fig. 4.9 shows a circular column with spirals i.e. the longitudinal bars are tie continuously with the help of a spiral reinforcement. The columns with helical or spiral as compared to links thus they increase the buckling resisance and ductility of the column.

4.13 ARRANGEMENT OF TRANSVERSE / LATERAL REINFORCEMENT

The arrangement of transverse reinforcement as per IS : 456 : 2000 (clause 26.5.3.2(b)) is as follows :

- If the longitudinal bars are not spaced more than 75 win or either side, the transverse reinforcement need only to go round corner and alternate bar for the purpose of providing effective lateral supports.

- If the longitudinal bars are spaced at a distance not exceeding 48 times the diameter of the bar, the ties are effectively tied in two directions, additional longitudinal bars in between these bars need to be tied in one direction by open ties.

- Where the longitudinal reinforcing bars in a compression member are placed in more than one row, effective lateral support to the longitudinal bars in

the inner rows may be assumed to have been provided if :

- ➢ Transverse reinforcement is provided for the outer most row in accordance with 2.

- ➢ No bar of the inner row is closer to the nearest compression face than three mimes the diameter of the largest bar in the inner row.

- Where the longitudinal bars in a compression member are grouped (not in contact) and each group adequately tied with transverse reinforcement in accordance with the codal provisions, the transverse reinforcement for the compression member as a whole may be provided on the assumption that each group is a single longitudinal bar for the purpose of determining the pitch and diameter of the transverse reinforcement in accordance with the codal provisions. The diameter of such transverse reinforcement should not exceed 20 mm.

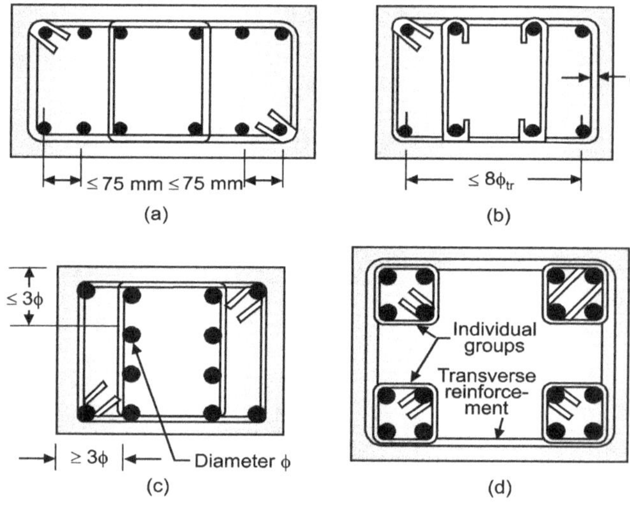

Fig. 4.10

4.14 IS SPECIFICATION REGARDING COLUMN

(Refer, cause 26.5.3, Pg. No. 48, IS 456 : 2000)

1. Longitudinal Reinforcement : The longitudinal reinforcement consist of steel bars placed longitudinally in a column. It is also called main steel.

(a) Area of Longitudinal Reinforcement :

$$A_{sc\ min} = 0.80\ \% \text{ of } A_g$$

$$A_{sc\ max} = 6\% \text{ of } A_g$$

where, A_g = Total gross cross sectional area of column

A_{sc} = Cross sectional are of longitudinal reinforcement

Note : The use of 6% r/f may involve practical difficulties in placing and compacting concrete. Hence % of steel shall not exceed 4%.

(b) Minimum Number of Longitudinal Bars :

   (i) For rectangular column = 4 No's.

   (ii) For circular column = 6 No's.

(c) Minimum Diameter of Longitudinal Bar :

   The diameter of the longitudinal bars shall not be less than 12 mm. ($\therefore$ $\phi \nless$ 12 mm).

(d) Spacing of Longitudinal Reinforcement / Bar

   (i) The minimum clear distance between two main bars shall not be less than nominal size of aggregate plus 5 mm

   (clear spacing $\nless$ Nominal size of aggregate

   + 5 mm)

   (ii) The maximum spacing of longitudinal bars shall not be more that 300 mm along periphery of the column.

(e) Nominal Cover to Longitudinal Bar :

   Nominal cover should not be less than

   (i) 40 mm

   (ii) Largest dia. of longitudinal bar   } whichever is more

Note : In case of column width is 200 mm or less and whose reinforcing bars does not exceed 12 mm, a nominal cover of 25 mm may be used.

(f) Lap Length :

   Lap length is the length of the overlap of bar required to safely transfer stress from one bar to another. If two different diameter bar are to be lapped, a lap length based on smaller diameter.

•   Lap length in column should be 45 D.

•   Lap length in beam / slab should be 69D.

   Where, D = Diameter of bar in mm.

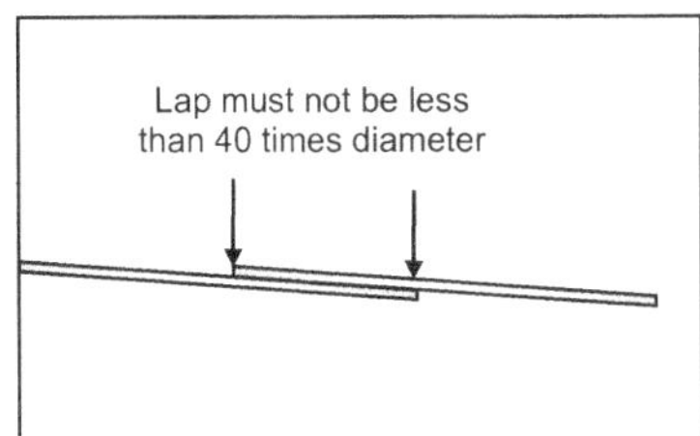

Fig. 4.11

2. Lateral / Transverse Reinforcement

The lateral or transverse reinforcement in a column maybe provided in the form of lateral ties or spiral or helical r/f. It is also called as links, ring, hoop r/f , binders etc.

(a) Diameter of Lateral Ties / Spiral or Helical Reinforcement

The diameter of lateral ties should not be less than,

   (i) $\phi = \dfrac{1}{4} \times$ largest diameter of longitudinal bar   } Provide whichever is more

   (ii) ϕ = 6 mm

(b) Pitch / Spacing of Lateral Ties

   The pitch or spacing of the lateral ties (links) should not be greater than.

   (i) Least lateral dimensions of the column

   (ii) 16 times diameter of smallest diameter of longitudinal bar   } whichever is less

   (iii) 300 mm

(c) Pitch of Helical Reinforcement

   The pitch of the helical turns should not be more than,

   (i) 75 mm

   (ii) $\dfrac{1}{6}$ th of the diameter of core of concrete   } less

The pitch of the helical turns should not be less than :

   (i) 25 mm

   (ii) 3 times the diameter of steel bar forming the helix.

(d) Arrangement of Transverse Reinforcement

The arrangement of transverse reinforcement as per IS : 456 : 2000 (clause 26.5.3.2(b)) is as follows :

•   If the longitudinal bars are not spaced more than 75 win or either side, the transverse reinforcement need only to go round corner and alternate bar for the purpose of providing effective lateral supports.

•   If the longitudinal bars are spaced at a distance not exceeding 48 times the diameter of the bar, the ties are effectively tied in two directions, additional longitudinal bars in between these bars need to be tied in one direction by open ties.

•   Where the longitudinal reinforcing bars in a compression member are placed in more than one row, effective lateral support to the longitudinal bars in the inner rows may be assumed to have been provided if :

- ➢ Transverse reinforcement is provided for the outer most row in accordance with 2.
- ➢ No bar of the inner row is closer to the nearest compression face than three mimes the diameter of the largest bar in the inner row.

- Where the longitudinal bars in a compression member are grouped (not in contact) and each group adequately tied with transverse reinforcement in accordance with the codal provisions, the transverse reinforcement for the compression member as a whole may be provided on the assumption that each group is a single longitudinal bar for the purpose of determining the pitch and diameter of the transverse reinforcement in accordance with the codal provisions. The diameter of such transverse reinforcement should not exceed 20 mm.

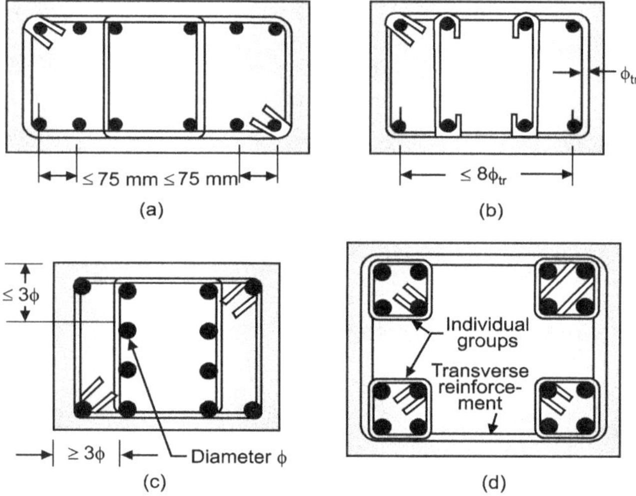

Fig. 4.12

4.15 SLENDERNESS LIMITS FOR COLUMN

The unsupported length between end restraints shall not exceed 60 times the least lateral dimension of a column.

If in any given plane, one end of a column is unrestrained, its unsupported length l, shall not exceed $\dfrac{100\, b^2}{D}$.

where,   b = width of that cross-section, and

D = depth of cross-section measured in the plane under consideration.

4.16 MINIMUM ECCENTRICITY

(Refer clause 25.4 and 39.3, Pg. No. 42, IS 456 : 2000)

All columns shall be designed for minimum eccentricity.

i.e. $e_{min} = \dfrac{\text{Unsupported length}}{500} + \dfrac{\text{lateral dimension}}{30}$

Subject to minimum of 20 mm.

The unsupported length (l) and lateral dimension can be ether 'b' or the depth 'D' of the column depending on the axis of bending.

Therefore,

$\therefore \qquad e_{min} = \dfrac{l_x}{500} + \dfrac{D}{30}$ or 20 mm, whichever is more

$\therefore \qquad e_{min} = \dfrac{l_y}{500} + \dfrac{b}{30}$ or 20 mm, whichever is more

and $\therefore\ e_{max} \not> 0.05\, D$ or $0.05\, b$

4.17 PERMISSIBLE STRESSES IN R.C.C. COLUMN

1. Permissible Stresses in Concrete :

(Refer Table – 21, Pg. No 81, IS 456 : 2000)

Grade of Concrete	M-10	M-15	M-20	M-25	M-30	M-35	M-40	M-45	M-50
Permissible Stress in Compression 'σ_{cc}' N/mm²	2.5	4.0	5.0	6.0	8.0	9.0	10.0	11.0	12.0

2. Permissible Stresses in Steel

(Refer Table 22, Pg. No. 82, IS 456 : 2000)

Sr. No.	Grade	Permissible compressive Stress in Longitudinal Steel Bars in N/mm²
1.	Mild steel (Fe-250)	130
2.	For HYSD Steel (Fe 415)	190

4.18 PERMISSIBLE LOADS IN COMPRESSION MEMBERS

(Refer Clause B-3, Pg. No. 81, IS 456 : 2000)

1. Short Column

(i) Short Column with Lateral Ties

The axial load 'P' on pedestal or short column reinforced with longitudinal bars and lateral ties is given by the following equations :

$\therefore \qquad \mathbf{P = \sigma_{cc} \cdot A_c + \sigma_{sc} \cdot A_{sc}}$

where, σ_{cc} = Permissible stress in concrete in direct compression

A_c = Cross sectional area of concrete excluding steel

σ_{sc} = Permissible compressive stress for column bars

A_{sc} = Cross-sectional area of the longitudinal steel

Note : The minimum eccentricity does not exceed 0.05 times the lateral dimensions.

(ii) Short Column with Helical Reinforcement

The permissible load for columns with helical reinforcement satisfy the requirement that the ratio of volume of helical reinforcement to the volume of core shall not be less than $0.36 \, (A_g / A_c - 1) \cdot (f_{ck} / f_y)$, shall be 1.05 times the permissible load for similar member with lateral ties or rings.

$$\therefore \qquad P = 1.05 \, (\sigma_{cc} \cdot A_c + \sigma_{sc} \cdot A_{sc})$$

2. Long Column :

When the ratio of the effective length and the least lateral dimension of a column exceeds 12, the column will be considered as long column. In the design of such columns considering the factor of buckling, lower value of working stresses in steel and concrete is adopted, by multiplying the general working stresses by the reduction coefficient C_r.

So for long column,

Safe stress in concrete = $C_r \times$ corresponding safe stress for short column

and, safe stress in steel = $C_r \times$ Corresponding safe stress for short column.

$\therefore$ The reduction coefficient can be obtained by the following formula,

$$\therefore \qquad C_r = 1.25 - \frac{l_{eff}}{48 \, b}$$

For more exact calculation

$$\therefore \qquad C_r = 1.25 - \frac{l_{eff.}}{160 \, I_{min}}$$

where, C_r = Reduction coefficient

$l_{eff.}$ = Effective length of column, and

b = Least lateral dimension of column, for column with helical r/f, 'b' is the diameter of core

I_{min} = Least moment of inertia

SOLVED EXAMPLES

Type I : Analysis – Axially Loaded Column

Example 4.1 : *A short column 400 mm $\times$ 400 mm in section is reinforced with 8 bars of 20 mm diameter. Find the safe load on the column by elastic theory. Take M-20 and Fe-415.*

Solution : Given : $b = D = 400$ mm

$$A_{sc} = 8 \times \frac{\pi}{4} \times 20^2$$

$$\therefore \qquad A_{sc} = 2513.27 \text{ mm}^2$$

$$\text{M-20}; \; \sigma_{cc} = 5 \text{ N/mm}^2$$

$$\text{Fe-415}; \; \sigma_{sc} = 190 \text{ N/mm}^2$$

1. Area of concrete (A_c) :

$$A_c = A_g - A_{sc} = (400 \times 400) - 2513.27$$

$$\therefore \qquad A_c = 157486.73 \text{ mm}^2$$

2. Calculation of load carrying capacity of column (P) :

$$P = \sigma_{cc} \cdot A_c + \sigma_{sc} \cdot A_{sc}$$

... Assume column carries lateral ties

$$P = (5 \times 157486.73) + (190 \times 2513.27)$$

$$\therefore \qquad P = 1264.95 \text{ kN}$$

Example 4.2 : *A reinforced concrete column 4 m long effectively held in position but not restrained against rotation at both ends, 400 mm in diameter is reinforced with 6 bars of 22 mm diameter. Take, M-20 concrete and Fe-415 steel.*

Solution : Given : $D = 400$ mm,

$$L = 4m,$$

$$l_{eff} = 1.00L = 4000 \text{ mm}$$

$$\text{Hence,} \quad \frac{l_{eff}}{D} = \frac{4000}{400} = 10 < 12, \text{ thus the column is short.}$$

$$A_{sc} = \sigma \times \frac{\pi}{4} \times 22^2 = 2280.80 \text{ mm}^2$$

$$\sigma_{cc} = 5 \text{ N/mm}^2, \; \sigma_{sc} = 190 \text{ N/mm}^2$$

1. Area of concrete (A_c) :

$$A_c = A_g - A_{sc}$$

$$= \left(\frac{\pi}{4} \times 400^2 \right) - 2280.80$$

$$\therefore \qquad A_c = 123382.91 \text{ mm}^2$$

2. Calculation of safe load carried by column (P) :

$$P = \sigma_{cc} \cdot A_c + \sigma_{sc} \cdot A_{sc}$$

$$= (5 \times 123382.91) + (190 \times 2280.80)$$

$$\therefore \qquad P = 1050.266 \text{ kN}$$

Example 4.3 : *In above question take, length of column is 8m, what will be the safe load the column can carry?*

Solution : Given : $L = 8.00$ m

$$l_{eff} = 1.00 \, L = 8000 \text{ mm}$$

$$A_{sc} = 2280.80 \text{ mm}^2$$

$$\sigma_{cc} = 5 \text{ N/mm}^2$$

$$\sigma_{sc} = 190 \text{ N/mm}^2$$

$$\text{Hence,} \quad \frac{l_{eff}}{D} = \frac{8000}{400} = 20 > 12$$

thus, the column is 'long column'

$$\therefore \quad \text{Reduction factor} = C_r = \left(1.25 - \frac{l_{eff}}{48b} \right)$$

$$= \left(1.25 - \frac{8000}{48 \times 400} \right)$$

$$\therefore \qquad C_r = 0.833$$

$\therefore$ Safe load on column (P) :

$$P = C_r \, [\sigma_{cc} \cdot A_c + \sigma_{sc} \cdot A_{sc}]$$

$$P = 0.833 \times 1050.266$$

$$\therefore \quad \textbf{P = 874.87 kN}$$

Example 4.4 : *A column 3.00 m long of 300 mm × 450 mm in section is reinforced with 8 bars of 16 mm diameter. The column is effectively held in position and restrained against rotation at both end. Find the safe load on column.*

Take M-15 and Fe-250 grades material.

Solution : Given :

$$b = 300 \text{ mm}, D = 450 \text{ mm}$$

$$L = 3.00 \text{ m}, l_{eff} = 0.65 \text{ L} = 1950 \text{ mm}$$

$$A_{sc} = 8 \times \frac{\pi}{4} \times 16^2 = 1608.50 \text{ mm}^2$$

$$\therefore \quad \sigma_{cc} = 4 \text{ N/mm}^2, \sigma_{sc} = 130 \text{ N/mm}^2$$

Hence, $\dfrac{l_{eff}}{b} = \dfrac{1950}{300} = 6.5 < 12$

Thus, the column is a short column

1. Area of concrete (A_c) :

$$A_c = A_g - A_{sc} = (b \times D) - A_{sc}$$

$$A_c = (300 \times 450) - 1608.50$$

$$\therefore \quad \textbf{A}_c \textbf{ = 133391.50 mm}^2$$

2. Safe load on column (P) :

$$P = \sigma_{cc} \cdot A_c + \sigma_{sc} \cdot A_{sc}$$

$$= (4 \times 133391.50) + (130 \times 1608.50)$$

$$\therefore \quad \textbf{P = 742.67 kN}$$

Example 4.5 : *Determine the safe L.C.C. of column 450 mm in diameter reinforced with 1.5% of longitudinal r/f. The column has an unsupported length of 3.2 m. Use M-25, Fe-415. Assume effective length to be unsupported length.*

Solution : Given :

$$D = 450 \text{ mm},$$

$$L = l_{eff} = 3.2 \text{ m}$$

$$A_{sc} = 1.5 \% \text{ of } A_g = \frac{1.5}{100} \times \left(\frac{\pi}{4} \times 450^2\right)$$

$$\therefore \quad \textbf{A}_{sc} \textbf{ = 2382.65 mm}^2$$

$$\sigma_{cc} = 6.0 \text{ N/mm}^2$$

$$\sigma_{sc} = 190 \text{ N/mm}^2$$

1. Area of concrete (A_c) :

$$A_c = A_g - A_{sc} = \left(\frac{\pi}{4} \times 450^2\right) - 2385.65$$

$$\therefore \quad \textbf{A}_c \textbf{ = 156657.48 mm}^2$$

2. Safe L.C.C. of column (P) :

$$P = 1.05 (\sigma_{cc} \cdot A_c + \sigma_{sc} \cdot A_{sc})$$

... for spiral columns, 5% increase in strength

$$\therefore \quad P = 1.05 (6 \times 156657.48 + 190 \times 2385.65)$$

$$\therefore \quad \textbf{P = 1462.88 kN}$$

Type II : Design – Axially Loaded Column

Example 4.6 : *Calculate the area of steel required for a R.C. column 350 mm × 450 mm in cross section to carry an axial load of 1000 kN. Assume grade of concrete M-20 and steel of grade, Fe-415.*

Solution : Given :

$$b = 350 \text{ mm}, D = 450 \text{ mm}$$

$$P = 1000 \text{ kN}, \sigma_{cc} = 5 \text{ N/mm}^2$$

$$\sigma_{sc} = 190 \text{ N/mm}^2$$

1. Calculation of gross area (A_g) :

$$A_g = b \times D = 350 \times 450$$

$$\therefore \quad A_g = 157500.00 \text{ mm}^2$$

2. Calculation of Area of steel (A_{sc}) :

$$P = \sigma_{cc} \cdot A_c + \sigma_{sc} \cdot A_{sc}$$

$$1000 \times 10^3 = 5 \times (157500 - A_{sc}) + 190 \times A_{sc}$$

$$\therefore \quad \textbf{A}_{sc} \textbf{ = 1148.65 mm}^2 \ngtr 0.8 \% \text{ of } A_g$$

Assume, diameter of longitudinal bar, $\phi = 18$ mm

$$\therefore \text{ No. of bars required} = \frac{A_{sc}}{\left(\frac{\pi}{4}\right) \times \phi^2} = \frac{1148.65}{\left(\frac{\pi}{4}\right) \times 18^2}$$

$$= \textbf{4.51} \approx \textbf{6 No's}$$

$$\therefore \textbf{ Provide, 6 – 18 mm } \phi \textbf{ longitudinal bars > 0.8 \% } \textbf{A}_g$$

3. Design of lateral ties :

(a) Diameter of lateral ties :

(i) $\phi = \dfrac{1}{4} \times$ largest dia. of longitudinal bar

$$\quad\quad = \dfrac{1}{4} \times 16 = 4 \text{ mm}$$

$\left.\begin{array}{l} \\ \\ \\ \\ \end{array}\right\}$ ... More

(ii) $\phi = 6$ mm

$$\therefore \quad \phi = \textbf{6 mm}$$

(b) Spacing (s) :

(i) S = least lateral dimension

$$\quad\quad = 350 \text{ mm}$$

(ii) S = 16 × smallest dia. of longitudinal bar

$$\quad\quad = 16 \times 16 = 256 \text{ mm}$$

$\left.\begin{array}{l} \\ \\ \\ \\ \end{array}\right\}$ ... Less

(iii) S = 300 mm

$$\therefore \quad \textbf{S = 256} \approx \textbf{250 mm}$$

$$\therefore \quad \textbf{Provide, 6 mm } \phi \textbf{ ties @ 250 mm c/c.}$$

Example 4.7 : *Design the reinforcement in a column of a 450 mm × 600 mm, subject to an axial load of 2000 kN under service load due to dead load and live loads. The column has an unsupported length of 3.00 m and is restrained in both directions. Use M-20 and Fe-415 grades material.*

Solution : Given :

$$b = 450 \text{ mm}, D = 600 \text{ mm}$$
$$P = 2000 \text{ kN}, l = 3.00 \text{ mm}$$
$$\sigma_{cc} = 5 \text{ N/mm2}, \sigma_{sc} = 190 \text{ N/mm}^2$$

Effective length $= 0.65\, l = 0.65 \times 3000$

$\therefore \quad l_{eff} = 1950.00 \text{ mm}$

1. Type of column :

Slenderness ratio $= \lambda = \dfrac{l_{eff}}{\text{least lateral dimension}} = \dfrac{1950}{450}$

$\therefore \quad \lambda = 4.33 < 12$

$\therefore$ Hence, the column may be designed as a short column.

2. Check for minimum eccentricity :

$$e_{x \cdot min} = \frac{l}{500} + \frac{D}{30} = \frac{3000}{500} + \frac{600}{30}$$
$$= 26.00 \text{ mm} > 20 \text{ mm}$$

and $\quad e_{max} = 0.05\, D = 0.05 \times 600 = 30 \text{ mm}$

$\therefore \ e_{x \cdot min} < e_{max}$ ... Okay

Also, $\quad e_{y,min} = \dfrac{l}{500} + \dfrac{b}{30} = \dfrac{3000}{500} + \dfrac{450}{30}$

$$= 21.00 \text{ m} > 20 \text{ mm}$$

$e_{max} = 0.05\, b = 0.05 \times 450 = 22.5 \text{ mm}$

$e_{y.min} < e_{max}$ ... Okay

3. Calculation of longitudinal reinforcement (A_{sc}) :

$$P = \sigma_{cc} \cdot A_c + \sigma_{sc} \cdot A_{sc}$$
$$= \sigma_{cc} \cdot (A_g - A_{sc}) + \sigma_{sc} \cdot A_{sc}$$
$$2000 \times 10^3 = 5 \times (450 \times 600 - A_{sc}) + 190 \times A_{sc}$$
$$\therefore \quad \mathbf{A_{sc} = 3513.51 \ mm^2} > 0.8 \text{ \% of } A_g$$

Assume, diameter of longitudinal bar $= \phi = 22$ mm

$\therefore$ No. of bars required $= \dfrac{A_{sc}}{\left(\dfrac{\pi}{4} \times \phi^2\right)} = \dfrac{3513.51}{\left(\dfrac{\pi}{4} \times 22^2\right)} = 9.24$

$$\approx 10 \text{ No's}$$

$\therefore$ **Provide, 10 – 22 mm ϕ, as a longitudinal reinforcement.**

4. Design of lateral ties :

(a) Diameter of lateral ties :

(i) $\quad \phi = \dfrac{1}{4} \times 22 = 5.50$ m

(ii) $\quad \phi = 6$ mm   ... More

$\therefore \quad \phi = \mathbf{6 \ mm}$

(b) Spacing / pitch if lateral ties (s) :

(i) $\quad$ S $=$ least lateral dimension

$\quad\quad\quad = 450$ mm

(ii) $\quad$ S $= 16 \times$ smallest dia. of longitudinal bar

$\quad\quad\quad = 16 \times 22 = 352$ mm   ... Less

(iii) $\quad$ S $= 300$ mm

$\therefore \quad$ S $= 300$ mm

Provide, 6mm ϕ ties @ 300 mm C/C

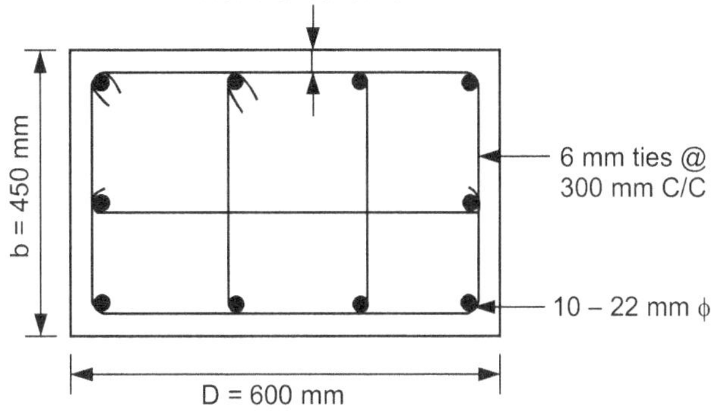

Fig. 4.13

Example 4.8 : *Design a column to carry an axial load of 800 kN. Design the column as per I.S. code. Permissible stresses in concrete and steel are 5 N/mm² and 130 N/mm².*

Solution : Given :

$$P = 800 \text{ kN}, \sigma_{cc} = 5 \text{ N/mm}^2$$
$$\sigma_{sc} = 130 \text{ N/mm}^2$$

Assume, Axially loaded short column

1. Assume % of longitudinal reinforcement :

$$A_{sc} = 1 \text{ \% of } A_g = 0.01\, A_g \quad\quad ... \text{(a)}$$

2. Area of concrete (A_c) :

$$A_c = A_g - A_{sc} = A_g - 0.01\, A_g \quad ... \text{from (a)}$$
$$\therefore \quad A_c = 0.99\, A_g \quad\quad\quad\quad ... \text{(b)}$$

3. Calculation of gross area (A_g) :

$$P = \sigma_{cc} \cdot A_c + \sigma_{sc} \cdot A_{sc}$$

From (a) and (b)

$$800 \times 10^3 = 5 \times 0.99\, A_g + 0.01\, A_g \times 130$$
$$\therefore \quad A_g = 128000 \text{ mm}^2$$

Assume, square column,

$\therefore$ Each side of column

$$= b = D = \sqrt{A_g} = \sqrt{128000}$$
$$\therefore \quad b = D = 357.77$$

$\therefore$ Provide, b $\approx$ D $\approx$ 360 mm

Hence, Provide, size of column 400 mm × 400 mm

4. Area of longitudinal r/f (A_{sc}) :

From (a)

$$A_{sc} = 0.01 \times (360 \times 360)$$

$$\therefore \quad A_{sc} = 1296.00 \text{ mm}^2 > 0.8 \text{ % of } A_g$$

Assume, diameter of longitudinal bar $= \phi = 16$ mm

$$\therefore \quad \text{No. of bars} = \frac{A_{sc}}{\left(\frac{\pi}{4} \times 16^2\right)} = 6.44 \approx 8 \text{ No's}$$

∴   Provide 8-16 mm ϕ as a longitudinal r/f.

5. Design of lateral ties :

(a) Diameter of lateral ties :

(i) $\phi = \dfrac{1}{4} \times$ diameter of main bar

$\qquad = \dfrac{1}{4} \times 16 = 4$ mm $\qquad$ ... More

(ii) $\phi = 6$ mm

∴ $\phi = \textbf{6 mm}$

(b) Spacing / Pitch of lateral ties :

(i)   S = least lateral dimension

$\qquad = 360$ mm

(ii) S = $16 \times$ dia. of main bar $\qquad$ ... Less

$\qquad = 16 \times 16 = 256$ mm

(iii) S = 300 mm

∴   **S = 256 $\approx$ 250 mm**

∴   **provide 6 mm ϕ ties @ 250 mm c/c**

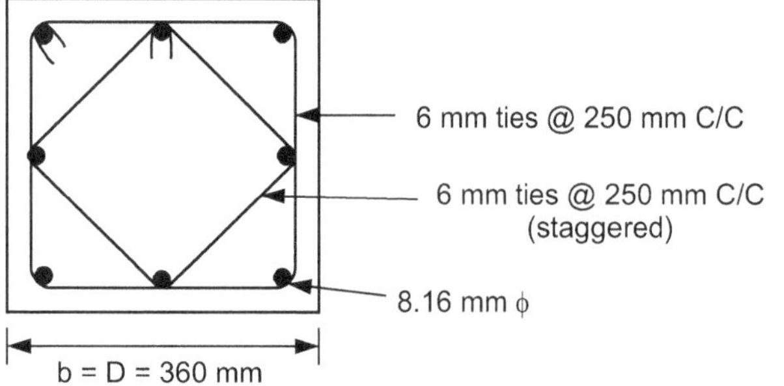

Fig. 4.14

Example 4.9 : *Design the reinforcement in a spiral column of 450 mm diameter subjected to service load of 1200 kN. The column has an unsupported length of 3.4 m. Use M-25, Fe-415.*

Solution : Given :

$$D = 450 \text{ mm}, \sigma_{cc} = 6 \text{ N/mm}^2$$
$$P = 1200 \text{ kN}, \sigma_{sc} = 190 \text{ N/mm}^2$$
$$l = 3.40 \text{ m}$$

Assume, effective length to be equal to unsupported length.

$$\therefore \quad l_{eff} = l = 3.40 \text{ m} = 3400 \text{ mm}$$

1. Type of column :

$$\text{slenderness ratio} = \frac{l_{eff}}{\text{least lateral dimension}}$$

$$= \frac{3400}{450} = 7.56 < 12$$

∴   The column is designed as a 'short column'.

2. Check for minimum eccentricity :

$$l_{min} = \frac{l}{500} + \frac{D}{30} = \frac{3400}{500} + \frac{450}{30}$$

$$= 21.00 > 20 \text{ mm}$$

and $\qquad e_{max} = 0.05 \times D$

$$= 0.05 \times 450$$

$$= 22.5 \text{ mm}$$

$$\therefore \quad e_{max} > e_{min} \qquad \text{... Okay}$$

So, the code formula for axially compressed short columns may be used.

3. Calculation of longitudinal reinforcement :

For spiral columns, 5% increase in strength is allowed.

$$\therefore \quad P = 1.05 \,(\sigma_{cc} \cdot A_c + \sigma_{sc} \cdot A_{sc})$$

$$P = 1.05 \,[\sigma_{cc} \times (A_g - A_{sc}) + \sigma_{sc} \cdot A_{sc}]$$

$$1200 \times 10^3 = 1.05 \left[6 \times \left(\frac{\pi}{4} \times 450^2 - A_{sc}\right) + 190 \times A_{sc}\right]$$

$$\therefore \quad A_{sc} = 1024.99 \text{ mm}^2 \not> 0.8 \text{ % of } A_g$$

$$\therefore \quad A_{sc} = 0.8 \text{ % of } A_g$$

$$A_{sc} = 1272.34 \text{ mm}^2$$

Assume diameter of bar $= \phi = \textbf{18 mm}$

$$\therefore \quad \text{No. of bars} = \frac{A_{sc}}{\left(\frac{\pi}{4} \times \phi^2\right)}$$

$$= \frac{1272.34}{\left(\frac{\pi}{4} \times 18^2\right)}$$

$$= 5.00$$

$$\approx 6 \text{ No's}$$

∴ Provide, 6-16 mm ϕ, as a longitudinal r/f

4. Design of lateral ties :

(a) Diameter of lateral ties :

(i) $\phi = \dfrac{1}{4} \times$ Dia. of main bar

$\qquad = \dfrac{1}{4} \times 16 = 4$ mm $\qquad$ ... More

(ii) $\phi = 6$ mm

∴ $\phi = \textbf{6 mm}$

(b) Pitch of spiral reinforcement :

(i) $\quad$ S $= 75$ mm

(ii) $\quad$ S $= \dfrac{1}{6} \times$ core diameter

$\qquad = \dfrac{1}{6} \times (450 - 2 \times 40)$ $\quad$ ... Less

$\qquad = 61.67$

∴ $\quad$ S $= 61.67 \approx 60$ mm > 25 mm

∴ **Provide, 6mm ϕ spiral @ 60 mm c/c pitch**

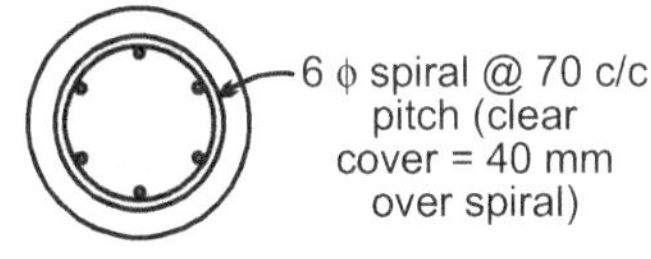

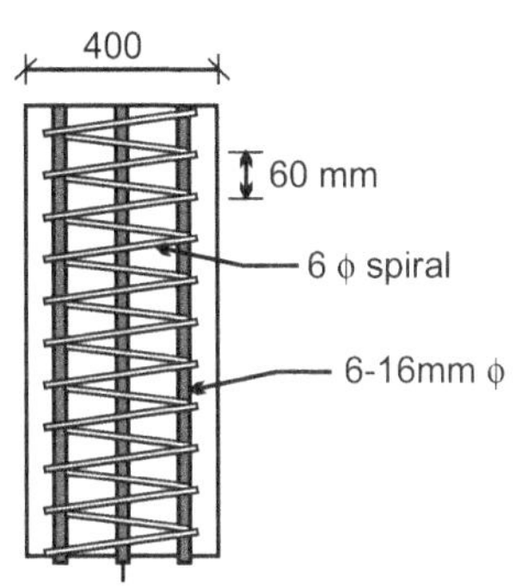

Fig. 4.15

Example 4.10 : *Design a rectangular column to carry an axial load of 900 kN. The length of column is 3.2 m, the column is effectively held in position at both ends but not restrained against rotation at one end. Take M-20 and Fe-250.*

Solution : Given :

$\qquad$ P $= 900$ kN,

$\qquad l = 3.2$ m

$\qquad \sigma_{cc} = 5$ N/mm^2

$\qquad \sigma_{sc} = 130$ N/mm^2

Effective length=

$\qquad l_{eff} = 0.80\, l = 0.80 \times 3200$

∴ $\qquad l_{eff} = 2560.00$ mm

Assume, Axially loaded short column.

1. Assume % of longitudinal reinforcement :

∴ $\qquad A_{sc} = 1$ % of $A_g = 0.1 \times A_g$ $\quad$ ... (a)

2. Area of concrete (A_c) :

∴ $\qquad A_c = A_g - A_{sc} = A_g - 0.01\, A_g$

∴ $\qquad A_c = 0.99\, A_g$

3. Calculation of gross area (A_g) :

$\qquad$ P $= \sigma_{cc} \cdot A_c + \sigma_{sc} \cdot A_{sc}$

$900 \times 10^3 = 5 \times 0.99\, A_g + 130 \times 0.01 \times A_g$

∴ $\qquad A_g = 144000$ mm^2

Provide, size of rectangular column is (400 × 450 mm)

4. Check :

(i) Type of column :

Slenderness ratio $= \lambda = \dfrac{l_{eff}}{\text{least lateral dimension}}$

$\qquad \lambda = \dfrac{2560}{400} = 6.4 < 12$

∴ $\quad$ The column is 'short column'

(ii) Minimum eccentricity :

$\qquad e_{min} = \dfrac{l}{500} + \dfrac{b}{30} = \dfrac{3200}{500} + \dfrac{400}{30} = 19.73$

$\qquad \approx 20$ mm

$\qquad e_{max} = 0.05 \times b = 0.05 \times 400 = 20$ mm

$\qquad e_{min} = e_{max}$ $\qquad$ ... Okay safe.

∴ $\quad$ Assumption is correct.

5. Calculation of longitudinal reinforcement :

from (a),

$\qquad A_{sc} = 0.01\, A_g = 0.01 \times (400 \times 450)$

∴ $\qquad A_{sc} = 1800$ mm$^2 > 0.8$ % of A_g

Assume, diameter of longitudinal bar $= \phi = 18$ mm

∴ No. of bars $= \dfrac{A_{sc}}{\left(\dfrac{\pi}{4} \times \phi^2\right)} = \dfrac{1800}{\left(\dfrac{\pi}{4} \times 18^2\right)} = 7.07$

$\qquad \approx 8$ No's

∴ **Provide, 8 - 18 mm ϕ as a longitudinal r/f.**

6. Design of lateral ties :

(a) Diameter of lateral ties :

(i) $\quad \phi = \dfrac{1}{4} \times$ dia. of main bar

$\qquad = \dfrac{1}{4} \times 18$ $\quad$ ... More

$\qquad = 4.5$ mm

(ii) $\quad \phi = 6$mm

∴ $\qquad \phi = $ **6mm**

(b) spacing (s) :

(i) $\quad$ S $=$ least lateral dimension

$\qquad = 400$ mm

(ii) $\quad$ S $= 16 \times \phi = 16 \times 18$ $\quad$ ... Less

$\qquad = 288$ mm

(iii) $\quad$ S $= 300$ mm

∴ $\qquad$ **S $= 288 \approx 275$ mm**

∴ **Provide, 6mm ϕ @ 275 mm c/c, lateral ties.**

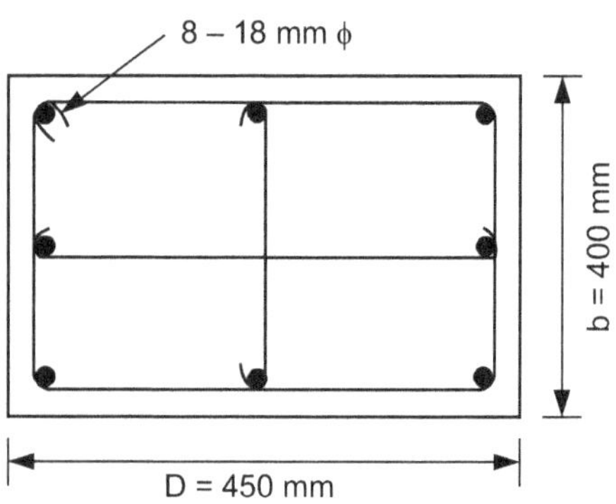

Fig. 4.16

Example 4.11 : *A column 400 mm × 400 mm × 6 m long has to support a load of 875 kN. Find the necessary reinforcement for the column. Use M-20 concrete. For a short column the safe stresses in concrete and steel are 5 N/mm² and 130 N/mm² respectively.*

Solution : Given :

$$b = D = 400 \text{ mm}$$
$$l = l_{eff} = 6 \text{ m}$$
$$P = 875 \text{ kN}$$
$$\sigma_{cc} = 5 \text{ N/mm}^2$$
$$\sigma_{sc} = 130 \text{ N/mm}^2$$

1. Type of column :

Slenderness ratio $= \lambda = \dfrac{l_{eff}}{b} = \dfrac{6000}{400}$

$\therefore \qquad \lambda = 15 > 12$

$\therefore$ The column is designed as 'Long Column'.

Reduction Coefficient (C_r) :

$$C_r = 1.25 - \frac{l_{eff}}{48b} = 1.25 - \frac{6000}{48 \times 400}$$
$$C_r = 0.938$$

2. Area of concrete (A_c) :

$$A_c = A_g - A_{sc} = 400 \times 400 - A_{sc}$$
$\therefore \qquad A_c = (160000 - A_{sc})$

3. Area of longitudinal reinforcement (A_{sc}) :

$$P = C_r \cdot (\sigma_{cc} \cdot A_c + \sigma_{sc} \cdot A_{sc})$$
$$875 \times 10^3 = 0.938 \, [5 \times (160000 - A_{sc}) + 130 \times A_{sc}]$$
$\therefore \qquad A_{sc} = 1062.69 \text{ mm}^2 \not> 0.8\% \text{ of } A_s$

$\therefore$ Provide,

$$A_{sc} = A_{sc} \text{ min} = 0.8 \text{ \% of } A_g$$
$$A_{sc} = \frac{0.8}{100} \times (400 \times 400)$$
$\therefore \qquad A_{sc} = 1280.00 \text{ mm}^2$

$\therefore$ Assume, diameter of longitudinal bar $= \phi = 16$ mm

$\therefore$ No. of bars $= \dfrac{A_{sc}}{\left(\dfrac{\pi}{4} \times \phi^2\right)} = \dfrac{1280}{\left(\dfrac{\pi}{4} \times 16^2\right)} = 6.37$

$$\approx 8 \text{ No's.}$$

$\therefore$ **Provide, 8-16 mmφ, as a longitudinal r/f**

4. Design of lateral reinforcement :

(a) Diameter of lateral ties (ϕ) :

(i) $\qquad \phi = \dfrac{1}{4} \times$ dia. of main bar

$\qquad\qquad = \dfrac{1}{4} \times 16 = 4$ mm   ⎫

(ii) $\qquad \phi = 6$ mm   ⎬ ... More

$\therefore \qquad$ **φ = 6 mm**   ⎭

(b) Pitch . Spacing (S) :

(i) $\qquad S =$ least lateral dimension   ⎫

(ii) $\qquad S = 16 \times \phi = 16 \times 16$ ⎬ ... Less

$\qquad\qquad = 256$ mm

(iii) $\qquad S = 300$ mm   ⎭

$\therefore \qquad$ **S = 256 ≈ 250 mm**

$\therefore$ **Provide, 6 mm φ lateral ties @ 250 mm C/C**

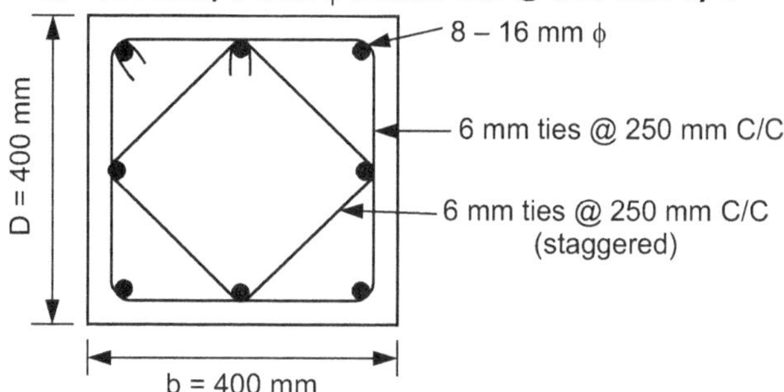

Fig. 4.17

Type III : Column Subjected to Combined Axial Load and Bending : (Clause B-4, Pg. no. 83, IS 456 : 2000)

A. Design Based on Uncracked Section :

A member subjected to axial load and bending (due to eccentricity of load, monolithic construction, lateral forces, etc.) shall be considered safe provided the following conditions are satisfied :

(a) $\qquad \dfrac{\sigma_{cc, \, cal}}{\sigma_{cc}} + \dfrac{\sigma_{cbc, cal}}{\sigma_{cbc}} \leq 1$

where,

$\qquad \sigma_{cc} =$ Permissible axial compressive stress in concrete

$\qquad \sigma_{cc, \, cal} =$ calculated direct compressive stress in concrete

$\qquad \sigma_{cbc} =$ Permissible bending compressive stress in concrete

$\qquad \sigma_{cbc, \, cal} =$ Calculated bending compressive stress in concrete

(b) The resultant tension in concrete is not greater than 35% and 25% of the resultant compression for biaxial and uniaxial bending respectively, or does not exceed three-fourths, the 7 days modulus of rupture of concrete.

Notes : 1. $\sigma_{cc,\,cal} = \dfrac{P}{A_c + 1.5\,m \cdot A_{sc}}$ for columns with ties

where,

$$P = \text{Axial load on column}$$
$$A_c = \text{Area of concrete}$$
$$A_{sc} = \text{Area of longitudinal r/f}$$
$$m = \text{modular ratio}$$

2. $\sigma_{cbc,\,cal} = \dfrac{M}{Z}$

where,

$$M = \text{Moment}$$
$$Z = \text{Modulus of section}$$

(In case of sections subject to moments in two directions, the stress shall be calculated separately and added algebraically.)

B. Design Based on Cracked Section :

If the above conditions are not satisfied, the stresses in concrete and steel shall be calculated by the theory of cracked section in which the tensile resistance of concrete is ignored. If the calculated stresses are within the permissible stress may be assumed to be safe.

Note : The maximum stress in concrete and steel may be found from tables and charts based on the cracked section theory or directly by determining the no-stress line which should satisfy the following requirements.

(a) The direct load should be equal to the algebric sum of the forces on concrete and steel.

(b) The moment of the external loads about any other reference lines should be equal to the algebric sum of the moment of the forces in concrete (ignoring the tensile force in concrete) and steel about the same line, and

(c) The moment of external loads about any reference lines should be equal to the algebric sum of the moment of the forces in concrete (ignoring the tensile force in concrete) and steel about the same line.

Example 4.12 : *A R.C. column 400 mm × 400 mm is reinforced with 4 bars of 25 mm diameter, placed at a cover of 50 mm to the centre of steel bars. Determine the maximum and minimum stresses in concrete if the column is subjected to a load of 400 kN at an eccentricity of 50 mm about one of the axes. Also, check whether the section is safe or not.*
Use M-15 concrete, taking m = 19.

Solution : Given :

$$b = D = 400 \text{ mm}$$
$$P = 400 \text{ kN}$$
$$e = 50 \text{ mm}$$
$$m = 19$$
$$A_{sc} = 4 \times \frac{\pi}{4} \times 25^2 = 1963.50 \text{ mm}^2$$

For M-15

$$\sigma_{cc} = 4 \text{ N/mm}^2$$
$$\sigma_{cbc} = 5 \text{ N/mm}^2$$

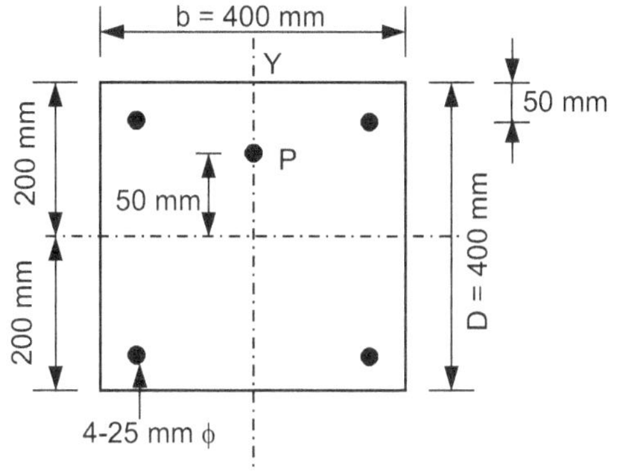

Fig. 4.18

1. **Equivalent area of concrete (A_e) :**

$$A_e = A_c + 1.5\,m \cdot A_{sc}$$
$$= (A_g - A_{sc}) + 1.5\,m \cdot A_{sc}$$
$$\therefore \quad A_e = A_g + A_{sc}\,(1.5m - 1)$$
$$= (400 \times 400) + 1963.50 \times (1.5 \times 19 - 1)$$
$$\therefore \quad A_e = 213993.25 \text{ mm}^2$$

2. **Equivalent moment of Inertia about the centroidal axis X-X (I_e) :**

$$I_e = \frac{bD^3}{12} + (1.5\,m - 1) \cdot A_{sc}\left(\frac{D}{2} - d_c\right)^2$$
$$= \frac{400 \times 400^3}{12}$$
$$+ (1.5 \times 19 - 1) \times 1963.50 \times (200 - 50)^2$$
$$\therefore \quad I_e = 3348.25 \times 10^6 \text{ mm}^4$$

3. **Calculated direct compressive stress in concrete ($\sigma_{cc,\,cal}$) :**

$$\sigma_{cc,\,cal} = \frac{P}{A_e} = \frac{P}{(A_c + 1.5\,m \cdot A_{sc})}$$
$$= \frac{400 \times 10^3}{213996.25}$$
$$\therefore \quad \sigma_{cc,cal} = 1.87 \text{ N/mm}^2$$

4. **Calculated bending compressive stress in concrete ($\sigma_{cbc,cal}$):**

$$\sigma_{cbc,\,cal} = \frac{M}{Z} = \frac{P \cdot e}{(I_e/y)} = \frac{P \cdot e}{I_e} \times \frac{D}{2}$$

$$= \frac{400 \times 10^3 \times 50}{3348.25 \times 10^6} \times \frac{400}{2}$$

$$\therefore \quad \sigma_{cbc,cal} = 1.19 \text{ N/mm}^2$$

5. **Maximum and minimum stresses in concrete :**

Max. stress $= \sigma_{max} = \sigma_{cc,cal} + \sigma_{cbc,cal}$

$$= 1.87 + 1.19$$

$$\therefore \quad \sigma_{max} = 3.06 \text{ N/mm}^2$$

Min. stress $= \sigma_{min} = \sigma_{cc,\,cal} - \sigma_{cbc,\,cal}$

$$= 1.87 - 1.19$$

$$\therefore \quad \sigma_{min} = 0.68 \text{ N/mm}^2$$

6. **Check for section :**

$$\frac{\sigma_{cc,cal}}{\sigma_{cc}} + \frac{\sigma_{cbc,cal}}{\sigma_{cbc}} \leq 1$$

$$\frac{1.87}{4} + \frac{1.19}{5} < 1$$

$$0.71 < 1 \qquad \text{... column section is Safe.}$$

Example 4.13 : *A rectangular reinforced concrete section 400 mm wide and 600 mm deep is reinforced with 6 bars of 20 mm diameter placed at a cover of 40 mm from the top edge and 6 similar bars at the same cover from the bottom edge. Determine the maximum thrust (load) on the section, which can be applied at a distance of 80 mm, from the centre line, if the compressive stress in concrete is not to exceed 7 N/mm². Take, M-20 concrete.*

Solution : Given :

$$b = 400 \text{ mm}, \ D = 600 \text{ mm}$$

$$e = 80 \text{ mm}, \ d_c = 40 + \frac{\phi}{2} = 50 \text{ mm}$$

$$A_{sc} = 12 \times \frac{\pi}{4} \times 20^2$$

$$\therefore \quad A_{sc} = 3769.91 \text{ mm}^2$$

For M-20,

$$\sigma_{cc} = 5 \text{ N/mm}^2$$

$$\sigma_{cbc} = 7 \text{ N/mm}^2$$

Permissible compressive stress in concrete = 7 N/mm²

$$\therefore \quad \text{Modular ratio} = m = \frac{280}{3 \times \sigma_{cbc}} = \frac{280}{3 \times 7}$$

$$\therefore \quad m = 13.33$$

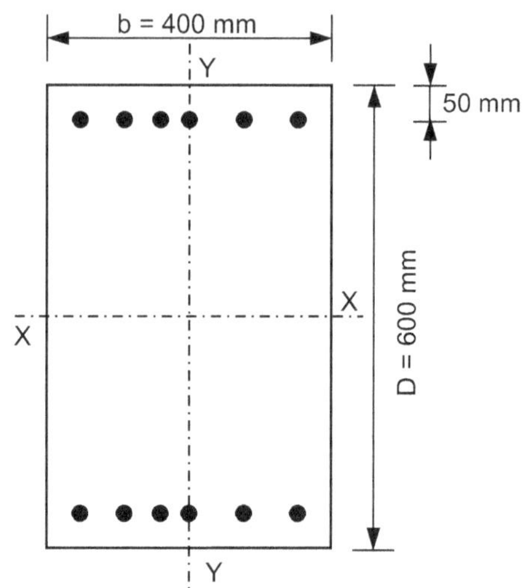

Fig. 4.19

1. **Equivalent area of concrete (A_e) :**

$$A_e = A_c + 1.5 \, m \cdot A_{sc}$$

$$= (A_g - A_{sc}) + 1.5 \, m \cdot A_{sc}$$

$$A_e = A_g + A_{sc} (1.5 \, m - 1)$$

$$= (400 \times 600) + 3769.91 \times (1.5 \times 13.33 - 1)$$

$$\therefore \quad A_e = 311609.44 \text{ mm}^2$$

2. **Equivalent moment of Inertia about the centroidal axis X – X (I_e) :**

$$\therefore \quad I_e = \frac{bD^3}{12} + (1.5 \, m - 1) \cdot A_{sc} \cdot \left(\frac{D}{2} - d_c\right)^2$$

$$= \frac{400 \times 600^3}{12} + (1.5 \times 13.33 - 1)$$

$$\times 3769.91 \times (300 - 50)^2$$

$$\therefore \quad I_e = 11675.59 \times 10^6 \text{ mm}^4$$

3. **Permissible compressive stress in concrete (σ_{max}) :**

$$\sigma_{max} = \sigma_{cc,\,cal} + \sigma_{cbc,\,cal}$$

$$7 = \frac{P}{A_e} + \frac{M}{Z}$$

$$7 = \frac{P}{A_e} + \frac{P \cdot e}{I_e} \times \frac{D}{2}$$

$$7 = \frac{P}{311609.44} + \frac{P \times 80}{11675.59 \times 10^6} \times \frac{600}{2}$$

$$\therefore \quad \textbf{P = 1329.61 kN}$$

Example 4.14 : *An R.C.C. column 300 mm × 300 mm is reinforced with 4 bars of 25 mm diameter placed at an effective cover of 50 mm. Find the eccentricity, the line of thrust may pass along the YY axis without causing tension in concrete. Take, m = 18.*

Solution : Given :

$$b = D = 300 \text{ mm}$$

$$d_c = 50 \text{ mm}$$

$$m = 18$$

$$A_{sc} = 4 \times \frac{\pi}{4} \times 25^2$$

$$\therefore \quad A_{sc} = 1963.50 \text{ mm}^2$$

1. Equivalent area of concrete (A_e) :

$$A_e = A_c + 1.5 \, m \cdot A_{sc}$$

$$A_e = A_g + A_{sc} (1.5 \, m - 1)$$

$$= (300 \times 300) + 1963.50 \times (1.5 \times 18 - 1)$$

$$\therefore \quad A_e = 141051.00 \text{ mm}^2$$

2. Equivalent moment of Inertia about centroid axis x-x (I_e) :

$$I_e = \frac{bD^3}{12} + (1.5 \, m - 1) \cdot A_{sc} \cdot \left(\frac{D}{2} - d_c\right)^2$$

$$= \frac{300 \times 300^3}{12} + (1.5 \times 18 - 1)$$

$$\times 1963.50 \, (150 - 50)^2$$

$$\therefore \quad I_e = 1185.51 \times 10^6 \text{ mm}^4$$

3. If tension in concrete should be just avoided the direct stress and bending stress should be equal for the section :

$$\sigma_{cc, cal} = \sigma_{cbc, cal}$$

$$\frac{P}{A_e} = \frac{M}{Z}$$

$$\therefore \quad \frac{P}{A_e} = \frac{P \cdot e}{I_e} \times \frac{D}{2}$$

$$\frac{1}{141051.00} = \frac{e}{1185.51 \times 10^6} \times \frac{300}{2}$$

$$\therefore \quad e = 56.03 \text{ mm}$$

Example 4.15 : *A circular column is subjected to an axial load of 500 kN and bending moment of 25 kN.m. Taking the permissible stresses, σ_{cc} = 4 N/mm², σ_{cbc} = 5 N/mm², check whether the section is safe or not. Take, m = 19. Diameter = 450 mm, Reinforcement : A_{sc} = 10 – 20 mm ϕ, effective cover = 50 mm*

Solution : Given :

$$D = 450 \text{ mm}, \, d_c = 50 \text{ mm}$$

$$P = 500 \text{ kN}, \, M = 25 \text{ kN.m}$$

$$m = 19,$$

$$\sigma_{cc} = 4 \text{ N/mm}^2,$$

$$\sigma_{cbc} = 5 \text{ N/mm}^2$$

$$A_{sc} = 10 \times \frac{\pi}{4} \times 20^2$$

$$\therefore \quad A_{sc} = 3141.59 \text{ mm}^2$$

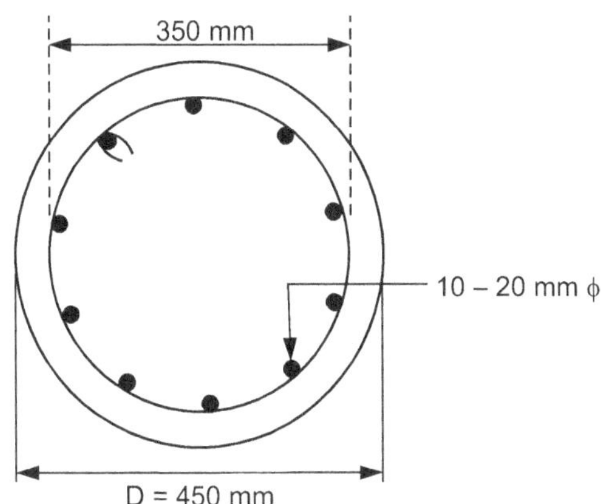

Fig. 4.20

1. Equivalent area of concrete (A_e) :

$$A_e = A_c + 1.5 \, m \cdot A_{sc}$$

$$= A_g + A_{sc} (1.5 \, m - 1)$$

$$A_e = \left(\frac{\pi}{4} \times 450^2\right) + 3141.59 \times (1.5 \times 19 - 1)$$

$$\therefore \quad A_e = 245436.85 \text{ mm}^2$$

2. Equivalent moment of Inertia (I_e) :

Before finding the moment of inertia of the section, let us find the moment of inertia of the steel ring. Polar moment of inertia (I_p) of ring is,

$$I_p = \Sigma \, \delta \, A_{sc} \cdot r^2 = \frac{Asc \cdot d^2}{4}$$

$$\therefore \quad I_{XX} = I_{YY} = \frac{T_p}{2} = \frac{A_{sc} \cdot d^2}{8}$$

$$\therefore \quad I_e = \frac{\pi}{64} \, (450)^4 + (1.5 \times 19 - 1)$$

$$\times \frac{3141.59 \times 350^2}{8}$$

$$I_e = 3335.79 \times 10^6 \text{ mm}^4$$

3. Calculated direct compressive stress in concrete ($\sigma_{cc, cal}$) :

$$\sigma_{cc, cal} = \frac{P}{A_e} = \frac{500 \times 10^3}{245436.85} = 2.04 \text{ N/mm}^2$$

4. Calculated bending compressive stress in concrete ($\sigma_{cbc, cal}$) :

$$\sigma_{cbc, cal} = \frac{M}{Z} = \frac{M}{I_e} \times \frac{D}{2}$$

$$= \frac{25 \times 10^6}{3335.79 \times 10^6} \times \frac{450}{2}$$

$$\sigma_{cbc, cal} = 1.69 \text{ N/mm}^2$$

5. Check for section :

$$\frac{\sigma_{cc, cal}}{\sigma_{cc}} + \frac{\sigma_{cbc, cal}}{\sigma_{cbc}} \leq 1.00$$

$$\frac{2.04}{4} + \frac{1.69}{5} < 1.00$$

$$0.85 < 1.00$$

... Section is Safe.

Example 4.16 : *A R.C. section 250 mm × 350 mm is reinforced with 4 bars of 18 mm φ arranged as shown in Fig. 4.21. The section is subjected to*

　(i) axial load = 200 kN

　(ii) bending moment of 3 kN.m about x-axis

　(iii) bending moment of 2 kN.m about y-axis

　Taking σ_{cc} = 4 N/mm², σ_{cbc} = 5 N/mm², find whether the section is safe or not. Take m = 19.

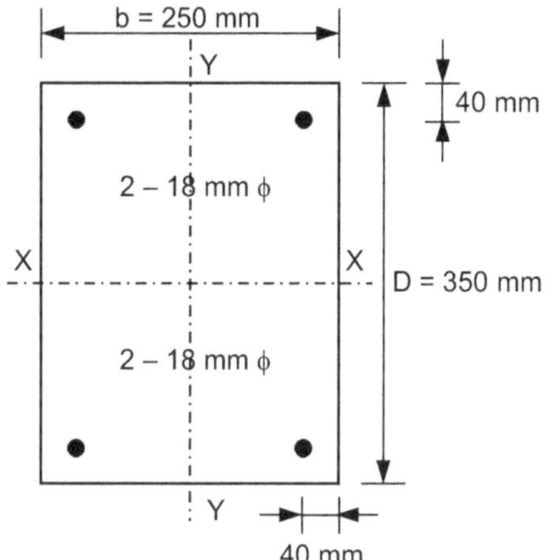

Fig. 4.21

Solution :

$$b = 250 \text{ mm}, D = 350 \text{ mm}$$
$$P = 200 \text{ kN}, m = 19$$
$$M_x = 3 \text{ kN.m}, M_y = 2 \text{ kN.m}$$
$$\sigma_{cc} = 4 \text{ N/mm}^2, \sigma_{cbc} = 5 \text{ N/mm}^2$$
$$A_{sc} = 4 \times \left(\frac{\pi}{4} \times 18^2\right) = 1017.88 \text{ mm}^2$$

1. Equivalent area of concrete (A_e) :

$$A_e = A_c + 1.5\, m \cdot A_{sc}$$
$$A_e = A_g + A_{sc} \times (1.5\, m - 1)$$
$$\therefore \quad A_e = (250 \times 350) + 1017.88 \times (1.5 \times 19 - 1)$$
$$\therefore \quad A_e = 115491.70 \text{ mm}^2$$

2. Equivalent moment of Inertia (I_e) :

$$I_{e,x} = \frac{bD^3}{12} + (1.5\, m - 1) \cdot A_{sc} \cdot \left(\frac{D}{2} - dc\right)^2$$
$$= \frac{250 \times 350^3}{12} + (1.5 \times 19 - 1)$$
$$\times 1017.88 \times (175 - 40)^2$$
$$\therefore \quad I_{e,x} = 1403.38 \times 10^6 \text{ mm}^4$$

$$I_{e,y} = \frac{Db^3}{12} + (1.5\, m - 1) \cdot A_{sc} \cdot \left(\frac{b}{2} - d_c\right)^2$$
$$= \frac{350 \times 250^3}{12} + (1.5 \times 19 - 1) \times 1017.88$$
$$\times (125 - 40)^2$$
$$\therefore \quad I_{e,y} = 657.97 \times 10^6$$

3. Calculated direct compressive stress in concrete ($\sigma_{cc,\,cal}$) :

$$\sigma_{cc,\,cal} = \frac{P}{A_e} = \frac{200 \times 10^3}{115491.70}$$
$$\therefore \quad \sigma_{cc,\,cal} = 1.73 \text{ N/mm}^2$$

4. Calculated bending compressive stress in concrete ($\sigma_{cbc,\,cal}$) :

$$(\sigma_{cbc,\,cal})_x = \frac{M_x}{Z_x} = \frac{M_x}{I_{e\cdot x}} \times \frac{D}{2}$$
$$= \frac{3 \times 10^6}{1403.38 \times 10^6} \times 175$$
$$\therefore \quad (\sigma_{cbc,\,cal})_x = 0.37 \text{ N/mm}^2$$
$$(\sigma_{cbc,\,cal})_y = \frac{M_y}{Z_y} = \frac{M_y}{I_{e,y}} \times \frac{b}{2}$$
$$= \frac{2 \times 10^6}{657.97 \times 10^6} \times 125$$
$$(\sigma_{cbc,\,cal})_y = 0.38 \text{ N/mm}^2$$
$$\therefore \quad \text{Total bending stress} = \sigma_{cbc,\,cal} = (\sigma_{cbc,\,cal})_x + (\sigma_{cbc,\,cal})_y$$
$$\therefore \quad \sigma_{cbc,\,cal} = 0.37 + 0.38$$
$$\sigma_{cbc,\,cal} = 0.75 \text{ N/mm}^2$$

5. Check for section :

$$\frac{\sigma_{cc,cal}}{\sigma_{cc}} + \frac{\sigma_{cbc,cal}}{\sigma_{cbc}} \leq 1.00$$
$$\frac{1.73}{4} + \frac{0.75}{5} < 1.00$$
$$0.58 < 1.00$$

... Okay section is Safe.

Example 4.17 : *Design a suitable column section to carry an axial load of 450 kN and bending moment of 8 kN·m. Take, M-15 concrete and M.S. reinforcement.*

Solution : Given :

$$P = 450 \text{ kN}, \sigma_{cc} = 4 \text{ N/mm}^2$$
$$M = 8 \text{ kN·m}, \sigma_{cbc} = 5 \text{ N/mm}^2$$
$$\therefore \text{Modular ratio} = \frac{280}{3 \times \sigma_{cbc}} = \frac{280}{3 \times 5}$$
$$\therefore \quad m = 18.67$$

Let, us Assume a section of 400 mm × 400 mm, reinforced symmetrically with 4.25 mmφ with effective cover 50 mm.

$$\therefore \quad A_{sc} = 4 \times \frac{\pi}{4} \times 25^2$$
$$\therefore \quad A_{sc} = 1963.50 \text{ mm}^2$$

1. Equivalent area of concrete (A_e) :

$$A_e = A_c + 1.5\, m \cdot A_{sc}$$
$$A_e = A_g + A_{sc} + (1.5\, m - 1)$$
$$\therefore \quad A_e = (400 \times 400) + 1963.50 \times (1.5 \times 18.67 - 1)$$
$$\therefore \quad A_e = 213024.32 \text{ mm}^2$$

2. Equivalent moment of Inertia (I_e) :

$$I_e = \frac{bD^3}{12} + (1.5\ m - 1) \cdot A_{sc} \cdot \left(\frac{D}{2} - d_c\right)^2$$

$$= \frac{400 \times 400^3}{12} + (1.5 \times 18.67 - 1)$$

$$\times 1963.50 \times (200 - 50)^2$$

$$\therefore \quad I_e = 3326.38 \times 10^6 \ mm^4$$

3. Calculated direct compressive stress in concrete ($\sigma_{cc,cal}$) :

$$\sigma_{cc,\ cal} = \frac{P}{A_e} = \frac{450 \times 10^3}{213024.32}$$

$$\therefore \quad \sigma_{cc,\ cal} = 2.11 \ N/mm^2$$

4. Calculated bending compressive stress in concrete ($\sigma_{cbc,cal}$) :

$$\sigma_{cbc,cal} = \frac{M}{Z} = \frac{M}{I_e} \times \frac{D}{2}$$

$$= \frac{8 \times 10^6}{3326.38 \times 106} \times \frac{400}{2}$$

$$\therefore \quad \sigma_{cbc,cal} = 0.48 \ N/mm^2$$

5. Check for section :

$$\frac{\sigma_{cc,cal}}{\sigma_{cc}} + \frac{\sigma_{cbc,cal}}{\sigma_{cbc}} \leq 1.00$$

$$\frac{2.11}{4} + \frac{0.48}{5} < 1.00$$

$$0.62 < 1.00$$

... Okay ... section is Safe.

(B) FOOTING

4.19 INTRODUCTION

- The foundation is the lowermost part of the structure which transfers the load from structure to soil below it. The ground on which foundation rests is called sub-grade. The foundations should be designed, so that the stresses on any part of soil should be within permissible limit.

- The settlement of soil depends upon the different factors such as loading intensity, type of soil, etc. The settlement of the structure should be uniform to avoid differential settlement or rotation. This can be achieved by providing area of footing in such a way that the intensity of soil reaction is same under all the footings of a structure. To avoid the tilting of the structure, a foundation should be designed in such a way that the centre of gravity of loads should coincide with centre of gravity of footing area.

4.20 CLASSIFICATION OF FOOTING

Foundations may be broadly classified under two heads :

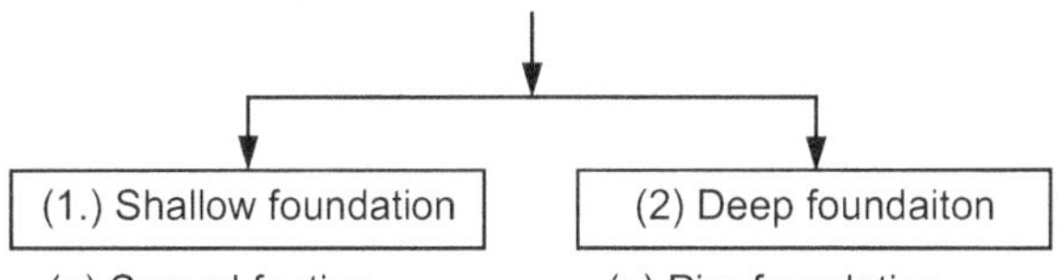

(a) Spread footing   (a) Pier foundation
(b) Combined footing   (b) Pile foundation
(c) Strap / cantilever footing   (c) Well / caisson foundation
(d) Mat / raft foundation

1. Shallow Foundations

- If the depth of foundation is less than width of the foundation it is known as shallow foundation. These type of foundations are used when supporting soil or rock layer carries the load very near to the ground surface.

(i) Spread Footing

- A spread footing is used to transfer the load from the structure to the sub soil over a larger area. Spread footing is classified as

   (a) Strip Footing : These types of footings are provided under light structures such as walls etc.

   (b) Isolated Footing : These footings are generally provided for individual columns. These footings can be square, rectangular, circular or sloped type footing.

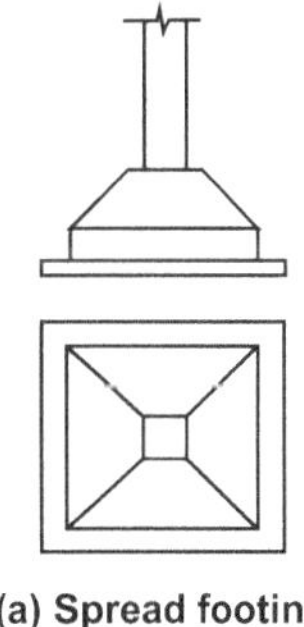

(a) Spread footing

Fig. 4.22

(ii) Combined Footing

- When two columns are very closer and carrying very heavy load and soil bearing capacity is low, these types of footings are used. In this case area of footing of adjacent columns are overlapping each other or column is near to property line in such cases combined footings are used. These footing may be rectangular or trapezoidal.

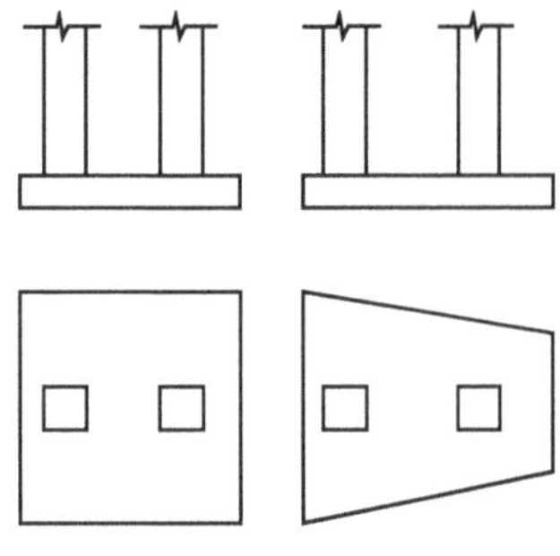

(b) Combined footing

Fig. 4.22

(iii) Strap or Cantilever Footing

- A strap footing consist of spread footing of two columns connected by a strap beam.

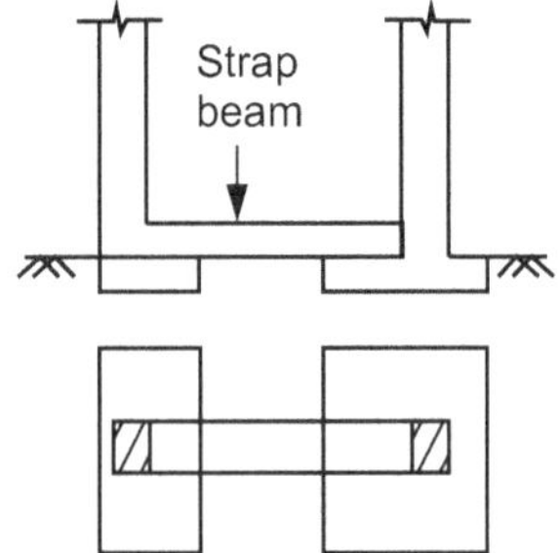

Fig. 4.22 (c) : Strap or cantilever footing

The strap beam does not remain in contact with soil, and thus does not transfer any pressure to the soil. Such a footing is generally used to combine the footing of outer column to the adjacent column footing so that footing of outer column does not extend beyond property line.

(iv) Mat or Raft Foundation :

- It consist of thick reinforced concrete slab covering the entire area under a supporting several columns and walls. It is used in condition where the soil bearing capacity is low and building loads are heavy.

- When spread footing area cover more than one half (i.e. 50%) of the area it may prove more economical to use mat / raft foundation.

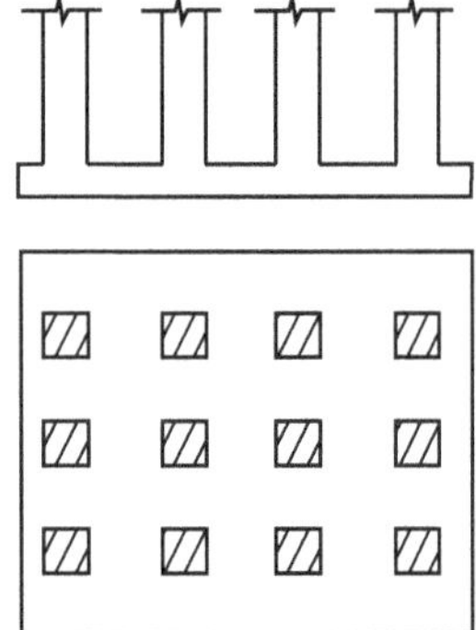

(d) Mat / raft foundation

Fig. 4.22

2. Deep Foundation :

- If the depth of foundation is more than width of the foundation is known as deep foundation.

- A deep foundation is a type of foundation that transfer building loads to the earth farther down from the surface than a shallow foundation does to a subsurface layer or a range of depths.

(i) Pile Foundation :

- Pile foundation is a deep foundation used where the top soil is very low. Piles transfer the load to a lower stratum of greater bearing capacity, by way of end bearing, or to the intermediate soil through skin friction.

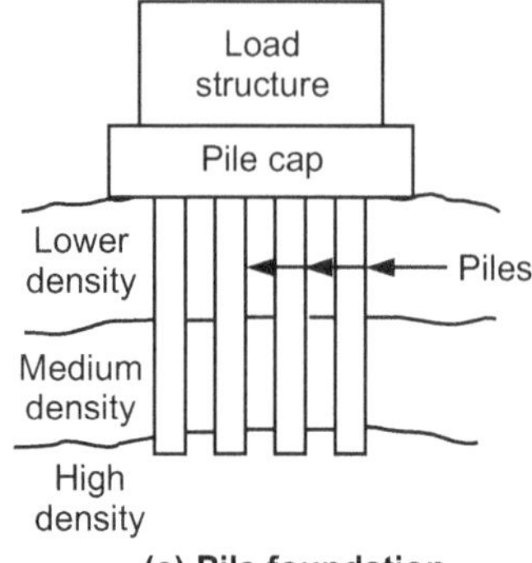

(e) Pile foundation

Fig. 4.22 (e)

4.21 PURPOSE / FUNCTIONS OF FOUNDATION

- To distribute the load of the structure over a large bearing area so as to bring the intensity of load within the safe bearing capacity of soil.

- To load the bearing surface at a uniform rate to avoid differential settlement.

- To prevent the lateral movement of supporting material.

- To attain a level and firm bed for building operations.

- To increase the stability of the structure as a whole.

4.22 I.S. SPECIFICATION REGARDING DESIGN OF FOOTING

1. **Minimum Nominal Cover :** (Clause 26.4.2.2)

- The minimum nominal cover for the footings should not be less than 50 mm.

- The actual cover may be more depending on the presence of harmful chemicals, minerals, water table or environmental exposure condition given in (Table 16, 16-A, Pg. No. 47, IS 456 : 2000)

2. **Thickness at the Edge of Footing :** (Clause 34.1.2, IS 456)

- In reinforced and plain concrete footings, the thickness at the edge shall be not less than 150 mm for footings, nor less than 300 mm for pile cap above piles.

- In the case of plain concrete pedestal, the angle between the plane passing through the bottom edge of the pedestal and the corresponding junction edge of the column with pedestal and the horizontal plane shall be governed by the expression :

$$\tan \alpha \nleq 0.9 \sqrt{\frac{100\, q_o}{f_{ck}} + 1}$$

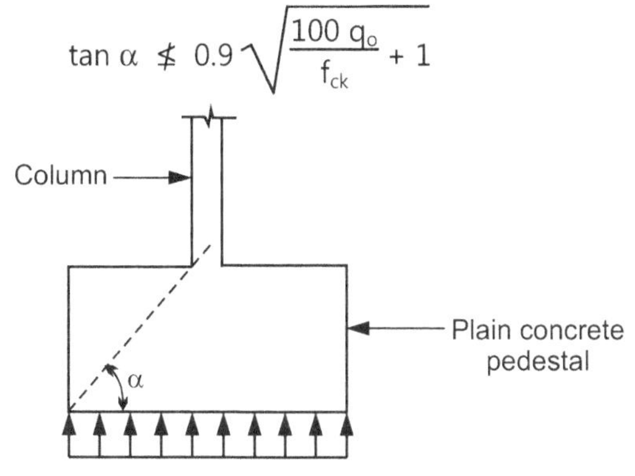

Fig. 4.23

where, q_o = calculated max. bearing pressure (MPa)

f_{ck} = Characteristic strength of concrete at 28 days in (MPa)

3. **Bending Moments :** (Clause 34.2 , Pg. No. 64 , IS 456 : 2000)

- The critical section of maximum bending moment for the purpose of designing an isolated concrete footing which supports a column, pedestal or wall shall be ;

(i) At the face of the column, pedestal or wall for footing supporting a concrete column, pedestal or reinforced concrete wall.

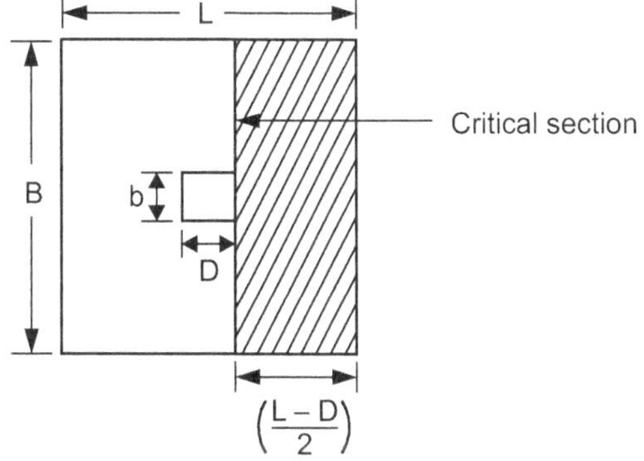

$$\left(\frac{L - D}{2}\right)$$

Fig. 4.24

(ii) Halfway between the centre line and the edge of the wall, for footing under masonry wall.

4. **Shear Force :**

- The shear strength of footings is governed by the more severe of the following two conditions :

(i) **One Way Shear :** (Clause 34.2.4, Pg. No. 65, IS 456 : 2000)

- The footing acting essentially as a wide beam, with a potential diagonal crack extending in a plane across the entire width, the critical section at a distance equal to the effective depth of footing.

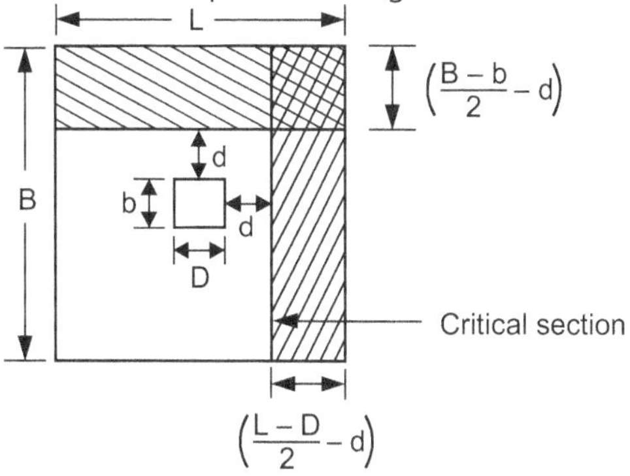

$$\left(\frac{B - b}{2} - d\right)$$

$$\left(\frac{L - D}{2} - d\right)$$

Fig. 4.25

(ii) **Two Way Shear :** (Clause 31.6.1, 34.2.4, IS 456)

- Two way action of the footing, with potential diagonal cracking along the surface of truncated cone or pyramid around the concentrated load, in this case, the footing shall be designed for the critical section for shear shall be at a distance of 'd/2' along the periphery of the column.

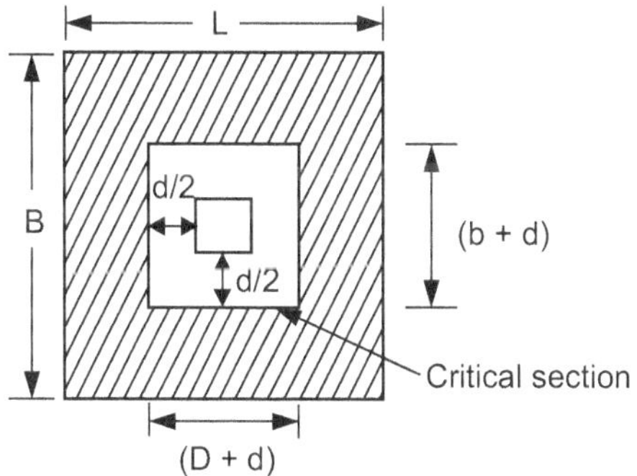

Fig. 4.26

Example 4.18 : *Design an isolated footing of uniform thickness of a R.C. column carrying a load of 500 kN and having size 500 mm × 500 mm. The safe bearing capacity of soil may be takes as 120 kN/m². Use M-20 and Fe-415 materials.*

Solution : Given :

column size, b = D = 500 mm

Axial load = P = 500 kN

S.B.C. of soil = 120 kN/m²

M-20; σ_{cbc} = 7.0 N/mm²

Fe-415; σ_{st} = 230 N/mm²

1. Calculation of design constant :

(i) Modular ratio (m) :

$$m = \frac{280}{3 \times \sigma_{cbc}} = \frac{280}{3 \times 7}$$

$\therefore$ $m = 13.33$

(ii) N.A. depth factor (K) :

$$K = \frac{m \cdot \sigma_{cbc}}{m \cdot \sigma_{cbc} + \sigma_{st}}$$

$\therefore$ $K = 0.29$

(iii) lever arm factor (j) :

$$j = 1 - \frac{K}{3} = 1 - \frac{0.29}{3}$$

$\therefore$ $j = 0.90$

(iv) Moment resisting factor (Q) :

$$Q = \frac{1}{2} \times \sigma_{cbc} \times j \times k$$

$\therefore$ **Q = 0.91**

2. Calculation of size of footing ($L_f \times B_f$) :

Assume, self wt. of footing = 10% of P = 50 kN

$\therefore$ Total load = 500 + 50 = 550 kN

$$A_{f\ required} = \frac{\text{Total load}}{\text{S.B.C. of soil}} = \frac{550}{120}$$

$\therefore$ $A_{f\ req.} = 4.58\ m^2$

Assume, square footing,

Each side of footing = $L_f = B_f = \sqrt{A_f} = \sqrt{4.58}$

$\therefore$ **$L_f = B_f = 2.14 \approx 2.20$ m**

$\therefore$ **Provide, size of footing ($L_f \times B_f$) i.e. 2.20 m $\times$ 2.20 m**

3. Calculation of depth of footing (D_f) :

(i)   Net upward pressure = p_o

$$= \frac{\text{Axial load excluding self wt.}}{A_{f\ provided}}$$

$\therefore$ $$P_o = \frac{500}{2.20 \times 2.20}$$

$\therefore$ **$P_o = 103.31\ kN/m^2$**

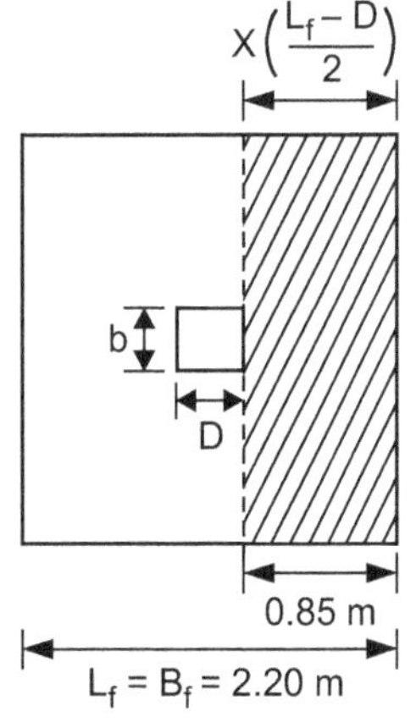

Fig. 4.27 : Critical section for moment

(ii) The maximum bending moment acts at the face of column (x-x) :

$\therefore$ $$M = \frac{(P_0.B) \times 0.85^2}{2}$$

$$= \frac{103.31 \times 2.20 \times 0.85^2}{2}$$

$\therefore$ **M = 82.11 kN.m**

(iii) Check for depth / flexure / bending

By equating (M) and Resisting moment

$$M = Qbd^2$$

$$82.11 \times 10^6 = 0.91 \times 2200 \times d^2$$

$$d_{req} = 202.52\ mm$$

$\therefore$ provide, d = 240 mm, c = 60 mm

$D_f = 300$ mm

4. Calculation of area of steel (A_{st}) :

$$M = T \cdot (j \cdot d)$$

$$A_{st} = \frac{M}{\sigma_{st} \cdot j \cdot d}$$

$$= \frac{82.11 \times 10^6}{230 \times 0.90 \times 240}$$

Assume, diameter of bar = ϕ = 10 mm

$\therefore$ No. of bars = $\dfrac{A_{st}}{\left(\dfrac{\pi}{4} \times \phi^2\right)}$

$$= 21.04 \approx 22\ No's.$$

$\therefore$ **Provide, 22-10 mm ϕ uniformly distributed in each direction**

5. Check for shear :

A. Check for one way shear :

Critical section is acting at distance 'd' from the face of the column

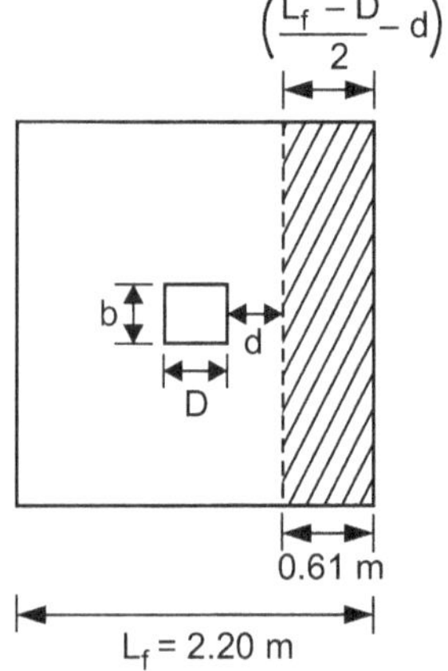

Fig. 4.28 : Critical section for one way shear

(i)   Nominal shear stress (τ_v) :

$$\therefore \ \tau_v = \frac{V}{B_o} = \frac{P_o \times \text{shaded area}}{B_f \times d}$$

$$\tau_v = \frac{103.31 \times (0.61 \times 2.20)}{2.20 \times 0.24} \times 10^{-3}$$

$$\therefore \qquad \tau_v = \mathbf{0.26 \ N/mm^2}$$

(ii)   Design shear strength of concrete $(\tau_c \cdot k)$

(Refer Table-19, clause 40.2.1.1, IS 456 : 2000)

$$Pt_{max} = \frac{A_{st}}{bd} \times 100$$

$$= \frac{(22 \times \frac{\pi}{4} \times 10^2)}{2200 \times 240} \times 100$$

$$\therefore \qquad Pt_{max} = 0.33 \ \%$$

	Pt_{max}	τ_c
0.08 {	0.25	0.36
	0.33	
	0.50	0.48
	0.25	0.12
	1	0.48

$$\therefore \qquad \tau_c = 0.36 + (0.08 \times 0.48)$$

$$\therefore \qquad \tau_c = 0.40 \ N/mm^2 \text{ and } K = 1$$

$$\therefore \qquad \tau_{c \cdot K} = \mathbf{0.40 \ N/mm^2}$$

(iii)   Compare τ_γ and $\tau_{c \cdot K}$:

$$0.26 \ < \ 0.40$$

$$\therefore \qquad \tau_c \ < \ \tau_{c \cdot K} \qquad \text{... Okay. Safe in one way shear}$$

B.   Check for two way shear / punching shear :

Critical section is acting at distance 'd/2' from the face of the column along periphery of the column.

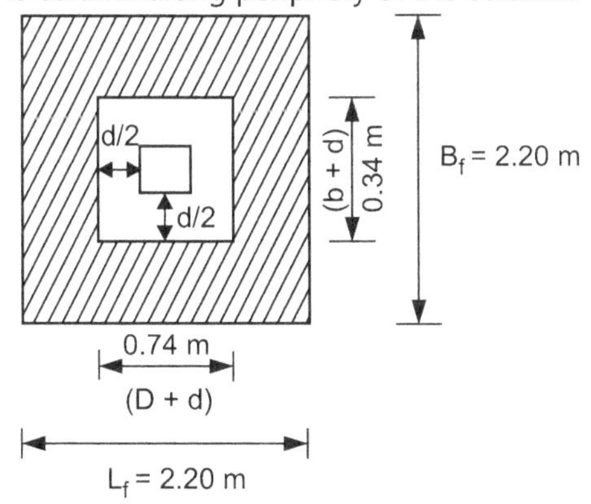

Fig. 4.29

(i)   Nominal shear stress (τ_v) :

$$\tau_v = \frac{V}{B_o d} = \frac{P_o \times \text{Shaded Area}}{\text{Perimeter} \times d}$$

$$\tau_v = \frac{103.31 \times [(2.20)^2 - (0.74)^2]}{(4 \times 0.74) \times 0.24} \times 10^{-3}$$

$$\therefore \qquad \tau_v = \mathbf{0.62 \ N/mm^2}$$

(ii)   Permissible shear stress of concrete $(\tau_c \cdot K_s)$:

$$\therefore \qquad K_s = (0.5 + B_c) = \left(0.5 + \frac{500}{500}\right)$$

$$= 1.5 > 1 \qquad \{\because K_s \not> 1\}$$

$$\therefore \quad \text{Hence, } K_s = 1$$

$$\text{and} \qquad \tau_c = 0.16 \sqrt{f_{ck}} = 0.16 \sqrt{20} = 0.72 \ N/mm^2$$

$$\therefore \qquad \tau_c \cdot K_s = \mathbf{0.72 \ N/mm^2}$$

(iii)   Compare τ_v and $\tau_c \cdot K_s$:

$$0.62 \ < \ 0.72$$

$$\therefore \qquad \tau_v \ < \ \tau_c \cdot K_s \quad \text{... Okay ... safe in two way shear}$$

6.   Check for development length :

$$Ld = \frac{\phi \ \sigma_s}{4 \times \tau_{bd}} = \frac{10 \times 230}{4 \times 0.9 \times 1.6} = 449.22 \ mm$$

$$\therefore \qquad Ld \approx 450 \ mm$$

Length of bars available $= \dfrac{1}{2} \ (B_f - b) - \text{cover}$

$$= \frac{1}{2} \times (2200 - 500) - 60$$

$$= 790 \ mm > Ld \qquad \text{... Okay ... safe.}$$

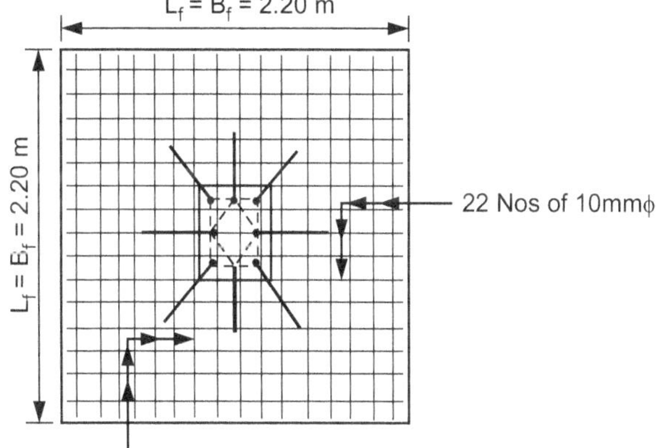

Fig. 4.30 : Isolated footing with uniform depth

Example 4.19 : *Design a rectangular isolated footing of uniform thickness for R.C. column bearing a vertical load of 750 kN and having size 400 mm × 600 mm. The S.B.C. of soil is 150 KPa. Use M-20 concrete and Fe-415 steel.*

Solution : Given :

Column size, b = 400 mm, D = 600 mm

Axial load = P = 750 kN

S.B.C. of soil = 150 kN/m^2

M-20; σ_{cbc} = 7.0 N/mm^2

Fe-415; σ_{st} = 230 N/mm^2

1. Calculation of design constant

(i) Modular ratio (m) :

$$m = \frac{280}{3 \times \sigma_{cbc}} = \frac{280}{3 \times 7.0} = 13.33$$

(ii) N.A. depth factor (K) :

$$K = \frac{m \cdot \sigma_{cbc}}{m \cdot \sigma_{cbc} + \sigma_{st}} = 0.29$$

(iii) lever arm factor (j) :

$$j = \left(1 - \frac{K}{3}\right) = \left(1 - \frac{0.29}{3}\right) = 0.90$$

(iv) Moment resisting factor (Q) :

$$Q = \frac{1}{2} \times \sigma_{cbc} \times j \times k = 0.91$$

2. Calculation of size of footing ($L_f \times B_f$) :

Assume, self wt. of footing = 10% of P = 75 kN

$\therefore$ Total load = 750 + 75 = 825 kN

$$A_{f\,required} = \frac{\text{Total load}}{\text{S.B.C. of soil}} = \frac{825}{150}$$

$\therefore$ **$A_{f\,required}$ = 5.50 m^2**

Let, ratio of B_f to $L_f = \dfrac{400}{600} = \dfrac{2}{3}$

$\therefore \qquad B_f = \dfrac{2}{3} \times L_f$

For rectangular footing,

$$L_f \times B_f = A_f$$

$$L_f \times \frac{2}{3} \times L_f = 5.50$$

$\therefore \qquad L_f$ = 2.87 m and B_f = 1.91 m

$\therefore$ Provide, L_f = 2.87 $\approx$ 3.00 m

$\qquad\qquad B_f$ = 1.91 $\approx$ 2.00 m

3. Calculation of depth of footing (D_f)

(i) Net upward pressure = P_o

$$= \frac{\text{Axial load excluding self wt.}}{A_f \text{ provided}}$$

$\therefore \qquad P_o = \dfrac{750}{(3 \times 2)}$

$\therefore \qquad P_o$ = 125.00 kN/m^2

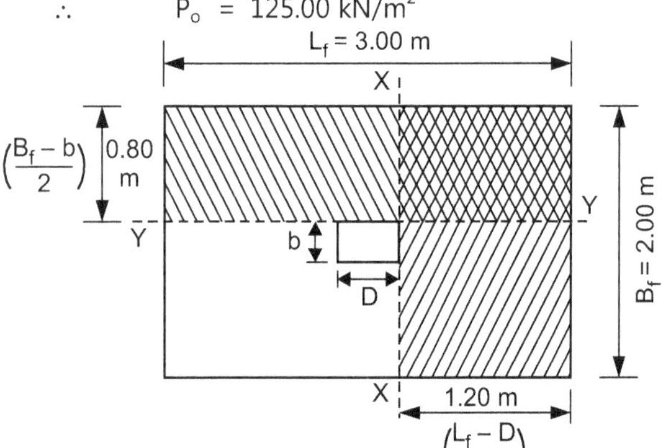

Fig. 4.31 : Critical section for moment

(ii) The maximum bending moment acts at the face of column (x − x)

$$M_1 = \frac{(P_0 \cdot B_f) \times 1.20^2}{2} = \frac{125 \times 2 \times 1.20^2}{2}$$

$\therefore \qquad$ **M_1 = 180.00 kN.m**

(iii) Check for depth / flexure / bending

by equating ,

$$M = Qbd^2$$

$$180 \times 10^6 = 0.91 \times 2000 \times d^2$$

$$d_{req} = 314.49$$

$\therefore$ **Provide , d $\approx$ 390 mm, c = 60 mm, D_f = 450 mm**

(iv) The maximum bending moment at the face of column at the face of column (y-y)

$$M_2 = \frac{(P_o \times L_f) \times 0.80^2}{2}$$

$$= \frac{125 \times 3 \times 0.80^2}{2}$$

$\therefore \qquad$ **M_2 = 120.00 kN.m** $< m_1$

$\qquad\qquad\qquad \therefore$ No need to check for M_2

4. Calculation of area of steel (A_{st}) :

(a) Along section x-x (i.e. r/f parallel to length)

$$A_{st_1} = \frac{M_1}{\sigma_{st} \times j \times d}$$

$$= \frac{180 \times 10^6}{230 \times 0.90 \times 390}$$

$\therefore \qquad A_{st_1}$ = 2229.65 mm^2

Assume, diameter of bar = ϕ = 12

No. of bars $= \dfrac{A_{st}}{(\pi/4 \times \phi^2)} = 19.71 \approx 20$ Nos.

$\therefore$ **Provide, 20-12 mm ϕ in x-direction (i.e. parallel to length)**

(b) Along section y-y (i.e. parallel to width)

$$A_{st_2} = \frac{M_2}{\sigma_{st} \times j \times d} = \frac{120 \times 10^3}{230 \times 0.90 \times 390}$$

$\therefore \qquad A_{st_2}$ = 1486.44 mm^2

Assume, diameter of bar = ϕ = 10 mm

$\therefore$ No. of bars $= \dfrac{A_{st}}{(\pi/4 \times \phi^2)} = 18.93 \approx 20$ Nos.

$\therefore$ **Provide, 20-10 mm ϕ in y-direction (i.e. parallel to width)**

5. Check for shear

(a) Check for one way shear

Critical section is acting at distance 'd' from the face of column.

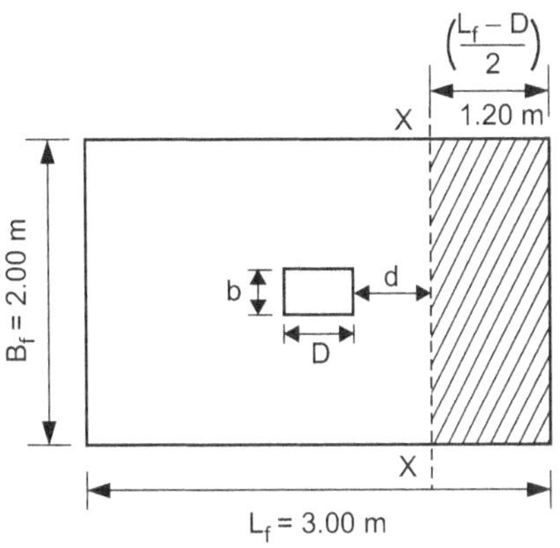

Fig. 4.32 : Critical section for one way shear

(i) Nominal shear stress (τ_v)

$$\tau_v = \frac{V}{B_o d} = \frac{P_o \times \text{shaded area}}{B_f \times d}$$

$$= \frac{125 \times (1.2 \times 2.00)}{2.00 \times 0.39} \times 10^{-3}$$

$$\therefore \quad \tau_v = \mathbf{0.38 \ N/mm^2}$$

(ii) Design shear strength of concrete $(\tau_{c.K})$

$$P_{t\,max} = \frac{A_{st}}{bd} \times 100 = \frac{\left(20 \times \dfrac{\pi}{4} \times 12^2\right)}{2000 \times 390} \times 100$$

$$\therefore \quad P_{t\,max} = 0.29 \ \%$$

	Pt_{max}	τ_c
0.04 {	0.25	0.36
	0.29	
	0.50	0.48

0.25	0.12
1	0.48

$$\therefore \quad \tau_c = 0.36 + (0.04 \times 0.48)$$

$$\therefore \quad \tau_c = 0.38 \ N/mm^2$$

$$\text{and} \quad k = 1.00$$

$$\therefore \quad \tau_c \cdot k = \mathbf{0.38 \ N/mm^2}$$

(iii) Compare τ_v and $\tau_{c.K}$

$$0.38 = 0.38$$

$$\tau_v = \tau_{c.k} \qquad \text{...okay safe in one way shear}$$

(B) Check for two way shear / punching shear

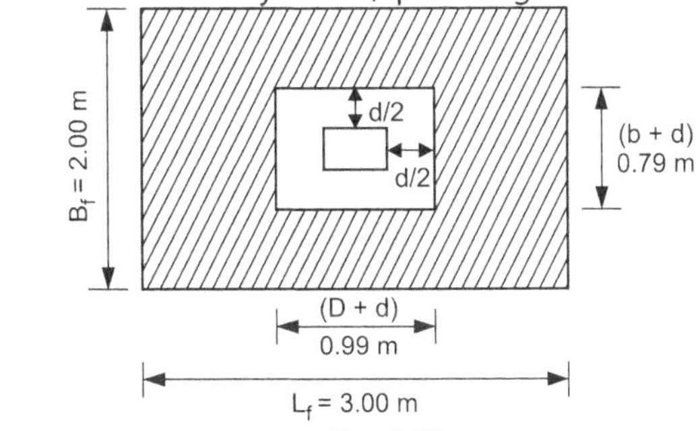

Fig. 4.33

(i) Nominal shear stress (τ_v)

$$\tau_v = \frac{V}{B_o d} = \frac{P_o \times \text{shaded area}}{\text{Perimeter} \times d}$$

$$\tau_v = \frac{125 \times [(3 \times 2) - (0.99 \times 0.79)]}{(2 \times 0.99 + 2 \times 0.79) \times 0.39} \times 10^{-3}$$

$$\therefore \quad \tau_v = \mathbf{0.47 \ N/mm^2}$$

(ii) Design shear stress of concrete $(\tau_{c.k_s})$

$$\therefore \quad k_s = (0.5 + \beta_c) = \left(0.5 + \frac{400}{600}\right) = 1.17 > 1$$

$$\{\therefore \ ks \not> 1\}$$

$$\therefore \ \text{Hence, } ks = 1$$

$$\text{and} \quad \tau_c = 0.16 \sqrt{f_{ck}} = 0.16 \sqrt{20} = 0.72 \ N/mm^2$$

$$\therefore \quad \tau_{c.k_s} = \mathbf{0.72 \ N/mm^2}$$

(iii) Compare (τ_v) and $(k_s \cdot \tau_c)$

$$0.47 < 0.72$$

$$\therefore \quad \tau_v < \tau_c.k_s \qquad \text{...Okay safe in two way shear}$$

6. Check for development length

$$L_d = \frac{\phi . \sigma_{st}}{4 \times \tau_{bd}} = \frac{12 \times 230}{4 \times 0.8 \times 1.6} = 539.06$$

$$\therefore \quad \mathbf{L_d \approx 540 \ mm}$$

Length of bars available $= \dfrac{1}{2} \times (B_f - b) - \text{cover}$

$$= \mathbf{740 \ mm > L_d} \qquad \text{...okay safe}$$

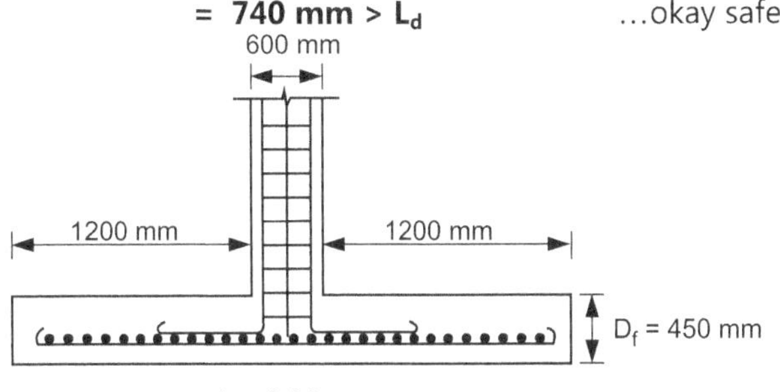

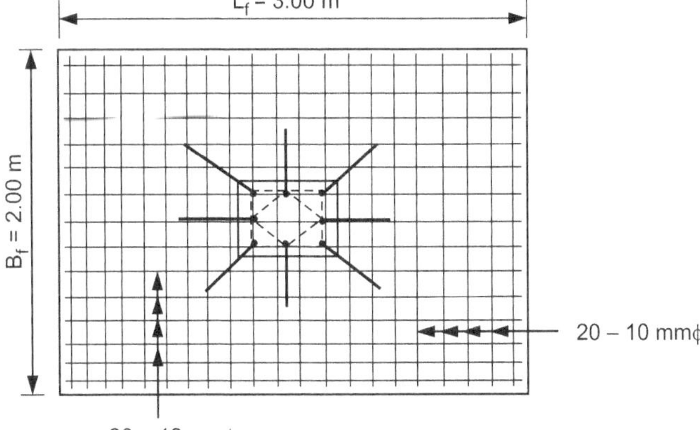

Fig. 4.34 : Isolated footing (rectangular) with uniform depth

EXERCISE

1. Define column, explain classification of column.
2. Differentiate between short column and long column.
3. Define : (i) Effective length (ii) Slenderness ratio (iii) Radius of gyration.

4. Write the functions of longitudinal reinforcement.

5. State the necessity of lateral reinforcement in column.

6. A short column 300 mm × 400 mm in section is reinforced with 6-12 mm ϕ. Determine the safe load on the column. Take M-15 and Fe-250 grades material.

7. Calculate the service load carried by circular column with helix of diameter 300 mm reinforced with 8-16 mm ϕ, the column of length 3.5 m, effectively held in position but not restrained against rotation at both ends.

8. Design a column carries a safe load of 600 kN. Take M-20 and Fe-500 grades material.

9. Design a circular column with spiral ties to carry an axial load of 400 kN. The length of column is 2.8 m. Take M-25, Fe-415.

10. Design a suitable column to carry an axial load of 600 kN and bending moment of 20 kN/m. Take, M-25 concrete.

11. Design an isolated footing for column 350 mm × 350 mm carries a live load of 1000 kN. The S.B.C. of soil is 120 kN/m^2. Use M-15 concrete and M. S. reinforcement. Sketch the details of the reinforcement.

12. A R.C. column section 400 mm × 500 mm is reinforced with 10 bars of 12 mm ϕ arranged, 5 bars on each side parallel to width of column with cover of 50 mm.

    The section is subjected to,

    (i)   Axial load = 500 kN

    (ii) Bending moment about x-axis, e_x = 50 mm

    (iii) Bending moment about y-axis, e_y = 30 mm

    Check whether section is safe or not. Take M-25 concrete.

13. Design a rectangular footing for column of size 300 × 600 mm carries a load of 1200 kN. The bearing capacity of soil is 150 kN/m^2. Use M-25 and Fe-415 grades material.

14. Design a sloped square footing for column 300 × 300 mm to carry a safe load of 700 kN. The S.B.C. of soil is 160 kPa. Use M-20 and Fe-415 grades material.

15. Design a sloped rectangular footing for column 350 mm × 450 mm to carry a load of 650 kN. The S.B.C. of soil is 105 kN/m^2. Draw the reinforcement detailing.

16. Define footing. Explain the different types of footing.

17. State the functions of foundation.

18. Explain different modes of failure in case of footing.

◈ ◈ ◈

5.1 INTRODUCTION

- Staircase is used to provide an access to different floors of same building. As staircase is the only means for communication at different floors so its location should be in such a way that it is easily accessible from all parts of the building. Staircase should be with sufficient light and properly ventilated for easy access.

5.2 DIFFERENT PARTS OF STAIRCASE

- **Tread :** The horizontal distance of a step is known as tread. Tread consists of going and noising as shown in Fig. 5.1.

- **Nosing :** This is the outer projecting part of a tread. If space available for stair is sufficiently available, then noising is not provided.

- **Going :** The net horizontal distance of a step is known as going.

- **Rise :** The vertical height of a step is known as rise.

- **Winders :** Steps used for changing the direction of the stair, are triangular in plan.

- **Flight :** The inclined slab of a stair is known as flight of a staircase. It consists of a series of steps.

- **Landing :** The horizontal slab of a stair is known as landing.

- **Line of Nosing (Pitch Line) :** This is a straight line touching the nosings of the various steps and parallel to the slope of the stair.

- **Headroom :** This is the vertical distance measured from the pitch line to the ceiling above the staircase, or any obstacle above it. The height of headroom should not be less than 2.00 m

- **Stringer Beam :** Girder is called as stringer beam. In case of Non RCC staircase where steel, wood or composite materials are used, the stringer beams are used. Sometimes only one stringer beam in the middle or sometimes two beams at the sides are used on which steps rest simply supported.

- **Spandrel :** The triangular space underneath of a staircase (when there is not another flight underneath).

- **Pitch :** The slope of the staircase measured as the ratio between the rise and going.

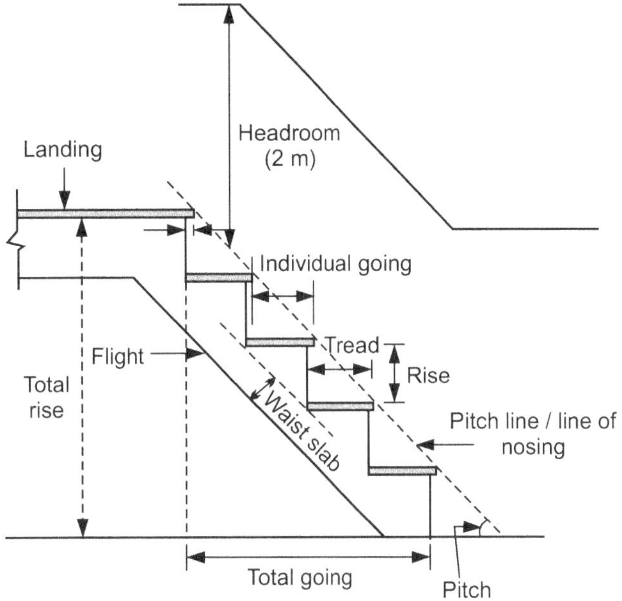

Fig. 5.1 : Different parts of flight

5.3 CLASSIFICATION OF STAIRS

Classification of stairs are as follows :

1. Straight fight stairs
2. Turning stairs
   (i) Quarter turn stairs
   (ii) Half turn stairs (open newel or open well stair)
   (iii) Dog-legged stairs
   (iv) Bifurcated stairs
3. Geometrical stair
   (i) Circular stairs
   (ii) Spiral stairs
   (iii) Helical stairs

1. **Straight Stair :** When staircase room available is narrow, then straight stairs are preferable. In this type, the steps are provided in the same direction. This type of stair consists of one flight without landing.

2. **Dog–Legged Stair :** It consists of two flights in the opposite direction. The distance between two flights is from 0 to 150 mm. Landing is provided at the level from which the direction of flight changes. Generally, dog-legged stairs are provided in residential building.

3. **Open–Newel Stair :** A rectangular space is provided between two flights and these two flights are

connected by landing. If floor to floor height is large, then a flight with less steps can also be provided between two flights. The ventilation purpose can be served by open space. Generally, in public or commercial buildings, these types of staircase are provided. This space sometimes may be used as light well.

4. **Geometrical Stair :** In this type of stair, the well between the flights is curved. Steps known as winders are provided in this curved portion with small width inner side for walking over the stair from inner edge.

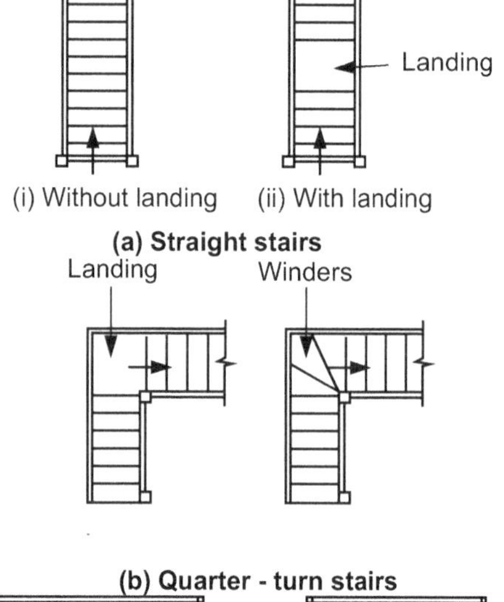

(i) Without landing　　(ii) With landing

(a) Straight stairs

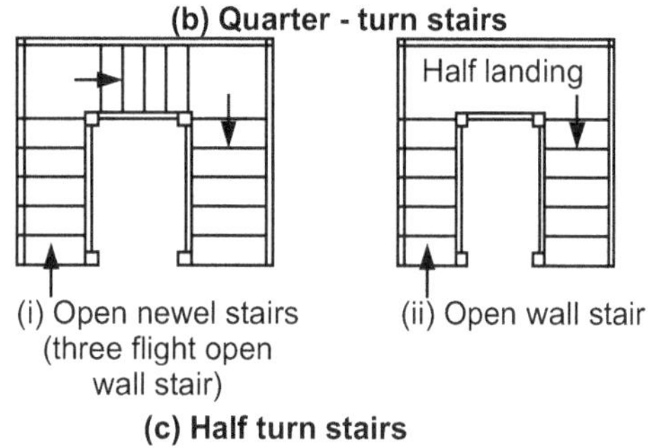

(b) Quarter - turn stairs

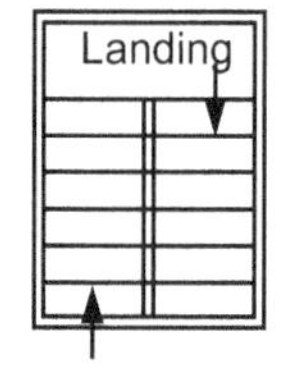

(i) Open newel stairs
(three flight open
wall stair)

(ii) Open wall stair

(c) Half turn stairs

(d) Dog.legged staircase

**(e) Circular stairs
(curved)**

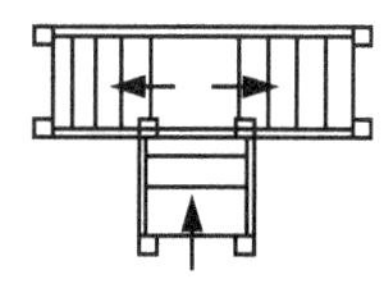

(f) Spiral stairs　　　　**(g) Bifurcated stairs**

Fig. 5.2 : Different types of stairs

5.4　GENERAL REQUIREMENT OF GOOD STAIR

To provide easy, quick and safe mode of communication between the floors. Following are the general guidelines or requirements which a stair should fulfill.

1. **Location :**
- It should be so located as to provide easy access to the occupants building.
- It should be so located that it is well lightened and ventilated from directly the exterior.

2. **Width of Stair :**
- It should be wide enough to carry the user without much crowd and inconvenience, it depends on location in building and type of building.
- In domestic or residential building the minimum width of stair should be 900 mm, while in public building in may vary in between 1.2 m to 1.8 m.

3. **Length of Flight :**
- The number of steps are not more than 12 and not less than 3.

4. **Pitch of Stair :**
- A comfortable slope is achieved when twise rise plus going is equal to 60 cm approximately pitch should however be limited to 30° to 45°.

5. **Head Room :**
- Clear space between tread and soffit of the flight should not be less 2.00 m in any case.

5.5　LOADS

5.5.1　Dead Load

- It consists of three loads :
1. **Self Weight of Waist Slab :** In case of flight, this is inclined slab so it is multiplied by $(\sqrt{R^2 + T^2})/T$ to the dead load of landing.
2. **Self Weight of Step :** Step is considered as a beam with depth equal to half the rise of step.
3. **Floor Finish :** Floor finish is considered on horizontal span. It is taken as 1 kN/m^2.

5.5.2　Live Load

- Live load is taken according to IS 875–1987 (part II). The live load is considered as 3000 N/m^2 on the buildings where there is no possibility of overcrowding, such as residential buildings, hotels, hostels, club etc. The live load on the buildings which are liable for overcrowding is taken as 4000 to 5000 N/m^2 such as public buildings, commercial complex etc.

5.5.3 Distribution of Loading on Stairs

(Distribution of loading is according to IS 456–2000, clause 33.2, page 63).

- In the case of stairs with open wells, where spans partly crossing at right angles occur, the load on areas common to any two such spans may be taken as one-half in each direction as shown in Fig. 5.3.

- Where flights or landing are embedded into walls for a length of not less than 110 mm and are designed to span in the direction of the flight, a 150 mm strip may be deducted from the loaded area and the effective breadth of the section increased by 75 mm for purposes of design. [See Fig. 5.4].

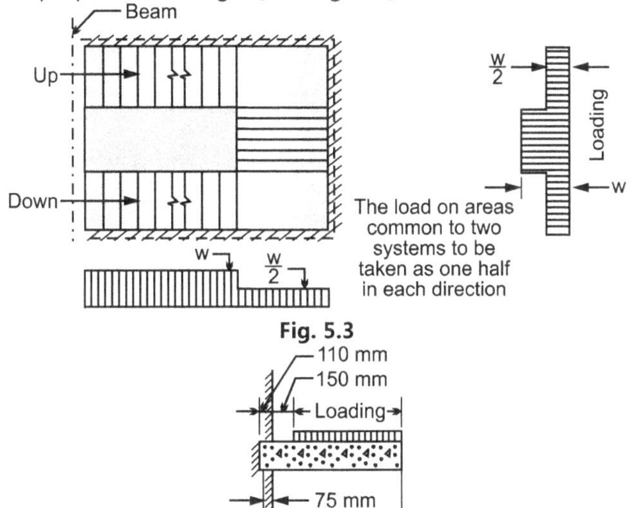

Fig. 5.3

Fig. 5.4

5.6 EFFECTIVE SPAN OF STAIRS

(IS 456–2000, clause 33.1, page 63)

- Stair slab may be divided structurally into two parts depending upon the direction in which the stair slab spans.
  1. Stair slab spanning horizontally.
  2. Stair slab spanning longitudinally.

5.6.1 Stair Slab Spanning Horizontally

- In this type of stair, slab is supported along the flight on each side. It may be supported by walls or by wall on one side and beam on the other side. In this type of stair, each step is designed as spanning horizontally considering equivalent to a rectangular beam with width b and effective depth equal to $\dfrac{D}{2}$.

- Width b is measured parallel to the slope of the stair. As slab is considered spanning parallel to step, so main

reinforcement is provided along the step and distribution reinforcement is provided along the flight.

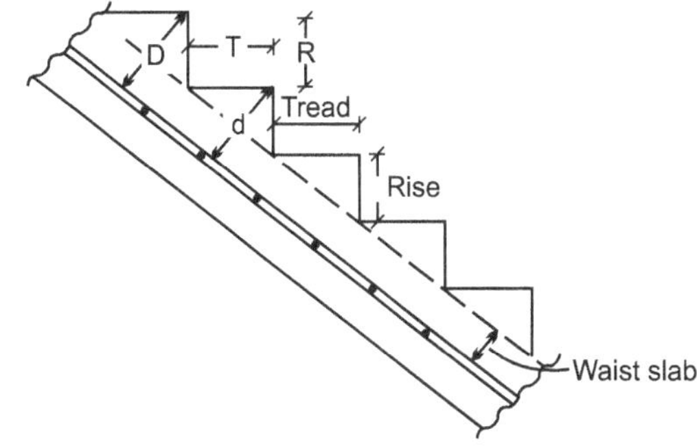

Fig. 5.5

5.6.2 Stair Slab Spanning Longitudinally

- In this type of stair, slab is supported at top and bottom of flight or landing and unsupported on sides. The effective span of slab in such cases is taken as follows :
  - ➤ Where spanning at top and bottom risers by beams spanning parallel with the risers, the distance centre-to-centre of beams.
  - ➤ Where spanning on to the edge of landing slab, which spans parallel with risers, a distance equal to the going of the stairs plus at each end either half the width of landing or one metre, whichever is smaller.

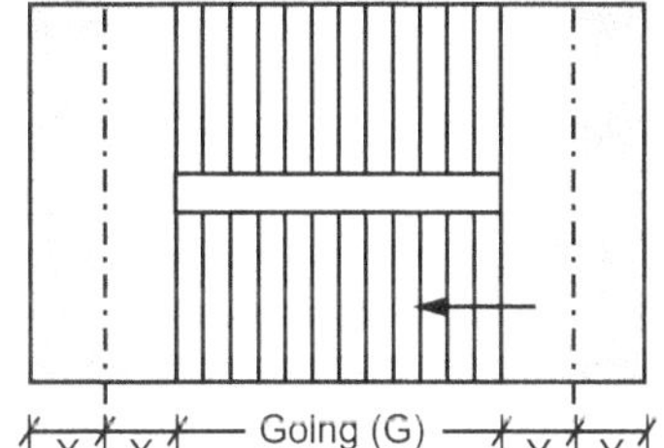

X	Y	Span in metres
< 1 m	< 1 m	G + X + Y
< 1 m	≥ 1 m	G + X + 1
≥ 1 m	< 1 m	G + 1 + Y
≥ 1 m	≥ 1 m	G + 1 + 1

Fig. 5.6 : Effective span

  - ➤ Where the landing slab spans in the same direction as the stairs, they shall be considered as acting together to form a single slab and the span determined as the distance centre–to–centre of the supporting beams or walls, the going is measured horizontally.

5.7 DESIGN PROCEDURE OF STAIRS

1. **Assume Rise (R)**
   - 150 mm to 180 mm (Residential Building),
   - 120 mm to 150 mm (Public Building)

2. **Height of One Flight**

   = Floor to floor height/2

   … Dog legged stairs

   = Floor to floor height/2.5

   … Open well stairs

3. **Number of Risers** = Height of one flight/Rise

4. **Number of Treads** = No. of risers – 1

5. **Width of Stair** = 0.8 m to 1 m (Residential Building) = 1.8 m to 2 m (Public Building)

6. **Width of Landing** = Width of stair

7. **Available Length** = Long span – 2(Width of landing)

8. **Tread (T)** = Available length / Number of tread

9. **Effective Span ($l_{eff.}$)** = c/c distance of wall = Width of wall + 2 (width of landing) + Available length

10. **Depth (thickness) of Waist Slab**

    (a) Effective depth available ($d_{available}$) = $\dfrac{l_{eff}}{20}$ to $\dfrac{l_{eff}}{25}$

    (b) Overall depth available ($d_{available}$) = $d_{available}$ + cover + (diameter of bar/2)

11. **Loads**

    (a) Self weight of the waist slab per unit horizontal area = $25 \times D_{available} \times \sqrt{[1 + (R/T)^2]}$ kN/m.

    (b) Weight of steps per unit horizontal area = $25 \dfrac{R}{2}$ kN/m

Note : The weight of step is calculated by treating the step to be equivalent to the horizontal slab of thickness equal to half the rise.

    (c) Self weight of landing = $25\, D_{available}$ (kN/m)

    (d) Assume L.L = 3 kN/m^2

    (e) Assume F.F. = 1 kN/m^2

    (f) Total load on Going portion = (Self weight of the waist slab + Weight of steps + LL + FF)

    (g) Total load on Landing = (Self weight of landing + LL + FF)

12. **Design Moment**

    (a) Draw Fig. loading on the flight and find out reaction at support, $\Sigma\, f_y = 0$.

    (b) Find out maximum moment where shear force is zero. (or direct take as $wl^2/8$)

    (c) Calculate depth required by equating maximum moment to resisting B.M.

    $$M = Qbd^2$$

    $$d_{req.} = \ldots\ldots \text{ mm} < d_{available} \qquad \ldots\text{Ok Safe.}$$

13. **Main Steel (A_{st})**

    $$M = T \times (j \times d)$$

    $$A_{st} = \dfrac{M}{\sigma_{st} \times j \times d}$$

    $$\therefore \quad A_{st} = \ldots\ldots \text{ mm}^2$$

    Assume, diameter of bar = ϕ = $\ldots\ldots$ mm

Spacing (s) :

(i) $\qquad s = \dfrac{1000 \times \text{Area of one bar}}{\text{total steel area}}$ $\left.\begin{array}{l}\\ \\ \\ \\ \\ \\ \end{array}\right\}$ whichever is less

(ii) $\qquad s = 3d$

(iii) $\qquad s = 300$ mm

14. **Distribution Steel**

(a) $\qquad A_{st} = 0.15\%\ A_g$

    $$= \left(\dfrac{0.15}{100}\right) b.\ D_{available} - \text{Mild steel (Fe250)}$$

    $$\ldots b = 1000 \text{ mm}$$

    $$= 0.12\% A_g$$

    $$= \left(\dfrac{0.12}{100}\right) b.\ D_{available} - \text{HYSD steel (Fe415 \& Fe500)}$$

(b) $\qquad$ Spacing $= \left[\dfrac{1000 \times \text{Area of one bar}}{\text{Total area of steel } (A_{st})}\right] < 5\ d_{available}$

    or 450 mm, whichever is less

15. **Summary of Design**

16. **Detailing**

SOLVED EXAMPLES

Type I : Dog Legged Staircase

Example 5.1 : *Design a dog-legged stair for a building in which the vertical distance between floors is 3.00 m. The stair hall measures 2.7 m × 5.20m. The live load may be taken as 3 kN/m². Use M-20 concrete, Fe-415 steel.*

Solution : Given :

Stair hall size = 2.70×5.20 m

Floor to floor height = 3.00 m

$\qquad$ L.L. = 3 kN/m^2

M-20; $\quad \sigma_{cbc} = 7$ N/mm^2

Fe-415 $\quad \sigma_{st} = 230$ N/mm^2

(I) Calculation of design constant

1. Modular ratio (m) :

    $$m = \dfrac{280}{3 \times \sigma_{cbc}} = \dfrac{280}{3 \times 7} = 13.33$$

2. N.A. depth factor (k) :

    $$k = \dfrac{m \cdot \sigma_{cbc}}{m \cdot \sigma_{cbc} + \sigma_{sy}} = 0.29$$

3. Lever arm factor (j) :

    $$j = \left(1 - \dfrac{k}{3}\right) = \left(1 - \dfrac{0.29}{3}\right) = 0.90$$

4. Moment resisting factor (Q) :

    $$Q = \dfrac{1}{2} \times \sigma_{cbc} \times j \times k = 0.91$$

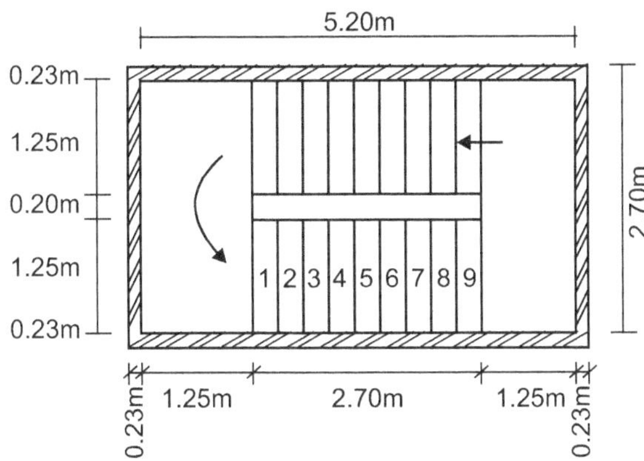

Fig. 5.7

(II) Calculation of dimensions

Fig. shows the plan of stair hall, floor to floor ht. is 3.00 m.

1. Ht. of each flight = $\dfrac{3.00}{2}$ = 1.50 m

2. Assume, Rise = R = 150 mm

3. No. of risers required = $\dfrac{1.50}{0.15}$ = 10 in each flight

4. No. of treads in each flight = 10 − 1 = 9 No's.

5. Assume, Tread = 300 mm

6. Going = space occupied by treads
   = 9 × 0.3 = 2.7 m

7. Assume, clearance = 0.2 m

8. Width of stairs = $\dfrac{2.7 - 0.2}{2}$ = 1.25 m

9. Width of landing = 1.25 m

(III) Effective span

Effective span = $l_{eff.}$ = 1.25 + 2.7 + 1.25 + 0.23

∴ $l_{eff.}$ = 5.43 m

(IV) Calculation of loading

Thickness of waist slab

$$= \frac{l}{20} \text{ to } \frac{l}{25} = \frac{5.43}{20} \text{ to } \frac{5.43}{25}$$

$$= 271.50 \text{ to } 217.20$$

∴   Assume,   D = 275 mm

                d = 250 mm

Consider, 1m width of flight

(i)   Live load = 3 kN/m

(ii) Assume F.F. = 1 kN/m

(iii) Wt. of steps = $\dfrac{1}{2} \times R \times 25$ = 1.88 kN/m

(iv) Self wt. of waist slab

$$= 25 \times D \times \frac{\sqrt{R^2 + T^2}}{T}$$

$$= 25 \times 0.275 \times \frac{\sqrt{0.15^2 + 0.300^2}}{0.300}$$

= 7.69 kN/m

∴   Total udl on going = w = (3 + 1 + 1.88 + 7.69)

∴   **w = 13.57 kN/m**

Note : The load 'w' on the landing portion will be 13.57 − 1.88 = 11.69 kN/m, since weight of steps will not come on it. However, a uniform value of 'w' has been adopted here.

(V) Calculation of maximum bending moment

$$\text{Max. B.M.} = M = \frac{wl^2}{8} = \frac{13.57 \times 5.43^2}{8}$$

∴   M = 50.01 kN.m

(VI) Check for depth

$$M = Qbd^2$$

$$50.01 \times 10^6 = 0.91 \times 1000 \times d^2$$

∴ $d_{req.}$ = 234.40 mm < $d_{provided}$ …Okay safe.

(VII) Calculation of Area of steel (A_{st}) :

$$M = T \times j \times d = \sigma_{st} \times A_{st} \times j \times d$$

∴ $A_{st} = \dfrac{M}{\sigma_{st} \times j \times d} = \dfrac{50.01 \times 10^6}{230 \times 0.90 \times 250}$

∴ A_{st} = 966.38 mm^2

and $A_{st\,min}$ = 0.12 % of Ag = $\dfrac{0.12}{100} \times 1000 \times 275$

∴ $A_{st\,min}$ = 330.00 mm^2

∴ Provide,   **A_{st} = 966.38 mm^2**

Assume, diameter of bar = ϕ = 10 mm

Spacing (s) :

(i)   s = $\dfrac{1000 \times (\pi/4 \times \phi^2)}{A_{st}}$ = 81.27 mm

(ii) s = 3d = 3 × 250 = 750 mm   } less

(iii) s = 300 mm

∴   s = 81.27 ≈ 80.00 mm

∴   **Provide 10 mm ϕ @ 80 mm c/c, as a main steel.**

(VIII) Distribution steel (A_{std}) :

∴ A_{std} = A_{stmin} = 330 mm^2

Assume diameter of bar = ϕ = 8 mm

Spacing (s) :

(i)   s = $\dfrac{1000 \times (\pi/4 \times \phi^2)}{A_{std}}$ = 152.32 mm

(ii) s = sd = 5 × 250 = 1250 mm   } less

(iii) s = 450 mm

∴   s = 152.32 ≈ 150.00 mm

∴   **Provide, 8mm ϕ @ 150 mm c/c as a distribution steel.**

(IX) Development length (Ld) :

$$L_d = \frac{\phi \cdot \sigma_s}{4 \times \tau_{bd}} = \frac{10 \times 230}{4 \times (0.8 \times 1.6)} = 449.22 \text{ mm}$$

∴   Provide,   **L_d = 450 mm**

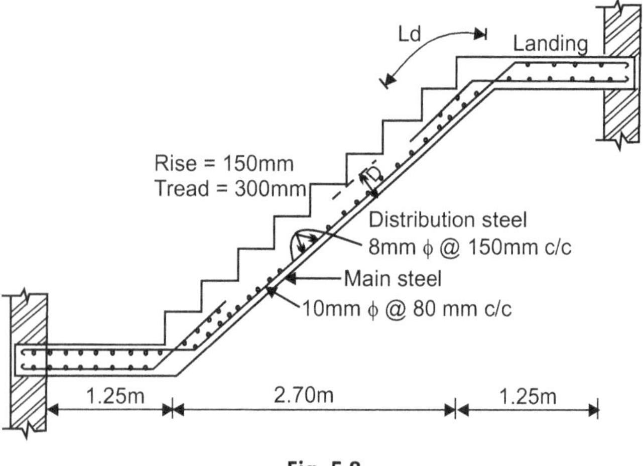

Fig. 5.8

Example 5.2 : *Design a dog-legged stair for a residential building in which the vertical distance between floor to floor is 3.1m. The space available for staircase is 2.00 m × 3.2 m. The live load may be taken as 2.75 kN/m². Use M-15 concrete Fe-250 steel.*

Solution :

Stair hall size = 2.00 m × 3.20 m

Floor to floor height = 3.00 m

L.L. = 2.75 kN/m²

M-15, σ_{cbc} = 5 N/mm²

Fe-250 ; σ_{st} = 130 N/mm²

(I) Calculation of design constant :

1. Modular ration (m) :

$$\therefore \quad m = \frac{280}{3 \times \sigma_{cbc}} = \frac{280}{3 \times 5} = 18.66$$

2. N.A. depth factor (k) :

$$\therefore \quad k = \frac{m \cdot \sigma_{cbc}}{m \cdot \sigma_{cbc} + \sigma_{st}} = 0.42$$

3. Lever arm factor (j) :

$$\therefore \quad j = \left(1 - \frac{k}{3}\right) = \left(1 - \frac{0.42}{3}\right) = 0.86$$

4. Moment resisting factor (Q) :

$$Q = \frac{1}{2} \times \sigma_{cbc} \times j \times k = 0.90$$

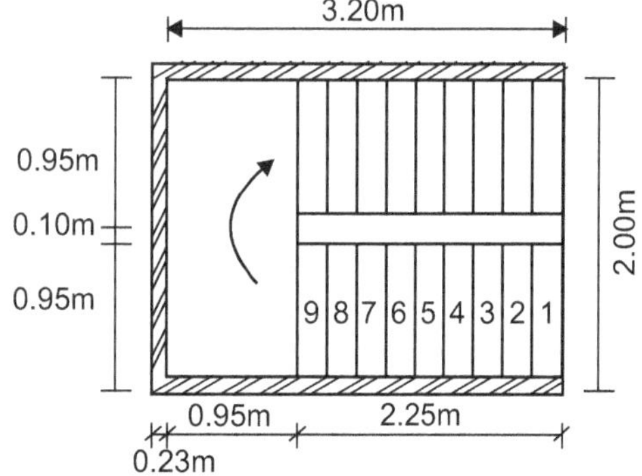

Fig. 5.9

(II) Calculation of dimensions

1. Floor to floor height = 3.10 m

2. Height of each flight = $\frac{3.10}{2}$ = 1.55 m

3. Assume, Rise = R = 155 m

4. No. of rises required = $\frac{1550}{155}$ = 10 Nos.

5. Assume clearance = 100 mm

6. Width of stair = $\left(\frac{2 - 0.10}{2}\right)$ = 0.95 m

7. Assume, width of landing = 0.95 m

8. Space available for going = 3.2 − 0.95 = 2.25 m

9. Consider, Tread = T = 250 mm

10. No. of Treads in each flight = $\frac{2.25}{0.25}$ = 9 Nos.

(III) Effective span ($l_{eff.}$) :

Effective span = $l_{eff.}$ = (0.95 + 2.25 + 0.23)

$$\therefore \quad l_{eff.} = 3.43 \text{ m}$$

(IV) Calculation of loading (w)

Thickness of waist slab

$$= \frac{l}{20} \text{ to } \frac{l}{25} = 171.50 \text{ to } 137.20$$

$\therefore$ Assume   D ≈ 150 mm

  d ≈ 125 mm

Consider '1m' width of flight

(i)   Live load = 2.75 kN/m

(ii) Assume, F.F. = 1.00 kN/m

(iii) Wt. of steps = $\left(\frac{R}{2} \times 25\right)$ = 1.94 kN/m

(iv) Self wt. of waist slab = $25 \times D \times \frac{\sqrt{R^2 + T^2}}{T}$ = 4.41 kN/m

$\therefore$ Total udl on going = w = (2.75 + 1.00 + 1.94 + 4.41)

$$\therefore \quad w = 10.10 \text{ kN/m}$$

Note : The load 'w' on the landing portion will be 10.10-1.94 = 8.16 kN/m, since the weight of steps will not come on it. However, a uniform value of 'w' has been adopted here.

(V) Calculation of maximum bending moment

$$\text{Max. B.M.} = M = \frac{wl^2}{8} = \frac{10.10 \times 3.43^2}{8}$$

$$\therefore \quad M = 14.85 \text{ kN.m}$$

(VI) Check for depth

By equating,   M = Qbd^2

$$14.85 \times 10^6 = 0.90 \times 1000 \times d^2$$

$$\therefore \quad d_{req.} = 128.45 \text{ mm} \nless d_{provided0}$$

…Not okay … unsafe

$\therefore$ Increasing depth,

$\therefore$ Provide, $D = 155$ mm

$\qquad d = 130$ mm

(VII) Calculation of area of steel (A_{st}) :

$$M = T \times (j \times d)$$

$$A_{st} = \frac{M}{\sigma_{st} \times j \times d} = \frac{14.85 \times 10^6}{130 \times 0.86 \times 130}$$

$\therefore \qquad A_{st} = 1021.74$ mm^2

and $\quad A_{st\,min} = 0.15$ % of $A_g = \dfrac{0.15}{100} \times 1000 \times 155$

$\therefore \qquad A_{st\,min} = 232.50$ mm^2

$\therefore$ **Provide, $A_{st} = 1021.74$ mm^2**

Assume, diameter of bar $= \phi = 10$ mm

Spacing (s) :

(i) $\quad s = \dfrac{1000 \times (\pi/4 \times \phi^2)}{A_{st}} = 76.87$ mm $\qquad$ less

(ii) $\quad s = 3d = 3 \times 130 = 390$ mm

(iii) $\quad s = 300$ mm

$\therefore \qquad\qquad s = 76.87 \approx 75$ mm

$\therefore$ **Provide 10 mm ϕ @ 75 mm c/c, as a main steel.**

(VIII) Distribution steel (A_{std}) :

$\therefore \qquad A_{std} = A_{st\,min} = 232.50$ mm^2

$\qquad$ diameter of bar $= \phi = 8$ mm

Assume, spacing s $= \dfrac{1000 \times (\pi/4 \times 8^2)}{232.50}$

$\qquad\qquad = 216.19$ mm $< 5d$ or 450 mm

$\therefore$ Provide, $\quad s = 216.19 \approx 200$ mm

$\therefore$ **Provide 8 mm ϕ @ 200 mm c/c, as a distribution steel.**

(IX) Development length :

$$L_d = \frac{\phi \cdot \sigma_s}{4 \times \tau_{bd}} = \frac{10 \times 130}{4 \times 0.8 \times 1.6}$$

$$= 253.91 \text{ mm}$$

$\therefore \qquad$ **$L_d = 250.00$ mm**

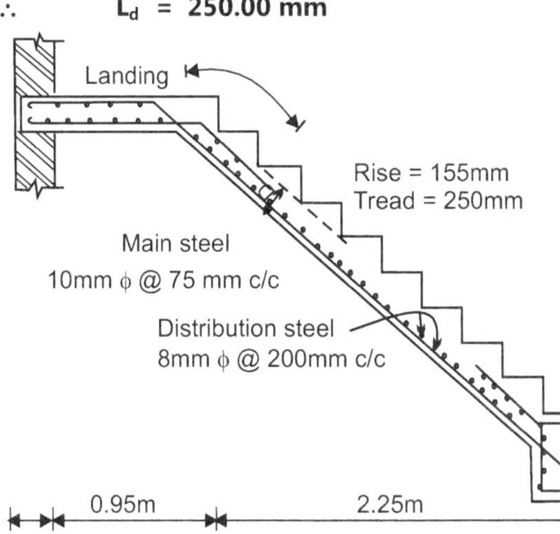

Fig. 5.10

Example 5.3 : *Design a dog-legged staircase for a public building. The columns are 230 mm $\times$ 230 mm. The floor to floor height of storey is 3.60 m. The live load on staircase is 3.5 kN/m^2 and floor finish of 0.75 kN/m^2. Use M-20 and Fe-415. The space available for staircase is 3.00 m $\times$ 5.20 m. Prepare working drawing showing details of reinforcement and dowel bars.*

Solution : Given :

$\quad$ Space available for staircase $= 3 \times 5.2$ m

$\quad$ Floor to floor height $= 3.60$ m

$\quad$ Live load $= 3.50$ kN/m^2

$\quad$ F.F. $= 0.75$ kN/m^2

$\quad$ M-20 ; $\sigma_{cbc} = 7$ N/mm^2

$\quad$ Fe-415; $\sigma_{st} = 230$ N/mm^2

(I) Calculation of design constant :

1. $\quad$ Modular ratio (m) :

$$m = \frac{280}{3 \times \sigma_{cbc}} = \frac{280}{3 \times 7} = 13.33$$

2. $\quad$ N.A. depth factor (K) :

$$K = \frac{m \cdot \sigma_{cbc}}{m \cdot \sigma_{cbc} + \sigma_{st}} = 0.29$$

3. $\quad$ lever arm factor (j) :

$$j = \left(1 - \frac{K}{3}\right) = \left(1 - \frac{0.29}{3}\right) = 0.90$$

4. $\quad$ Moment Resisting factor (Q) :

$$Q = \frac{1}{2} \times \sigma_{cbc} \times j \times k = 0.91$$

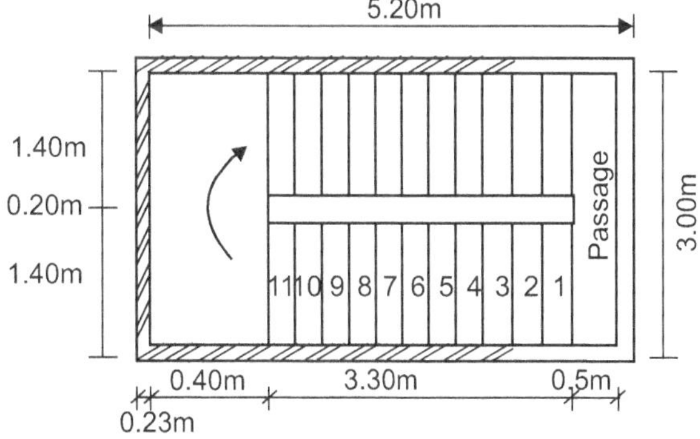

Fig. 5.11

(II) Calculation pf dimensions :

$\quad$ Fig. shows the plan of stair hall

1. $\quad$ Floor to floor height is $= 3.60$ m

2. $\quad$ Height of each flight $= \dfrac{3.60}{2} = 1.80$ m

3. $\quad$ Assume, Rise $= R = 150$ mm

4. $\quad$ No. of risers required $= \dfrac{1.80}{0.15} = 12$ No's.

5. No. of Treads in each flight = R-1 = 11 No's.

6. Assume, Tread = T = 300 mm

7. Going = 11×0.3 = 3.3 m

8. Assume, clearance = 0.20 m

9. Width of stairs = 1.40 m

10. Width of landing = 1.40 m

11. Distance available in addition to passage = 0.5 m

(III) Effective span (l_{eff}) :

Effective span = l_{eff} = 1.40 + 3.30 + 0.5 + 0.23

$\therefore$ l_{eff} **= 5.43 m**

(IV) Calculation of loading (w) :

Thickness of waist slab = D = $\dfrac{l}{20}$ to $\dfrac{l}{25}$

$\therefore$ **Assume, D = 275 mm**

               d = 250 mm

Consider, 1m width of flight,

  (i)   Live load = 3.50 kN/m

  (ii)   F.F. = 0.75 kN/m

  (iii) Wt. of steps = $\dfrac{R}{2} \times 25$ = 1.88 kN/m

  (iv) Self wt. of waist slab = $25 \times D \times \dfrac{\sqrt{R^2 + T^2}}{T}$

                       = 7.69 kN/m

Total udl on going = w = (3.50 + 0.75 + 1.88 + 7.69)

$\therefore$ **w = 13.82 kN/m**

Note : The load 'w' on the landing portion will be 13.82 − 1.88 = 11.94 kN/m, since weight of steps will not come on it. However, a uniform value of w has been adopted here.

(V) Calculation of maximum bending moment (M) :

Max. B.M. = M = $\dfrac{Wl^2}{8}$ = $\dfrac{13.82 \times 5.43^2}{8}$

$\therefore$ **M = 50.94 kN.m**

(VI) Check for depth :

By equating,

$\therefore$ M = Qbd^2

50.94×10^6 = $0.91 \times 1000 \times d^2$

d_{req} = 234.40 mm < $d_{provided}$ ... Okay safe.

(VII) Calculation of Area of steel (A_{st}) :

M = $T \times (j \times d)$

A_{st} = $\dfrac{M}{\sigma_{st} \times j \times d}$ = $\dfrac{50.94 \times 10^6}{230 \times 0.90 \times 250}$

$\therefore$ A_{st} = 984.35 mm^2

A_{stmin} = 0.12 % of A_g = $\dfrac{0.12}{100} \times 1000 \times 275$

$\therefore$ A_{stmin} = 330 mm^2

$\therefore$ **Provide, A_{st} = 984.35 mm^2**

Assume, diameter of bar = ϕ = 10 mm

- **Spacing (s) :**

(i)   S = $\dfrac{1000 \times \left(\dfrac{\pi}{4} \times \phi^2\right)}{A_{st}}$ = 79.79 mm

(ii)   S = 3d = 3×250 = 750 mm   } Less

(iii)   S = 300 mm

$\therefore$ Spacing = S = 75 mm

$\therefore$ **Provide, 10 mm ϕ @ 75 mm c/c, as a main steel.**

(VIII) Distribution steel (A_{std}) :

$\therefore$ A_{std} = A_{stmin} = 330 mm^2

Assume, diameter of bar = ϕ = 8 mm

- **Spacing (s) :**

(i)   S = $\dfrac{1000 \times \left(\dfrac{\pi}{4} \times \phi^2\right)}{A_{std}}$ = 152.32 mm

(ii)   S = 5d = 5×250 = 1250.00 mm   } Less

(iii)   S = 450 mm

$\therefore$ S = 152.32 $\approx$ 150 mm

$\therefore$ **Provide, 8 mm ϕ @ 150 mm C/C, as a distribution steel.**

(IX) Development length (L_d) :

L_d = $\dfrac{\phi \cdot \sigma_s}{4 \times \tau_{bd}}$ = $\dfrac{10 \times 230}{4 \times (0.8 \times 1.6)}$

     = 449.22 mm

$\therefore$ L_d $\approx$ **450 mm**

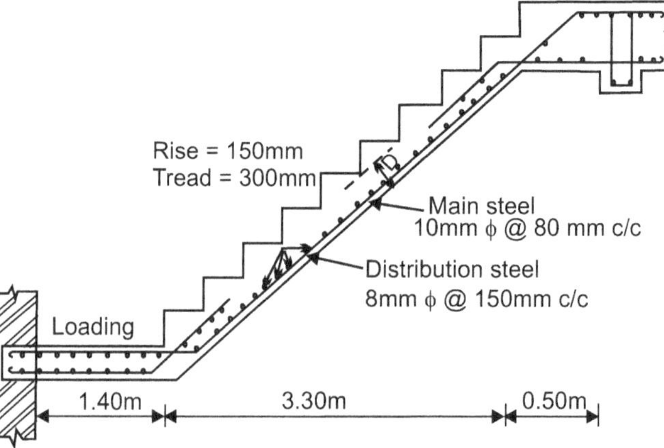

Fig. 5.12

Type II : Design of Open Well Staircase

Example 5.4 : *A stair which open well is having steps of size 280 × 150 mm. The arrangement of stair is as shown in Fig. Design the stairs for a live load of 3kN/m^2. Use M-20 concrete mix and Fe-415 steel. Sketch the details of reinforcement.*

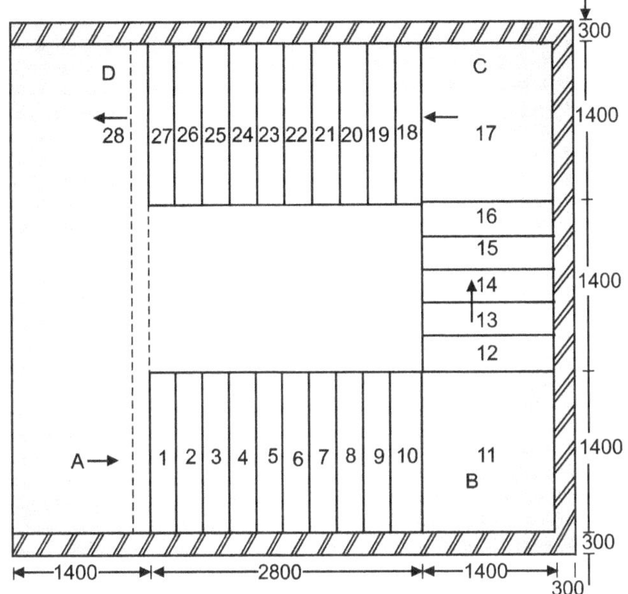

Fig. 5.13

Solution :

(I) Calculation of design constant :

1. Modular Ratio (m) :

$$m = \frac{280}{3 \times \sigma_{cbc}} = \frac{280}{3 \times 7} = 13.33$$

2. N.A. depth factor (K) :

$$k = \frac{m\,\sigma_{cbc}}{m \cdot \sigma_{cbc} + \sigma_{st}} = 0.29$$

3. Lever arm factor (j) :

$$j = \left(1 - \frac{K}{3}\right) = 0.90$$

4. Moment resisting factor (Q) :

$$Q = \frac{1}{2} \times \sigma_{cbc} \times j \times k = 0.91$$

(II) Effective span :

flight AB and CD,

$$l_{eff} = 2.8 + 1.4 + 0.15 + 0.15 = 4.5 \text{ m}$$

flight BC

$$l_{eff} = 1.4 + 1.4 + 1.4 + 0.15 + 0.15 = 4.5 \text{ m}$$

(III) Calculation of loading (w) :

Consider '1m' width of stair,

Thickness of waist slab $= \dfrac{l}{20}$ to $\dfrac{l}{25}$

$$= 225 \text{ to } 180$$

Provide, D = 225 mm

d = 200 mm

Consider '1m' width of flight,

(i)   Live load $= 3$ kN/m

(ii)   Assume, floor finish $= 1$ kN/m

(iii)   Wt. of steps $= \dfrac{R}{2} \times 25 = 1.88$ kN/m

(iv)   Self wt. of waist slab $= 25 \times D \times \dfrac{\sqrt{R^2 + T^2}}{T}$

$$= 6.38 \text{ kN/m}$$

∴   Total load on going $= w = 3 + 1 + 1.88 + 6.38$

∴ $w = 12.26$ kN/m

Self wt. of landing slab $= 25 \times D = 25 \times 0.225$

$$= 5.63 \text{ kN/m}$$

∴   Total load on landing $= 3 + 1 + 5.63$

$$= 9.63 \text{ kN/m}$$

- Flight AB and CD :

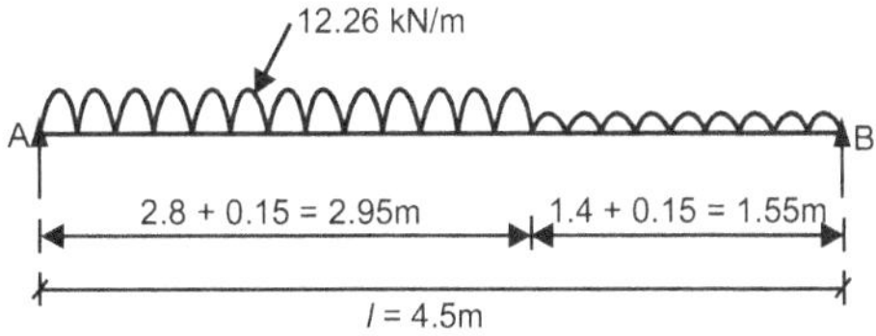

Fig. 5.14

- Flight BC :

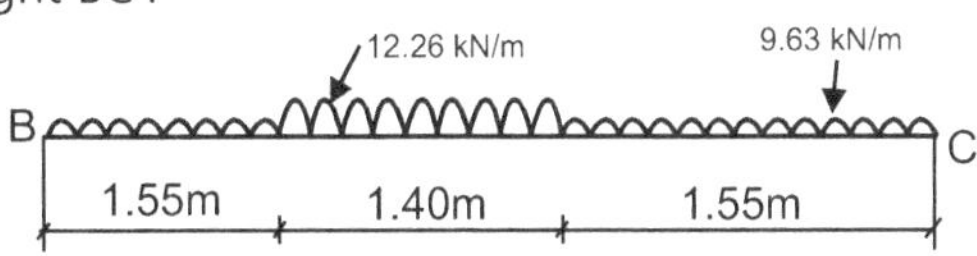

Fig. 5.15

Note : The load 'w' on the landing portion will be less, since, weight of steps will not come on it. However, a uniform value of 'w' has seen adopted here.

(IV) Calculation of maximum bending moment :

For flight AB, BC and CD :

$$\text{Max. B.M.} = M = \frac{wl^2}{8} = \frac{12.26 \times 4.5^2}{8}$$

∴ $M = 31.03$ kN.m

(V) Check for depth :

By equating,

$$M = Q\,bd^2$$

$$31.03 \times 10^6 = 0.91 \times 1000 \times d^2$$

∴ $d_{req} = 184.66$ mm $< d_{provided}$ ... Okay Safe.

(VI) Calculation of Area of steel (A_{st}) :

$$M = T \times (j \times d)$$

$$A_{st} = \frac{M}{\sigma_{st} \times j \times d} = \frac{31.03 \times 10^6}{230 \times 0.90 \times 200}$$

∴ $A_{st} = 749.52$ mm^2

and $A_{st\,min} = 0.12\%$ of $A_g = \dfrac{0.12}{100} \times 1000 \times 225$

∴ $A_{st\,min} = 270$ mm^2

∴   **Provide A_{st} = 749.52 mm^2**

Assume, diameter of bar $= \phi = 10$ mm

- **Spacing (s) :**

(i) $\quad S = \dfrac{1000 \times \left(\dfrac{\pi}{4} \times 10^2\right)}{A_{st}} = 104.79 \text{ mm}$

(ii) $\quad S = 3d = 3 \times 200 = 600 \text{ mm}$

(iii) $\quad S = 300 \text{ mm}$ ⎫ Less

$\therefore \quad S = 104.79 \approx 100 \text{ mm}$

∴ Provide, 10 mm φ @ 100 mm c/c, as a main steel

(VII) Distribution steel :

$\therefore \quad A_{std} = A_{stmin} = 270.00 \text{ mm}^2$

Assume, diameter of bar = φ = 8 mm

- **Spacing (s) :**

(i) $\quad S = \dfrac{1000 \times \left(\dfrac{\pi}{4} \times \phi^2\right)}{A_{std}}$

$\quad = 186.17 \text{ mm} < 5d \text{ or } 450 \text{ mm}$

$\therefore \quad S = 186.17 \approx 180.00 \text{ mm}$

∴ Provide, 8 mm φ @ 180.00 mm c/c as a distribution steel.

(VIII) Development length :

$$L_d = \dfrac{\phi \cdot \sigma_s}{4 \times \tau_{bd}} = \dfrac{10 \times 230}{4 \times (0.8 \times 1.6)} = 449.22$$

Provide, **L_d = 450 mm**

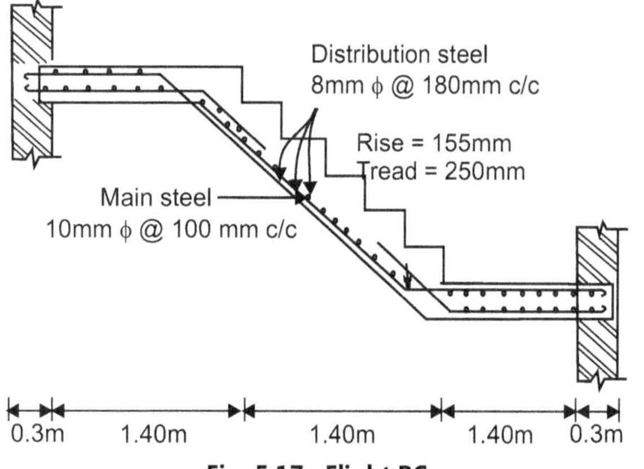

Fig. 5.16 : Flight AB and CD

Fig. 5.17 : Flight BC

Example 5.5 : *Design an open well staircase, the space available for staircase is 5.00 m × 3.55 m. The superimposed load on stair is 4.25 kN/m². Use M-25 concrete and Fe-415 steel. Sketch the reinforcement details. Floor to floor height is 3.60 m.*

Solution : Given :

Space available for staircase

$\quad = 5.00 \text{ m} \times 3.50 \text{ m}$

Superimposed load = (L.L + F.F) = 4.25 kN/m²

M-25; $\quad \sigma_{cbc} = 8.50 \text{ N/mm}^2$

Fe-415; $\quad \sigma_{st} = 230 \text{ N/mm}^2$

(I) Calculation of design constant :

1. Modular ratio (m) :

$$m = \dfrac{280}{3 \times \sigma_{cbc}} = \dfrac{280}{3 \times 8.5} = 10.98$$

2. N.A. depth factor (K) :

$$K = \dfrac{m \cdot \sigma_{cbc}}{m \cdot \sigma_{cbc} + \sigma_{st}} = 0.28$$

3. Lever arm factor (j) :

$$j = \left(1 - \dfrac{K}{3}\right) = \left(1 - \dfrac{0.28}{3}\right) = 0.91$$

4. Moment resisting factor (Q) :

$$Q = \dfrac{1}{2} \times \sigma_{cbc} \times j \times K = 1.08$$

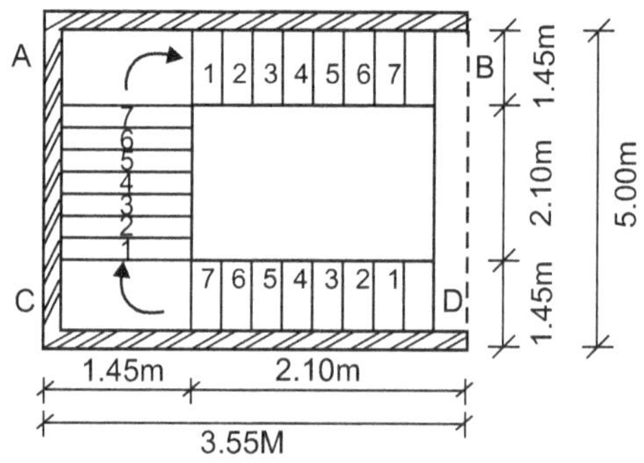

Fig. 5.18

(II) Calculation of dimensions :

1. Floor to floor height = 3.6 m

2. Consider Rise = R = 150 mm

3. No. of Risers = $\dfrac{3600}{150}$ = 24 Nos.

4. No. of Rise in each flight = $\dfrac{24}{3}$ = 8 Nos.

5. No. of treads in each flight = T = R − 1 = 8 − 1 = 7 No's.

6. Assume, Tread = T = 300 mm

7. Going = 7 × 0.3 = 2.1 m

8. Width of going = $\dfrac{5 - 2.1}{2}$ = 1.45 m

9. Width of landing = 1.45 m

(III) Effective span :

Flight AB and CD

$$l_{eff} = 1.45 + 2.10 + 0.23 = 3.78 \text{ m}$$

Flight AC

$$l_{eff} = 1.45 + 2.10 + 1.45 + 0.23 = 5.23 \text{ m}$$

(IV) Calculation of loading (w) :

Consider '1m' width of flight.

Thickness of waist slab $= \dfrac{l}{20}$ to $\dfrac{l}{25}$

$\therefore \qquad D = \dfrac{5230}{20}$ to $\dfrac{5230}{25}$

$\therefore \qquad D = 261.50$ to 209.20

$\therefore$ provide,

$$D = 250 \text{ mm}$$
$$d = 225 \text{ mm}$$

(i)   (Live load + floor finish) $= 4.25$ kN/m

(ii)   Self wt. of steps $= \dfrac{R}{2} \times 25 = 1.88$ kN/m

(iii)   Self wt. of waist slab $= 25 \times D \times \dfrac{\sqrt{R^2 + T^2}}{T}$

$$= 6.99 \text{ kN/m}$$

$\therefore$ Total load on going $= 4.25 + 1.88 + 6.99$

$\therefore \qquad\qquad\qquad w = 13.12$ kN/m

Self wt. of landing slab $= 25 \times D = 25 \times 0.250 = 6.25$ kN/m

$\therefore$ Total load on landing $= 4.25 + 6.25 = 10.5$ kN/m

- Flight AB and CD :

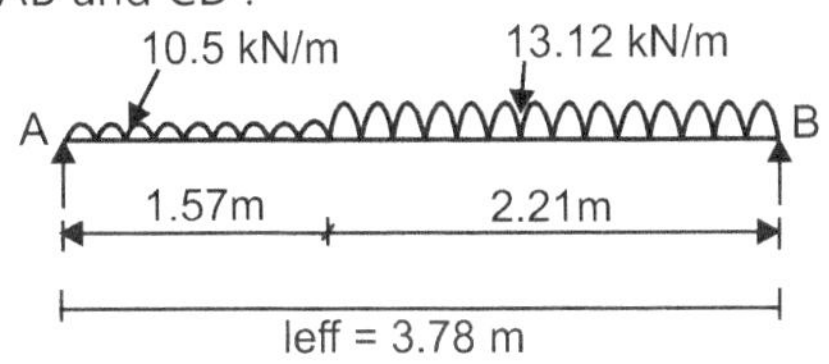

Fig. 5.19

- Flight AC :

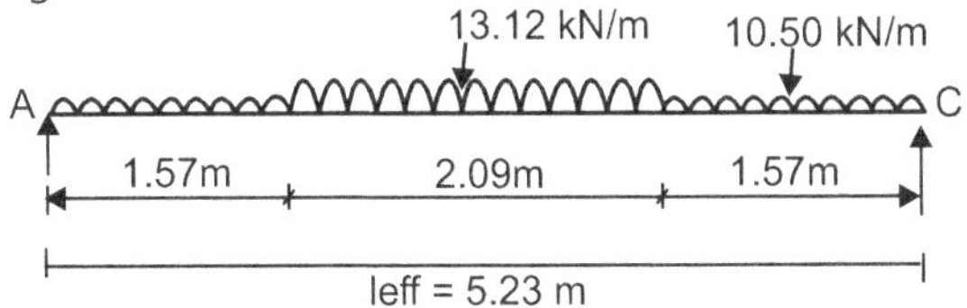

Fig. 5.20

Note : The load 'w' on the landing portion will be less, since, weight of steps will not come on it. However, a uniform value of w has been adopted here.

(V) Calculation of Bending moment maximum "

(i)   For flight AB and CD :

$$M_{AB} = \frac{wl^2}{8} = \frac{13.12 \times 3.78^2}{8}$$

$\therefore \qquad M_{AB} = 23.43$ kN.m

(ii)   For flight AC :

$$M_{AC} = \frac{wl^2}{8} = \frac{13.12 \times 5.23^2}{8}$$

$\therefore \qquad M_{AC} = 44.86$ kN.m

(VI) Check for depth :

By equating, M and resisting moment

(i)   Flight AB and CD

$$M = Qbd^2$$
$$23.43 \times 10^6 = 1.08 \times 1000 \times d^2$$

$\therefore \qquad d_{req} = 147.38$ mm $< d_{provided}$ ... Okay safe.

(ii)   Flight AC:

$$M = Qbd^2$$
$$44.86 \times 10^6 = 1.08 \times 1000 \times d^2$$

$\qquad\qquad d_{req} = 203.81$ mm $< d_{provided}$ ... Okay Safe.

(VII) Calculation of Area of steel (A_{st}) :

(i)   Flight AB and CD

$$A_{st} = \frac{M}{\sigma_{st} \times j \times d} = \frac{23.43 \times 10^6}{230 \times 0.91 \times 225}$$

$\therefore \qquad A_{st} = 497.53$ mm^2

and $A_{stmin} = 0.12\%$ of $A_g = \dfrac{0.12}{100} \times 1000 \times 250$

$\therefore \qquad A_{stmin} = 300$ mm^2

$\therefore$ Provide,

$$\mathbf{A_{st} = 497.53 \text{ mm}^2}$$

Assume, diameter of bar $= \phi = 10$ mm

- **Spacing (s) :**

(i) $S = \dfrac{1000 \times \left(\dfrac{\pi}{4} \times \phi^2\right)}{A_{st}}$

$$= 157.86 \text{ mm} < 3d \text{ or } 300 \text{ mm}$$

$\therefore \qquad S = 157.86 \approx 150$ mm

$\therefore$ Provide, 10 mm ϕ @ 150 mm c/c, as a main steel

(ii)   Flight AC :

$$A_{st} = \frac{M}{\sigma_{st} \times j \times d} = \frac{44.86 \times 10^6}{230 \times 0.91 \times 225}$$

$\therefore \qquad A_{st} = 952.59$ mm$^2 > A_{stmin}$

Assume, diameter of bar $= \phi = 12$ mm

- **Spacing (s) :**

(i) $S = \dfrac{1000 \times \left(\dfrac{\pi}{4} \times \phi^2\right)}{A_{st}}$

$$= 118.72 \text{ mm} < 3d \text{ or } 300 \text{ mm}$$

$\therefore$ Provide, 12 mm ϕ @ 110 mm c/c, as a main steel

(VIII) Distribution steel

$$\therefore \quad A_{std} = A_{stmin} = 300 \text{ mm}^2$$

Assume, diameter of bar = ϕ = 8 mm

$$\text{spacing} = S = \frac{1000 \times \left(\frac{\pi}{4} \times \phi^2\right)}{A_{std}}$$

$$= 167.55 \text{ mm} < 5d \text{ or } 450 \text{ mm}$$

$$\therefore \quad S = 167.55 \approx 150 \text{ mm}$$

∴ Provide, 8mm ϕ @ 150 mm c/c, as a distribution steel

(IX) Development length :

(i) Flight AB and CD :

$$L_d = \frac{\phi \cdot \sigma_s}{4 \times \tau_{bd}} = \frac{10 \times 230}{4 \times 0.9 \times 1.6} = 399.31 \text{ mm}$$

$$\therefore \quad \mathbf{L_d \approx 400 \text{ mm}}$$

(ii) Flight AC :

$$L_d = \frac{\phi \cdot \sigma_s}{4 \times \tau_{bd}} = \frac{12 \times 230}{4 \times 0.9 \times 1.6} = 479.17 \text{ mm}$$

$$\therefore \quad \mathbf{L_d \approx 480 \text{ mm}}$$

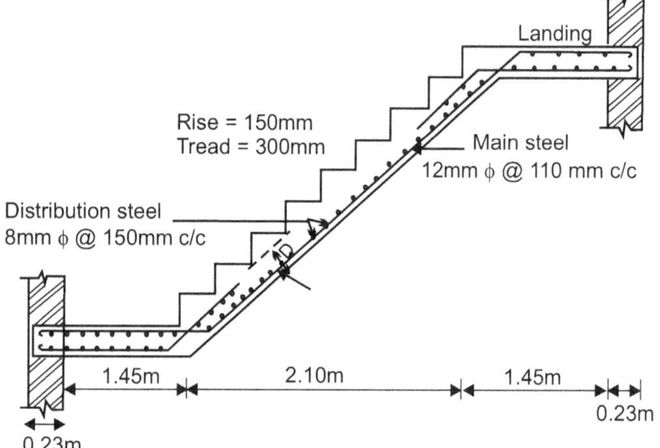

Fig. 5.21 : Flight AC

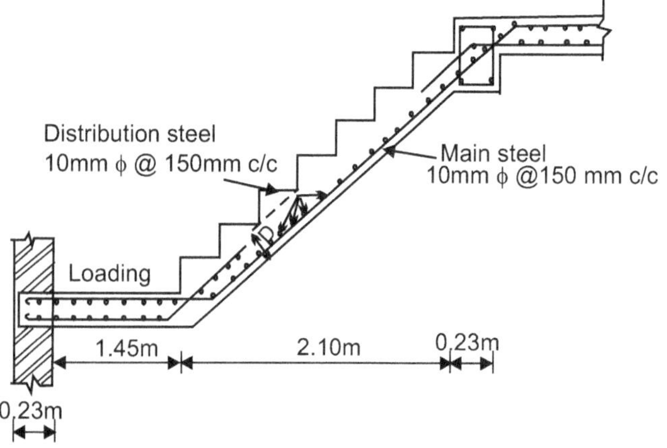

Fig. 5.22 : Flight AB and CD

EXERCISE

1. State and explain different parts of staircase.

2. Explain types of staircase.

3. State effective span of stair. Explain.

4. Explain the design procedure of design of dog-legged staircase.

5. Why development length is provided in staircase ?

6. Design a dog legged staircase for space 2.5m × 2.5m for residential building. The floor to floor height is 3.00 m. Take M-15 and Fe 250 grades material.

7. Design an open well staircase for a space 4m × 5.5m floor to floor ht is 4.1 m, Take M-25, Fe-415.

8. Design a suitable staircase for a three storied residential building. The size of a staircase room is 2.2 m × 4.25 m between centre of columns. The floor to floor height is 3.2 m. The live on stairs in 2.75 kN/m^2 and floor finish is 0.75 kN/m^2. Use M 20 grade of concrete and Fe 415.

9. Design a open–newel staircase for a multistoried building. The size of staircase room is 4.2 m × 5.2 m between centres of beams and walls. The floor to floor height is 4.2 m. The live load on stairs is 3 kN/m- and floor finish is 10 kN/m^2. Use M 20 grade of concrete and Fe 415.

10. Design a dog–legged staircase for a bungalow. The size of staircase room is 2.15 m × 3.4 m between centres of columns. The floor to floor height is 3.1 m. Use M 20 and Fe 415.

CHAPTER - 6
INTRODUCTION TO LIMIT STATE APPROACH

6.1 INTRODUCTION

- The philosophy of the limit state method of design is an advancement or improvement over traditional design philosophies. We know in working stress method, which is based on calculations on service working load condition only. In case of ultimate load method, which is based on calculations on ultimate load condition only and in limit state method aims for a comprehensive and rational solution to the design problem, by considering safety at ultimate loads and serviceability at working loads.

- In limit state method, a structure is designed to withstand all loads likely to act on it throughout its lifespan and also satisfy the serviceability requirement before failure can occur. The design provides a conditions that the structure will not become unfit for use for which it is meant or in other words the structure will not reach a limit state.

6.2 LIMIT STATE METHOD

(Refer Cl. No. 35.1, Pg. No. 67, IS 456 : 2000)

- The acceptable limit for the safety and serviceability requirements before failure occurs is called as 'limit state method'.

- The structure shall be designed on the basis of the most critical limit state and shall be checked for other limit states. For ensuring the above objective, the design should be based on characteristic values for material strengths and applied loads which take into account the variations in the material strengths and in the loads to be supported.

6.3 TYPES OF LIMIT STATE

(Refer Cl. No. 35.2, 35.3, Pg. No. 67,, IS 456 : 2000)

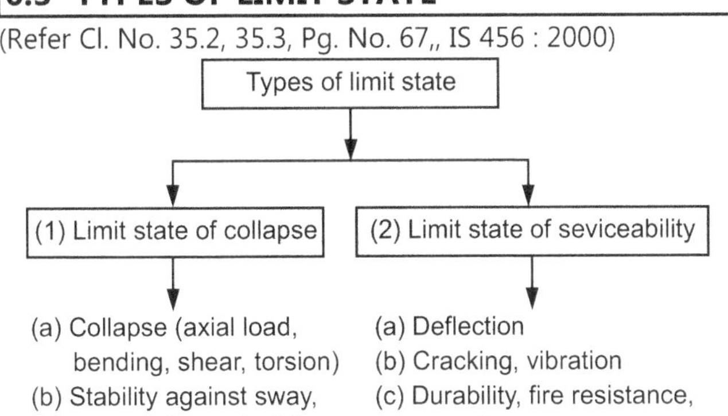

6.3.1 Limit State of Collapse

The resistance to bending, shear, torsion and axial load at every section shall not be less than the appropriate value at that section produced by the probable most unfavourable combination of loads on the structure using the appropriate partial safety factors.

Limit State of Strength Includes :

- Loss of equilibrium of the structure as a whole or its components.

- Loss of stability of the structure including the effect of sway, overturning and sliding.

- Failure of structure due to excessive deformation, rupture of the structure or its components.

- Cracks due to fatigue.

- Brittle fractures.

6.3.2 Limit State of Serviceability

Limit state of serviceability is related to the satisfactory performance of the structure at working load. The limit state is corresponds to deflection, cracking and durability.

(a) Limit State of Deflection

- The deflection of a structure or part of a structure there of shall affect the appearance or efficiency of the structure or finishes or partitions.

- A limit state provided by prescribing a limit for deflections or by providing limit to span-depth ratios of beam. The acceptable criteria of design for deflection is given in clause 23.2 and 23.2.1, IS 456 : 2000. The actual deflection should not be less than maximum allowable deflection.

(b) Limit State of Cracking

Cracking of concrete should also affect the appearance or durability of the structure, the acceptable limits of cracking would vary with the type of structure and environment. The actual width of crack will vary between wide is not possible. In general, the surface width of cracks should not exceed 0.3 mm.

To ensure that the crack width is not excessive, any of the two methods given below may be considered.

1. **Bar Spacing Control :** The rules of maximum spacing for bars in tension are given in clause 26.3 of IS 456 : 2000. This shall be used in all normal cases.

2. **Crack Width Calculation :** Where it is required to calculate the crack width, the crack width calculations may be done using formula given in Annex-F IS 456 : 2000.

(c) Limit State of Durability

Durability is defined as the ability of the structure to maintain its level of reliability and serviceability during its intended life span. It depends on its resistance to deterioration and the environment in which it is placed. The resistance of structure to weathering, chemical attack, abrasion, frost and fire depends largely upon its quality and constituent material. Depending upon these cause, they are divided into the following limit states

1. Limit state of fire resistance
2. Limit state of environmental and chemical actions.

Limit State of Serviceability Includes :

- Deformation within limit is acceptable if it exceeds certain limit it will cause discomfort to occupants excessive deformation may also cause damage to finishes and non structural members.
- Durability and corrosion and fire resistance.
- Vibration in the structural or any of its components causing discomfort to people or damages to the structure.
- Repairable damages or cracks due to fatigue.

6.4 CHARACTERISTIC STRENGTH

(Refer clause 36.1, Pg. No. 67, IS 456 : 2000)

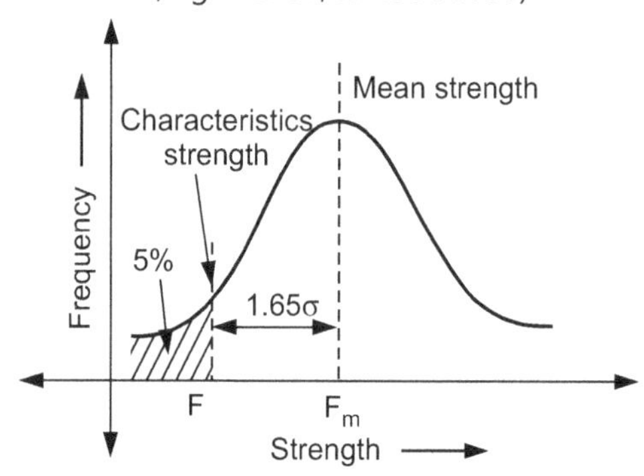

Fig. 6.1

- Characteristics strength of the material is the strength of material below which not more than 5% of the test results are expected to fall. It is denoted by 'f'.

Characteristics strength (f) = Mean strength − 1.65 × σ

$$\therefore \quad f = f_m - 1.65 \times \sigma$$

where, σ = standard deviation

value of 'σ' as per IS code

Grade	M-10, M-15	M-20, M-25	≥ M-30
'σ' N/mm²	3.50	4.00	5.00

6.4.1 Grades of Concrete

(Refer Table No. 2, Pg. No. 16, IS 456 : 2000)

Table 6.1

Group	Grade Designation	Specified Characteristic Compressive Strength of 150 mm cube at 28 days in N/mm²
(1)	(2)	(3)
Ordinary concrete	M10	10
	M15	15
	M20	20
Standard concrete	M25	25
	M30	30
	M35	35
	M40	40
	M45	45
	M50	50
	M55	55
High strength concrete	M60	60
	M65	65
	M70	70
	M75	75
	M80	80

Note :

- In the designation of concrete mix, M refers to the mix and the number to the specified compressive strength of 150 mm cube at 28 days, expressed in N/mm².
- The characteristic strength for steel (f_y) shall be assumed as minimum yield stress or 0.2 percent proof stress.

6.4.2 Grades of Steel Reinforcement

Table 6.2

Grade of Steel	Characteristics Min-yield Strength 'f_y' N/mm²
Fe – 250	250
Fe – 415	415
Fe – 500	500

6.5 CHARACTERISTICS LOAD

(Refer clause 36.2, Pg. No. 67, IS 456 : 2000)

- Characteristics load means the load on the structure which has a 95% probability of not being exceeded throughout the life of structure.
- Characteristics dead load is the weight of structure itself. Characteristics live load and wind load are taken

as per IS 875. Characteristics seismic load are taken as per IS 1893. It is denoted by 'F'.

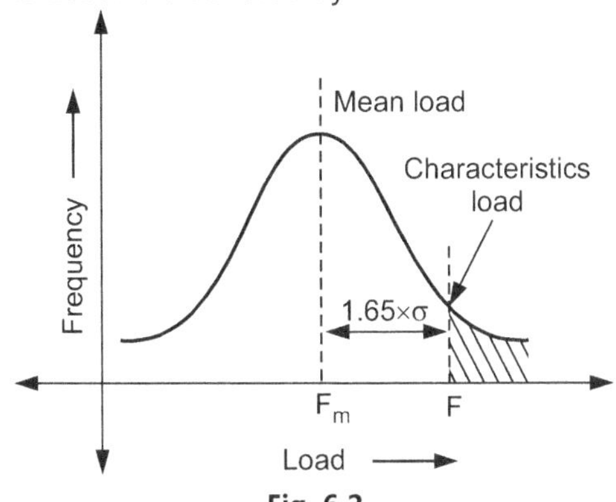

Fig. 6.2

Characteristic load (F) = Mean load + $1.65 \times \sigma$

∴ $$F = F_m + 1.65 \times \sigma$$

where, σ = standard deviation

6.6 PARTIAL SAFETY FACTOR

(Refer clause 36.4, Pg. No. 68, IS 456 : 2000)

- "It is the factor generally greater than unity by which either loads are multiplied or resistances / strengths are divided to obtain design values."

- The safety of the structure depends upon the two design factors i.e. load and material strength. These two different safety factors contribute partially to safety of the structure. Therefore, they are termed as partial safety factors. These are the factors, which when applied to loads and materials, give the design value.

(a) Partial Safety Factors for Material Strength : When assessing the strength of a structure or structural member for the limit state of collapse, the values of partial safety factor (γ_m) should be taken as clause 36.4.2.1; of IS : 456-2000, are given in Table 6.3.

Table 6.3

Material	Limit State of Collapse
Concrete	1.5
Steel	1.15

- A higher partial safety factor has been assigned to concrete. Compared to reinforcing steel, evidently because of the higher variability associated with it.

- For serviceability limit states; γ_m for concrete and steel is taken 1.0. A safety factor of unity is appropriate here, because the interest is in estimating the actual deflections and crack widths under the service loads.

(b) Partial Safety Factors for Loads : For limit state of collapse and limit state of serviceability, the partial safety factors (γ_f) for loads recommended by IS : 456 - 2000 are given in Table 6.4.

Table 6.4

Load Combination	Limit State of Collapse			Limit State of Serviceability		
	DL	LL	WL	DL	LL	WL
(1)	(2)	(3)	(4)	(5)	(6)	(7)
DL + LL	1.5	1.5	–	1.0	1.0	–
DL + WL	1.5 or 0.9*	–	1.5	1.0	–	1.0
DL + LL + WL	1.2	1.2	1.2	1.0	0.8	0.8

*   This value is to be considered when stability against overturning or stress reversal is critical.

Notes :

- While considering earthquake effects; substitute EL for WL.

- For the limit states of serviceability, the values of γ_f given in this table are applicable for short term effects. While assessing the long term effects due to creep the dead load and that part of the live load likely to be permanent may only be considered.

- The above partial safety factors shall normally be used. However, where the consequences of a structure attaining a limit state are of a serious nature such as huge loss of life and disruption of the economy, higher values of γ_m and γ_f than the above may be applied.

6.7 DESIGN VALUES

(Refer clause 36.3, Pg. No. 68, IS 456 : 2000)

The design values are obtained when partial safety factor are applied to the characteristics strength and loads.

6.7.1 Design Strength (f_d)

The ratio of characteristics strength and partial safety factor for material is called as design strength.

∴ Design strength (f_d) = $\dfrac{\text{Characteristics strength (f)}}{\text{Partial safety factor } (\gamma_m)}$

∴ $$f_d = \frac{f}{r_m}$$

6.7.2 Design Load (f_d)

It is the product of characteristics load and partial safety factor for load.

∴ Design load (F_d) = Characteristics load (F) × Partial safety factor (γ_f)

∴ $$F_d = F \times \gamma_f$$

6.8 ASSUMPTIONS IN LIMIT STATE OF COLLAPSE (FLEXURE)

(Refer clause 38.1, Pg. No. 69, IS 456 : 2000)

Design for the limit state of collapse in flexure shall be based on the assumptions given below :

1. Plane sections normal to the axis remain plane after bending.

2. The tensile strength of concrete is ignored.
3. There is perfect bond between steel and concrete right up to failure of structure.
4. For design purpose, the maximum compressive strength of concrete in the structure shall be assumed to be 0.67 times the characteristics strength, with application of partial safety factor.

$$\left\{ \text{Max. compressive stress in concrete} = \frac{0.67\, f_{ck}}{\gamma_m} = \frac{0.67\, f_{ck}}{1.5} = 0.446\, f_{ck} \right\}$$

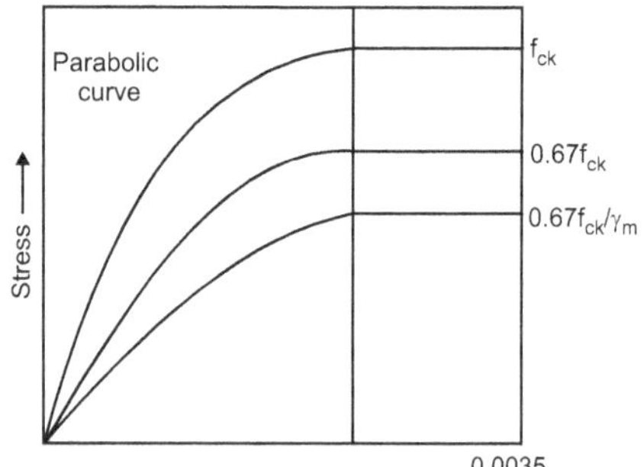

Fig. 6.3 : Stress-strain curve for concrete

5. For design purpose, the maximum tensile strength of steel shall be assumed to be characteristics strength with application of partial safety factor.

$$\left\{ \text{Max. tensile stress in steel} = \frac{f_y}{\gamma_m} = \frac{f_y}{1.15} = 0.87\, f_y \right\}$$

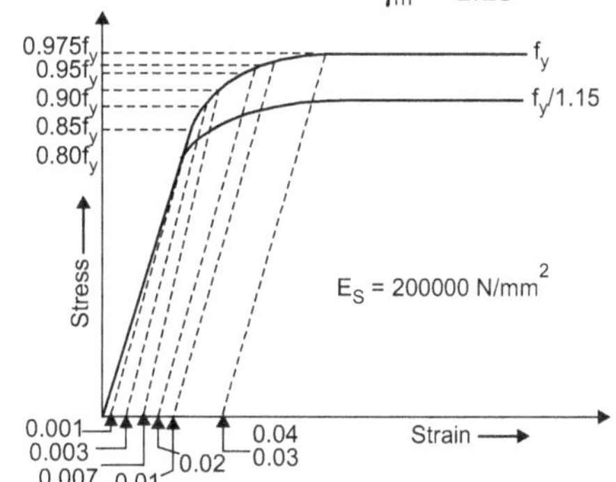

Fig. 6.4

Fig. 6.5

6. The maximum strain in concrete at the outermost compression fibre is taken as 0.0035 in bending.
7. The maximum strain in the tension reinforcement in the section at failure shall not be less than :

$$\frac{f_y}{1.15\, E_s} + 0.002$$

where f_y = Characteristics strength of steel

            E_s = Modulus of elasticity of steel

6.9 COVER ON RCC STRUCTURE

- To protect the reinforcement against fire.
- To provide cover against corrosion.
- To develop the sufficient bond strength along the surface area of steel bar.

Fig. 6.6

EXERCISE

1. State assumption is limit state of collapse – flexure.
2. Explain limit state method.
3. Why do we provide cover in RCC structure?
4. State and explain type of limit state.
5. Define :
   (i) Characteristics strength
   (ii) Characteristics load
   (iii) Design strength
   (iv) Design load
   (v) Partial safety factor
6. State the values of partial safety factor for material and load.
7. How to calculate design values in LSM as per IS.
8. State HOW, LSM is differ from WSM and ULM.
9. State grades of steel and concrete.
10. Explain : Characteristics strength in detail.
11. Explain the concept of under-reinforced balanced and over-reinforced section.

◈ ◈ ◈

CHAPTER - 7

LIMIT STATE OF COLLAPSE - FLEXURE (RECTANGULAR BEAM)

7.1 INTRODUCTION

- The introduction of beam are already discussed in Chapter 2. Reinforced concrete beam are structural elements that designed to carry transverse external loads. The loads causes bending moment, shear and torsion across their length. In this chapter we only study about bending moment on beam and provision to resist this bending moment.

Singly Reinforced Section :

- The beam in which steel reinforcement is placed in tensile zone only is called singly reinforced beam, and the steel bars provided in compression zone to hold the stirrups in position is called anchor bars.

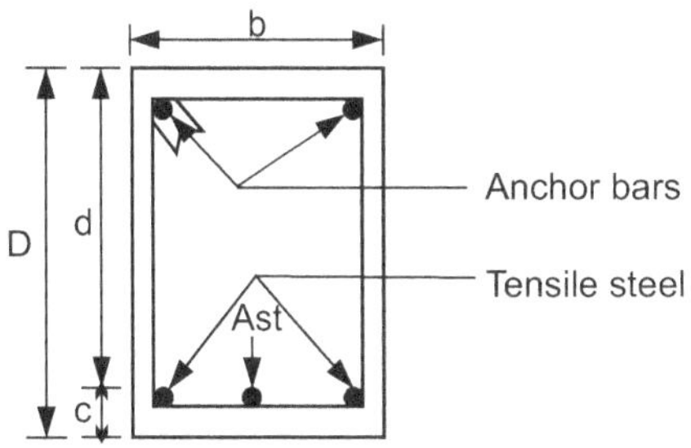

Fig. 7.1 : Singly reinforced beam

Doubly Reinforced Beam :

- The beam in which steel reinforcement is placed in tension as well as compression zone to assist concrete in taking compression is called doubly reinforced beam.

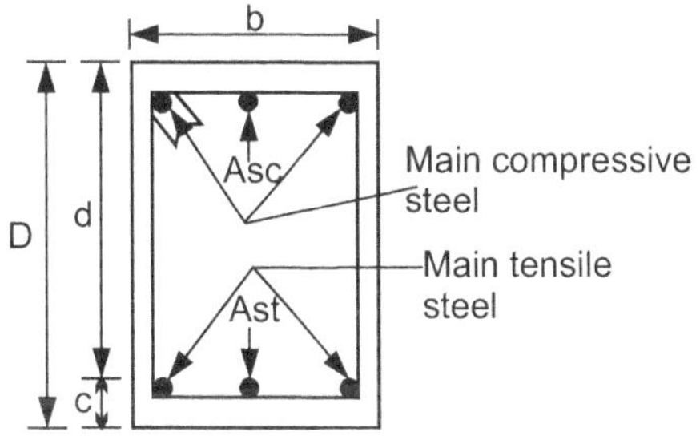

Fig. 7.2 : Doubly reinforced beam

(A) Singly Reinforced Rectangular Section (SRRS)

7.2 CONCEPT OF UNDER-REINFORCED, BALANCED AND OVER-REINFORCED SECTION

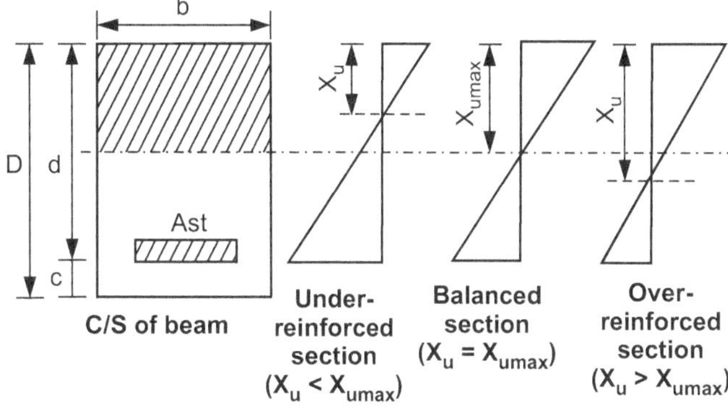

Fig. 7.3

1. Balanced Section

- Balanced section is the section in which the stress acting in concrete and that applied in steel section will attain permissible value simultaneously.

- The beam will fail due to both materials fail simultaneously. A sudden failure would occur with less alarming deflection.

- In this section same % of steel is provided that required for balanced section.

- This section is also called as best, economical, critical or engineering section.

- The actual N.A. (Neutral axis) is this section is called critical N.A. (i.e. $x_u = x_{umax}$).

2. Under Reinforced Section

- It is the section in which % of steel is less than required for balanced section.

- In case of under-reinforced tensile steel reaches maximum permissible stress in bending first. So the beam will fail due to overstress in steel.

- When the amount of steel is kept less than required for balanced section. The position of N.A. shifts upward to satisfy equilibrium condition. (i.e. $x_u < x_{umax}$).

- Every singly reinforced sections should be designed as under-reinforced section because this section gives enough warning before failure in the form of large deflection associated with well distributed cracks. Hence necessary precautions can be taken before collapse take place.

3. Over-Reinforced Section

- It is the section in which % of steel is more that required for balanced section.

- In case of over-reinforced concrete reaches their permissible stress level first so the beam will fail due to overstress in concrete.

- When the amount of steel is kept more than required for balanced section. The position of N.A. shifts downward to satisfy equilibrium condition. (i.e. $x_u > x_{umax}$)

- In case of over-reinforced section failure of section take place by crushing of concrete alone and therefore failure is sudden and explosive nature without any warning / signs. Therefore over-reinforced sections are not allowed in R.C.C. design.

- Hence when $x_u > x_{umax}$, the section should be redesigned or assume balanced section (i.e. put $x_u = x_{umax}$)

7.3 STRESS BLOCK PARAMETERS

- The stress strain behaviour of concrete under compression is generally obtained from cylinder or cube of concrete subjected to longitudinal compressive loading. Whereas the stress and strains are uniform for a cube (or for cylinder), they vary across the depth of a flexural member. The ratio between the maximum flexural compression in a beam and the crushing strength of a cube is likely to be different from unity. The IS code recommends the compressive strength of concrete in the structure equal to 0.67 times the characteristics strength.

- Fig. 7.4 (b) shows the strain diagram while Fig. 7.4 (c) shows the variation of stresses across the depth. In stress diagram concrete behaviour is parabolic from zero strain to 0.002 strain thereafter from 0.002 strain to $0.0035 \times$ strain the stress is constant.

Strain Diagram and Stress Block Diagram for Singly Reinforced Section.

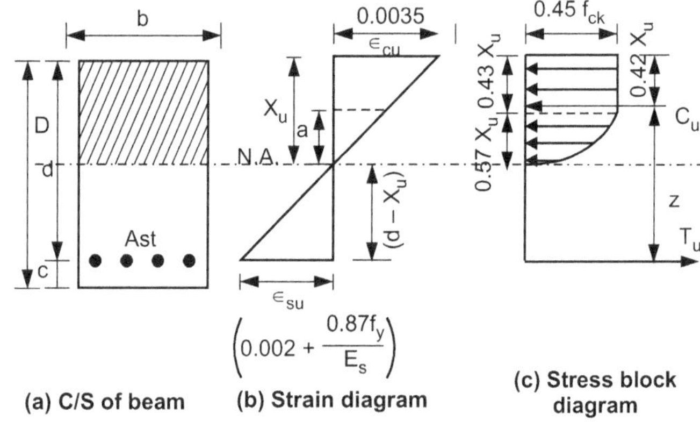

Fig. 7.4 : Stress strain diagram of SRRS

Where,

b = breadth of section

d = effective depth

c = effective cover

D = Overall depth

A_{st} = Area of tensile reinforcement

$N.A.$ = neutral axis

X_u = depth of N.A.

ε_{cu} = Max. strain in concrete

ε_{su} = Max. strain in steel

f_y = characteristics tensile stress in steel

f_{ck} = Characteristic compressive stress in concrete

E_s = Modulus of elasticity for steel

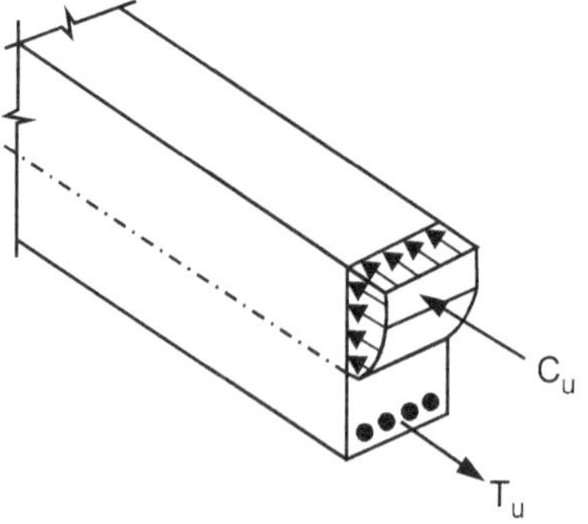

Fig. 7.5

1. T_u = Max. tensile force carried by steel

  ∴   **$T_u = 0.87\, f_y \cdot A_{st}$**

2. C_u = Max. compressive force carried by concrete

  ∴ C_u = (Area of rectangle + area of parabola) $\times$ width of s/c

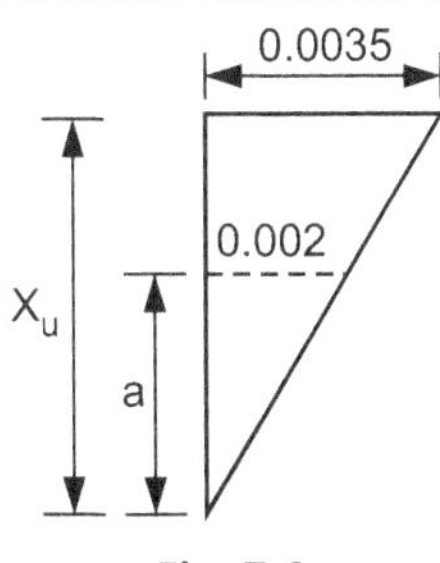

Fig. 7.6

From similar Δ,

$$\frac{a}{0.002} = \frac{x_u}{0.0035}$$

$$\therefore \quad a = 0.43\, x_u$$

$$\therefore \quad c_u = (0.45\, f_{ck} \times 0.43\, X_u) + \frac{2}{3}$$

$$\times 0.45 f_{ck} \times 0.57\, X_u$$

$$\therefore \quad \mathbf{c_u = 0.36\, f_{ck} \cdot X_u \cdot b}$$

3. $\quad z = $ lever arm

$$\therefore \quad z = (d - \bar{y})$$

$$\bar{y} = \frac{a_1 \cdot y_1 + a_2 \cdot y_2}{a_2 + a_2} = 0.42\, x_u$$

$$\therefore \quad \mathbf{z = (d - 0.42\, X_u)}$$

7.3.1   Depth of Neutral Axis

1. Actual Depth of N.A. (X_u) :

Equilibrium requirement in horizontal direction gives,

$$C_u = T_u$$

$$0.36\, f_{ck} \cdot x_u \cdot b = 0.87\, f_y \cdot A_{st}$$

$$\therefore \quad x_u = \frac{0.87\, f_y \cdot A_{st}}{0.36\, f_{ck} \cdot b}$$

2. Limiting / Critical Depth of N.A.

(Neutral axis depth for balanced section) :

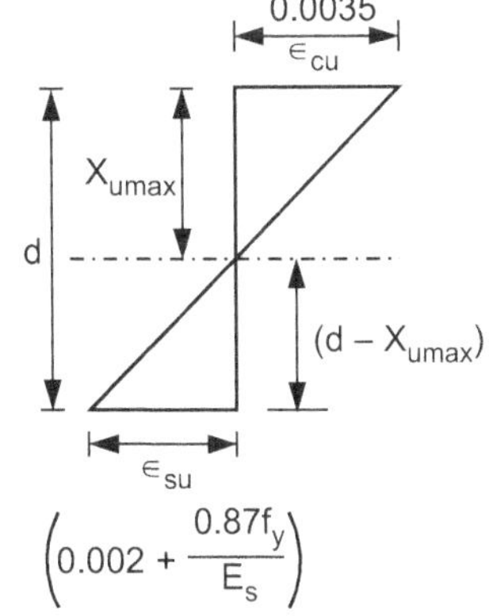

Fig. 7.7

From strain diagram for balanced s/c, from similar Δ,

$$\frac{x_{umax}}{(d - x_{umax})} = \frac{0.0035}{\left(0.002 + \dfrac{0.87\, f_y}{E_s}\right)}$$

(a) For $F_e - 250$ ($f_y = 250$ N/mm^2) :

$$\therefore \quad \frac{x_{umax}}{(d - x_{umax})} = \frac{0.0035}{\left(0.002 + \dfrac{0.87 \times 250}{2 \times 10^5}\right)} = 1.13$$

$$x_{umax} = 1.13\, d - 1.13\, x_{umax}$$

$$2.13\, x_{umax} = 1.13\, d$$

$$\therefore \quad \mathbf{x_{umax} = 0.53\, d}$$

(b) For Fe-415 ($f_y = 415$ N/mm^2)

$$\therefore \quad \frac{x_{umax}}{(d - x_{umax})} = \frac{0.0035}{\left(0.002 + \dfrac{0.87 \times 415}{2 \times 10^5}\right)} = 0.92$$

$$x_{umax} = 0.92\, d - 0.92\, x_{umax}$$

$$\therefore \quad \mathbf{x_{umax} = 0.48\, d}$$

(c) For Fe-500 ($f_y = 500$ N/mm^2)

$$\frac{x_{umax}}{(d - x_{umax})} = 0.84$$

$$\therefore \quad \mathbf{x_{umax} = 0.46\, d}$$

7.3.2   Moment of Resistance of SRRS

*   The flexural strength of RC section is also known as moment carrying capacity of the section. The compressive force 'C_u' in concrete and tensile force 'T_u' in steel are equal and opposite and are separated by distance ($d - 0.42\, x_u$), which is called as lever arm. Hence they form a couple. The couple moment is the moment of resistance and it is also called moment carrying capacity.

1. For Under Reinforced Section

Moment of resistance (M_u)

$$M_u = T_u \times z \qquad \text{...Tension failure}$$

$$\therefore \quad \mathbf{M_u = 0.87\, f_y \cdot A_{st} \times (d - 0.42\, x_u)}$$

2. For Over-Reinforced Section

$$M_u = C_u \times z \qquad \text{...Compression failure}$$

But, over-reinforced are not permitted in RCC design, hence assume balanced section.

3. For Balanced Section

$$M_u = T_u \times z$$

or $\qquad M_u = C_u \times z... \qquad\qquad$ Put $x_u = x_{umax}$

$$M_{umax} = 0.36\, f_{ck} \cdot x_{umax} \cdot b \times (d - 0.42 \times x_{umax})$$

(a) For Fe–250 ($x_{umax} = 0.53\, d$)

$$\therefore \quad M_{umax} = 0.36\, f_{ck} \cdot 0.53\, d \times b \times (d - 0.42 \times 0.53\, d)$$

$$\therefore \quad M_{umax} = 0.148\, f_{ck}\, bd^2$$

(b) For Fe-415 ($x_{umax} = 0.48\, d$) :

$$\therefore \quad M_{umax} = 0.36\, f_{ck}\, 0.48\, d \times b \times (d - 0.42 \times 0.48\, d)$$

$$\therefore \quad \mathbf{M_{umax} = 0.138\, f_{ck}\, bd^2}$$

(c) For Fe-500 (x_{umax} = 0.46 d) :

$$M_{u\,max} = 0.36\, f_{ck} \times 0.46\, d \times b \times (d - 0.42 \times 0.46d)$$

$$\therefore \quad \mathbf{M_{umax} = 0.133\, f_{ck}\, bd^2}$$

7.3.3 Area of Steel (A_{st})

- In design problems reinforcement is generally calculated for moment resisting capacity.

1. For Under-Reinforced Section

$$M_u = T_u \times z$$

$$M_u = 0.87\, f_y\, A_{st} \times (d - 0.42\, x_u)$$

$$M_u = 0.87\, f_y \cdot A_{st}\left(d - 0.42 \times \frac{0.87\, f_y \cdot A_{st}}{0.36\, f_{ck} \cdot b}\right)$$

$$M_u = 0.87\, f_y \cdot A_{st} \cdot d - 0.42 \times \frac{(0.87\, f_y \cdot A_{st})^2}{0.36\, f_{ck}\, b}$$

Dividing both sides by ($0.87\, f_y\, bd^2$) and re-arranging

$$\frac{f_y}{f_{ck}}\left(\frac{A_{st}}{bd}\right)^2 - \left(\frac{A_{st}}{bd}\right) + \frac{M_u}{0.87\, f_y\, bd^2} = 0$$

Quadratic equation in terms of $\left(\dfrac{A_{st}}{bd}\right)$ by solving we get,

$$\therefore \quad A_{st} = \frac{0.5\, f_{ck}}{f_y}\left[1 - \sqrt{1 - \frac{4.6\, M_u}{f_{ck}\, bd^2}}\right] \times bd$$

…used for balanced section also ($M_u = M_{umax}$)

2. For Balanced Section

- For balanced section equilibrium requirement of horizontal forces,

$$T_u = C_u$$

$$0.87\, f_y \cdot A_{st} = 0.36\, f_{ck} \cdot x_{umax} \cdot b$$

(a) For Fe-250 (x_{umax} = 0.53 d, f_y = 250 N/mm^2)

$$0.87\, f_y \cdot A_{st} = 0.36\, f_{ck} \times 0.53\, d \times b$$

$$\therefore \quad 100 \times \frac{A_{st}}{bd} = \frac{0.36 \times 0.53}{0.87 \times 250} \times f_{ck} \times 100$$

$$\therefore \quad \mathbf{Pt_{max} = 0.088\, f_{ck}\,\%}$$

(b) For Fe-415 (x_{umax} = 0.48 d, f_y = 415 N/mm^2)

$$0.87\, f_y \cdot A_{st} = 0.36\, f_{ck} \times 0.48\, d \times b$$

$$\therefore \quad 100 \times \frac{A_{st}}{bd} = \frac{0.36 \times 0.48}{0.87 \times 415} \times 100 \times f_{ck}$$

$$\therefore \quad \mathbf{Pt_{max} = 0.048\, f_{ck}\,\%}$$

(c) For Fe-500 (x_{umax} = 0.46 d, f_y = 500 N/mm^2)

$$0.87\, f_y \cdot A_{st} = 0.36\, f_{ck} \times 0.46\, d \times b$$

$$100 \times \frac{A_{st}}{bd} = \frac{0.36 \times 0.46}{0.87 \times 500} \times 100 \times f_{ck}$$

$$\therefore \quad \mathbf{P_{t\,max} = 0.038\, f_{ck}\,\%}$$

Grade of Steel	Critical Depth of N.A. 'X_{umax}'	Max. Moment of Resistance 'M_{umax}'	Max. % of Area of Steel 'Pt_{max}'
Fe-250	0.53 d	0.148 f_{ck} bd^2	0.088 f_{ck} %
Fe-415	0.48 d	0.138 f_{ck} bd^2	0.048 f_{ck} %
Fe-500	0.46 d	0.133 f_{ck} bd^2	0.038 f_{ck} %

7.4 I.S. SPECIFICATION REGARDING BEAM

7.4.1 Reinforcement

1. Minimum Reinforcement :

(Clause 26.5.1.1, Pg. No. 46, IS 456 : 2000)

$$A_{st\,min} = \frac{0.85}{f_y} \times bd$$

2. Maximum Reinforcement :

(Clause 26.5.1.1, 26.5.1.2, IS 456)

$$A_{st\,max} \not> 0.04\, bD$$

7.4.2 Cover the Reinforcement

(Clause 26.4, Pg. No. 46, IS 456 : 2000)

- Nominal cover is defined as the distance measured from the concrete surface (without plaster) to the nearest surface of the reinforcing bar.

- Specification regarding nominal cover to beam at different exposure condition are given in Table 16, Pg. 47, IS456 : 2000.

- But generally nominal cover to beam should not be less than 25 mm or largest diameter of longitudinal bar.

7.4.3 Spacing of Reinforcement

(Clause 26.3, Pg. No. 45, IS 456 : 2000)

1. Minimum horizontal spacing between individual bars
   - (i) Diameter of bar if diameter are equal
   - (ii) Largest diameter of bar　　whichever is
   - (iii) Nominal size of aggregate + 5 mm　　more
2. Maximum horizontal distance between tension bars
   - (i) For Fe-250; clear distance $\not> 300$ mm
   - (ii) For Fe-415; clear distance $\not> 180$ mm
   - (iii) For Fe-500; clear distance $\not> 150$ mm

7.4.4 Effective Spam

(Clause 22.2, Pg. No. 34, IS 456 : 2000)

(a) Simply Supported Beam / Slab
   - (i) $l_{eff.}$ = clear span + support width　　whichever
   - (ii) $l_{eff.}$ = clear span + effective depth　　is less

(b) Cantilever Beam / Slab

(i) $l_{eff.}$ = clear span + $\dfrac{1}{2}$ × support width

(ii) $l_{eff.}$ = clear span + $\dfrac{1}{2}$ ×effective depth

whichever is less

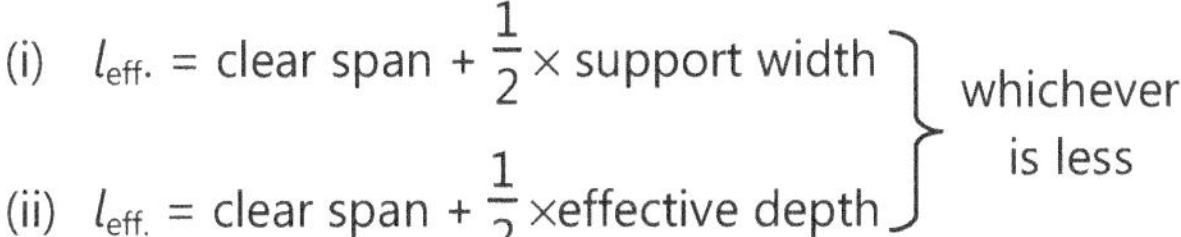

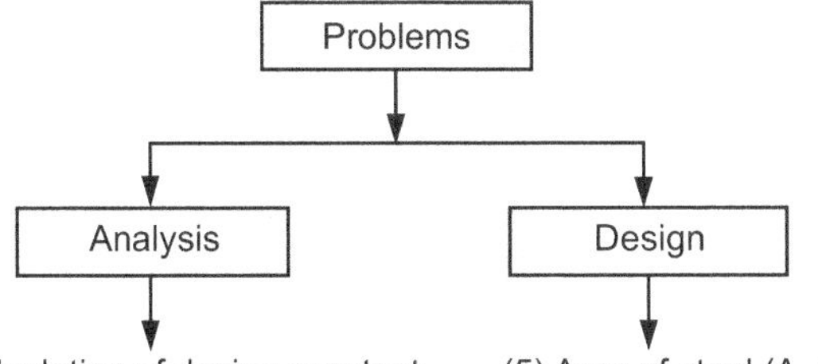

(1) Calculation of design constant

(2) Calculation of moment of resistance (M_u)

(3) Calculation of loading

(4) Calculation of stresses and strain in steel and concrete

(5) Area of steel (A_{st})

(6) Design of SRRS

SOLVED EXAMPLES

Type I : Calculation of Design Constants

Example 7.1 : *Determine the values of design parameters K_{umax}, R_{umax} and P_{tmax} for balanced section of grade M20 and Fe415.*

Solution : Given

$$f_{ck} = 20 \text{ MPa}$$
$$f_y = 415 \text{ MPa}$$

1. Calculation of k_{umax} : For balanced section

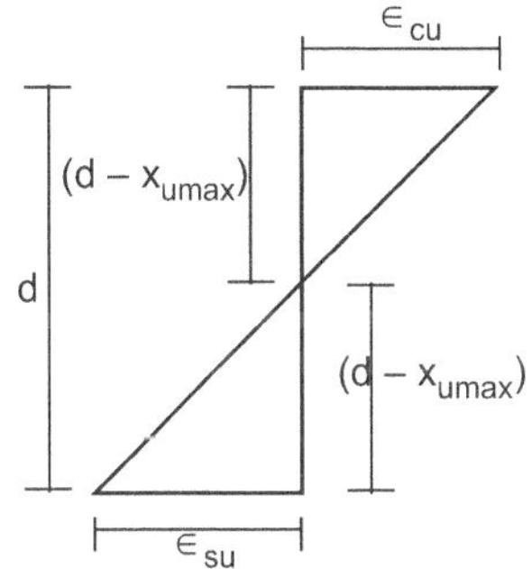

Fig. 7.8

From similar triangle,

$$\frac{x_{umax}}{d - x_{umax}} = \frac{\in_{cu}}{\in_{su}}$$

$$\frac{x_{umax}}{d - x_{umax}} = \frac{0.0035}{\left(0.002 + \dfrac{0.87\, f_y}{E_s}\right)}$$

$$\frac{x_{umax}}{d - x_{umax}} = \frac{0.0035}{\left(0.002 + \dfrac{0.87 \times 415}{2 \times 10^5}\right)} = 0.92$$

$$x_{umax} = 0.92\, d - 0.92\, x_{umax}$$

$\therefore$　　$1.92\, x_{umax} = 0.92\, d$

$\therefore$　　$x_{umax} = 0.48\, d$

$\therefore$　Design parameter $= k_{umax} = \dfrac{x_{umax}}{d}$

$\therefore$　　$K_{umax} = \dfrac{0.48d}{d}$

$\therefore$　　$\mathbf{K_{umax} = 0.48}$

2. Calculation of R_{umax}

$$M_u = C_u \times z \qquad \text{Put } (x_u = x_{umax})$$

$\therefore$　$M_{umax} = 0.36\, f_{ck}\, x_{umax} \cdot b\, (d - 0.42\, x_{umax})$

$M_{umax} = 0.36\, f_{ck} \times 0.48\, d \times b$
$\qquad\qquad \times (d - 0.42 \times 0.48d)$

$\therefore$　$\mathbf{M_{umax} = 0.138\, f_{ck}\, bd^2}$

$M_{umax} = 0.138 \times 20 \times bd^2$

$M_{umax} = 2.76\, bd^2$

$\therefore$　$R_{umax} = \dfrac{M_{umax}}{bd^2}$

$\therefore$　$R_{umax} = \dfrac{2.76\, bd^2}{bd^2}$

$\therefore$　$\mathbf{R_{umax} = 2.76\ N/mm^2}$

3. Calculation of P_{tmax}

$$T_u = C_u \qquad \dots \text{Put } x_u = x_{umax}$$

$0.87\, f_y\, A_{st} = 0.36\, f_{ck}\, x_{umax} \cdot b$

$0.87\, f_y\, A_{st} = 0.36\, f_{ck}\, 0.48 \cdot d \cdot b$

$\dfrac{A_{st}}{b.d} = \dfrac{0.36\, f_{ck}\, 0.48}{0.87\, f_y}$

$\dfrac{A_{st}}{b.d} = \dfrac{0.36 \times 0.48}{0.87 \times 415} \times f_{ck} \times \dfrac{100}{100}$

$\therefore$　$P_{t\,max} = 0.048\, f_{ck}$ %

$\therefore$　$\mathbf{P_{t\,max} = 0.96\ \%}$

Example 7.2 : *Determine the design constants R_u and P_t when neutral axis factor K_u is limited to 0.3. Assume M20 and Fe250 grades.*

Solution : Given

$$f_{ck} = 20 \text{ MPa}, \quad K_u = 0.3,$$
$$f_y = 250 \text{ MPa},$$

$\therefore$　$x_{ulim} = 0.3\, d$

1. Calculation of R_{ulim}

$$M_u = C_u \times z \qquad \dots x_u = x_{ulim}$$

$M_{ulim} = 0.36\, f_{ck}\, x_{ulim} \cdot b\, (d - 0.42\, x_{ulim})$

$M_{ulim} = 0.36 \times 20 \times 0.3d \times b\, (d - 0.42 \times 0.3\, d)$

$M_{ulim} = 0.36 \times 20 \times 0.3d \times b\, (0.87\, d)$

$$M_{ulim} = 1.88 \, bd^2$$

$$\therefore \quad R_{ulim} = \frac{M_{ulim}}{bd^2} = \frac{1.88 \, bd^2}{bd^2}$$

$$\therefore \quad \mathbf{R_{ulim} = 1.88 \ N/mm^2}$$

2. Calculation of P_{tlim}

$$T_u = C_u \qquad \ldots x_u = x_{ulim}$$

$$0.87 \, f_y \, A_{st} = 0.36 \, f_{ck} \, x_{ulim} \cdot b$$

$$0.87 \, 250 A_{st} = 0.36 \times 20 \times 0.3 \, d \times b$$

$$\frac{A_{st}}{b.d} = \frac{0.36 \times 20 \times 0.3}{0.87 \times 250} \times \frac{100}{100}$$

$$\frac{A_{st}}{b.d} = 0.99 \ \%$$

$$\therefore \quad \mathbf{P_{tlim} = 0.99 \ \%}$$

Example 7.3 : *Derive from first principles, the values of design parameters (constant $K_{u,\,max}$, $R_{u.max}$ and $P_{t.max}$ for a balanced section of concrete of grade M20 and steel of grade Fe 415.*

Given : $f_{ck} = 20 \ N/mm^2$, $f_y = 415 \ N/mm^2$, section balanced.

Required : $K_{u.max}$, $R_{u\,max}$ and $P_{t.\,max}$

Solution : From Equation

$$K_u = \frac{700}{1100 + 0.87 \, f_y}$$

$$K_{u\,max} = \frac{700}{1100 + 0.87 \times 415} = 0.48$$

$$R_{u.\,max} = 0.36 \, f_{ck} K_{u.max} \,(1 - 0.42 \, k_{u.max})$$

$$= 0.36 \times 20 \times 0.48 \,(1 - 0.42 \times 0.48)$$

$$\mathbf{R_{u.\,max} = 2.76 \ N/mm^2}$$

$$P_t = \frac{0.36 \, f_{ck} k_u}{0.87 \, f_y} = \frac{0.36 \times 20 \times 0.3}{0.87 \times 250}$$

$$\mathbf{P_t = 0.0096 \ i.e \ 0.96 \ \%}$$

Type II : Calculation of Moment of Resistance

Example 7.4 : *A beam 300 mm wide and 600 mm effective depth is reinforced with 4-16 mm ϕ. Find the ultimate moment of resistance of beam section. Use M20 and Fe415.*

Solution : Given

$$b = 300 \ mm, \ f_{ck} = 20 \ MPa, \ f_y = 415 \ MPa$$

$$A_{st} = 4 - 16 \ mm \ \phi = 4 \times \frac{\pi}{4}(16)^2 = 804.25 \ mm^2$$

$$d = 600 \ mm$$

1. Calculation of actual depth of Neutral Axis (x_u)

$$C_u = T_u$$

$$0.36 \, f_{ck} \cdot x_u \cdot b = 0.87 \, f_y \cdot A_{st}$$

$$\therefore \quad x_u = \frac{0.87 \times 415 \times 804.25}{0.36 \times 20 \times 300}$$

$$\therefore \quad \mathbf{x_u = 134.43 \ mm}$$

2. Calculation of critical depth of Neutral Axis (x_{umax})

$$x_{umax} = 0.48 \, d \qquad \ldots \text{for Fe415}$$

$$x_{umax} = 0.48 \times 600$$

$$\therefore \quad \mathbf{x_{umax} = 288 \ mm}$$

3. Compare x_u and x_{umax}

$$134.43 < 288$$

$$\therefore \quad x_u < x_{umax}$$

$$\ldots \text{section is under-reinforced section}$$

4. Calculation of ultimate moment of resistance (M_u)

$$\therefore \quad M_u = T_u \times z$$

$$M_u = 0.87 \, f_y \, A_{st} \,(d - 0.42 \, x_u)$$

$$M_u = 0.87 \times 415 \times 804.26 \times$$

$$(600 - 0.42 \times 134.43)$$

$$\mathbf{M_u = 157.83 \ kN\text{-}m}$$

Example 7.5 : *A reinforced concrete beam 300 mm wide is reinforced with 1435.82 mm^2 on tension side at effective depth 500 mm. Take M20 and Fe415. Determine :*

(1) Position of Neutral Axis (N.A.)

(2) Type of reinforced section

(3) Moment of resistance

Solution : Given

$$b = 300 \ mm, \ f_{ck} = 20 \ MPa, \ A_{st} = 1435.82 \ mm^2$$

$$d = 500 \ mm, \ f_y = 415 \ MPa$$

1. Calculation of position of Neutral Axis (N.A.) (x_u)

$$C_u = T_u; \quad x_u = \frac{0.87 \, f_y \cdot A_{st}}{0.36 \, f_{ck} \cdot b}$$

$$x_u = \frac{0.87 \times 415 \times 1435.82}{0.36 \times 20 \times 300}$$

$$\mathbf{x_u = 240.00 \ mm}$$

2. Type of reinforced section

$$x_{umax} = 0.48 \, d$$

$$= 0.48 \times 500$$

$$\mathbf{x_{umax} = 240 \ mm}$$

$$\therefore \quad \text{Compare } x_u \text{ and } x_{umax}$$

$$240 = 240$$

$$\therefore \quad x_u = x_{umax} \qquad \ldots \text{section is balanced section}$$

3. Calculation of moment of resistance (M_{umax})

$$M_{umax} = 0.138 \, f_{ck} \, bd^2$$

$$M_{umax} = 0.138 \times 20 \times 300 \times 500^2$$

$$\mathbf{M_{umax} = 207.00 \ kN\text{-}m}$$

Example 7.6 : *A singly reinforced rectangular beam of width 230 mm and 435 mm depth overall, reinforced with 3-20 mm ϕ with clear cover 25 mm. Find the maximum flexural strength beam can carry.*

Solution : Given

$b = 230$ mm,

$D = 435$ mm, $A_{st} = 3 - 20$ mm ϕ

$c = c' + \phi/2$, $A_{st} = 3 \times \dfrac{\pi}{4}(20)^2$

$\therefore$ $c = 25 + \dfrac{20}{2}$, $A_{st} = 942.48$ mm^2

$c = 35$ mm, $d = 400$ mm $\ldots d = D - C$

Assume, $f_{ck} = 20$ MPa, $f_y = 415$ MPa

1. Calculation of actual position of Neutral axis (x_u)

$$C_u = T_u$$

$$x_u = \frac{0.87 f_y \cdot A_{st}}{0.36\, f_{ck}\, b} = \frac{0.87 \times 415 \times 942.48}{0.36 \times 20 \times 230}$$

$$\mathbf{x_u = 205.48\ mm}$$

2. Calculation of limiting depth of Neutral Axis (x_{umax})

$$x_{umax} = 0.48\, d = 0.48 \times 400$$

$\therefore$ $\quad \mathbf{x_{umax} = 192\ mm}$

3. Compare x_u and x_{umax}

$$205.48 \; > \; 192$$

$$x_u \; > \; x_{umax}$$

…section is over-reinforced section

$\therefore$ But over-reinforced section is not allowed in R.C.C. design hence, assume balanced section.

$$M_u = C_u \times z = 0.138\, f_{ck}\, bd^2$$

$$M_u = 0.138 \times 20 \times 230 \times 400^2$$

$$\mathbf{M_u = 101.57\ kN.m}$$

Example 7.7 : *Find M_u for a S.R.R.S. beam having 230 mm width and 400 mm effective depth it is reinforced with 4 bars of 16 mm ϕ. Use M15 and Fe 415 Grades material. What will be the moment of resistance if 5-16 mm ϕ bars are used ? Will the moment of resistance further increase with in number of bars ?*

Solution : Given ;

$b = 230$ mm, $\quad f_{ck} = 15$ MPa

$d = 400$ mm, $\quad f_y = 415$ MPa

Case I : $A_{st} = \dfrac{\pi}{4}(16)^2 = 804.25$ mm^2

1. Calculation of actual depth of Neutral Axis (x_u)

$$C_u = T_u$$

$$x_u = \frac{0.87\, f_y\, A_{st}}{0.36\, f_{ck}\, b} = \frac{0.87 \times 415 \times 804.25}{0.36 \times 15 \times 230}$$

$\therefore$ $\quad \mathbf{x_u = 233.80\ mm}$

2. Calculation of critical depth of Neutral Axis (x_{umax})

$$x_{umax} = 0.48\, d = 0.48 \times 400$$

$$\mathbf{x_{umax} = 192\ mm}$$

3. Compare x_u and x_{umax}

$$233.80 \; > \; 192$$

$$x_u \; > \; x_{umax}$$

…section is over-reinforced section

$\therefore$ but, over-reinforced section is not allowed in R.C.C. design.

Hence, assume balanced section.

$\therefore$ $\quad M_{umax} = 0.138\, f_{ck} \cdot bd^2$

$$= 0.138 \times 15 \times 230 \times 400^2$$

$$\mathbf{M_{umax} = 76.18\ kN.m}$$

Case II : The given section is already over-reinforced section and its moment resisting capacity is **76.18 kN-m.**

$\therefore$ Further increase in number of bars, will not increase moment resisting capacity.

Example 7.8 : *Find the working moment of resistance SRRS having 230 mm $\times$ 500 mm overall section reinforced with 2-16 mm ϕ and 2-18 mmϕ on tensile zone, use M20 mix and mild steel reinforcement.*

Solution : Given : $b = 230$ mm, $D = 500$ mm, $f_{ck} = 20$ MPa,

$$A_{st} = 2 \times \frac{\pi}{4}(18^2 + 16^2) = 911.06\ \text{mm}^2$$

Assume, effective cover, $c = 40$ mm

$f_y = 250$ MPa

$\therefore$ $\quad d = D - C = 500 - 40$

$\quad d = 460$ mm

1. Calculation of actual depth of Neutral Axis (x_u)

$$C_u = T_u$$

$$x_u = \frac{0.87\, f_y\, A_{st}}{0.36\, f_{ck} \cdot b} = \frac{0.87 \times 250 \times 911.06}{0.36 \times 20 \times 230}$$

$\therefore$ $\quad \mathbf{x_u = 119.66\ mm}$

2. Calculation of critical / limiting depth of Neutral Axis (x_{umax})

$$x_{umax} = 0.53\, d = 0.53 \times 460$$

$\therefore$ $\quad \mathbf{X_{umax} = 243.8\ mm}$

3. Compare x_u and x_{umax}

$$119.66 \; < \; 243.8$$

$$x_u \; < \; x_{umax}$$

…section is under-reinforced section

4. Calculation of moment of resistance (M_u)

$$M_u = T_u \times z = 0.87\, f_y\, A_{st}\, (d - 0.42\, x_u)$$

$$M_u = 0.87 \times 250 \times 911.06 \times$$
$$(460 - 0.42 \times 119.66)$$

$$\therefore \quad M_u = 81.19 \text{ kN-m}$$

$$\therefore \quad \text{Safe moment of resistance} = M = \frac{M_u}{\gamma_f} = \frac{81.19}{1.5}$$

$$\therefore \quad \mathbf{M = 54.13 \text{ kN-m}}$$

Type III : Calculation of Loading (W_u)

Example 7.9 : *A R.C. beam of rectangular section 250 mm wide and 600 mm deep overall is reinforced on tension side by 4-20 mmϕ. Provided at 40 mm from soffit. The beam is subjected to mild exposure condition. Use M20 mix and HYSD Fe 415 bars. Determine –*

1. Ultimate moment of resistance

2. Safe udl the beam can carry over a simply supported span of 6 m.

Solution : Given data :

$$b = 250 \text{ mm},\ f_{ck} = 20 \text{ MPa},\ A_{st} = 4 - 20 \text{ mm}\phi,$$

$$D = 600 \text{ mm},\ f_y = 415 \text{ MPa},\ A_{st} = 4 \times \frac{\pi}{4}(20)^2$$

$$c = 40 \text{ mm},\ \therefore\ d = 560 \text{ mm},\ A_{st} = 1256.64 \text{ mm}^2$$

$$l_{eff} = 6\text{m}$$

(I) Calculation of ultimate moment of resistance (M_u)

(a) Calculation of actual depth of neutral axis (x_u)

$$C_u = T_u$$

$$x_u = \frac{0.87\, f_y\, A_{st}}{0.36\, f_{ck}\, b} = \frac{0.87 \times 415 \times 1256.64}{0.36 \times 20 \times 250}$$

$$\therefore \quad \mathbf{x_u = 252.06 \text{ mm}}$$

(b) Calculation of critical depth of Neutral Axis (x_{umax})

$$x_{umax} = 0.48\, d = 0.48 \times 560$$

$$\therefore \quad \mathbf{x_{umax} = 268.8 \text{ mm}}$$

(c) Compare x_u and x_{umax}

$$252.06 < 268.8$$

$$x_u < x_{umax}$$

…Section is under-reinforced section.

(d) Calculation of ultimate moment of resistance (M_u)

$$M_u = T_u \times z$$

$$M_u = 0.87\, f_y\, A_{st}\, (d - 0.42\, x_u)$$

$$M_u = 0.87 \times 415 \times 1256.64$$
$$(560 - 0.42 \times 252.06)$$

$$\therefore \quad \mathbf{M_u = 206.04 \text{ kN.m}} \qquad \text{...(1)}$$

(II) Calculation of safe udl (w) :

(a) Calculation of factored B.M. (BM_{max})

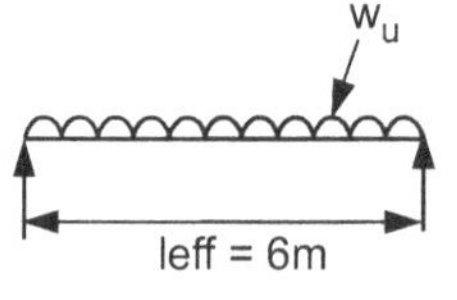

Fig. 7.9

$$BM_{max} = \frac{W_u \cdot l_{eff}^2}{8}$$

$$BM_{max} = \frac{W_u \cdot 6^2}{8}$$

$$BM_{max} = 4.5\, W_u \qquad \text{...(2)}$$

Equating equation (1) and (2)

$$BM_{max} = M_u$$

$$4.5\, W_u = 206.04$$

$$\therefore \quad \mathbf{W_u = 45.79 \text{ kN/m}} \qquad \text{...Factored udl}$$

$$\therefore \quad \text{Safe udl} = W = \frac{W_u}{\gamma_f} = \frac{45.79}{1.5}$$

$$\mathbf{W = 30.52 \text{ kN/m}}$$

Example 7.10 : *A simply supported beam of span 6 m is reinforced with 4-16mm ϕ on tension side. Find the safe concentrated load it can support at the centre in addition to its self weight. Size of beam is 300 mm $\times$ 500 mm. Take concrete M20 and steel Fe415 grade.*

Solution : Given :

$$b = 300 \text{ mm},\ f_{ck} = 20 \text{ MPa},\ A_{st} = 4\text{-}16\text{mm }\phi$$

$$D = 500 \text{ mm},\ f_y = 415 \text{ MPa},\ A_{st} = 4 \times \frac{\pi}{4}(16)^2$$

$$\text{Assume } C = 40 \text{ mm},\ d = 460 \text{ mm},\ A_{st} = 804.25 \text{ mm}^2$$

$$l_{eff} = 6\text{m},$$

(I) Calculation of moment of resistance (M_u)

(a) Calculation of actual depth of N.A. (x_u)

$$C_u = T_u$$

$$x_u = \frac{0.87\, f_y \cdot A_{st}}{0.36\, f_{ck} \cdot b} = \frac{0.87 \times 415 \times 804.25}{0.36 \times 20 \times 300}$$

$$\therefore \quad \mathbf{x_u = 134.43 \text{ mm}}$$

(b) Calculation of critical depth of N.A. (x_{umax})

$$x_{umax} = 0.48\, d = 0.48 \times 460$$

$$\therefore \quad \mathbf{x_{umax} = 220.8 \text{ mm}}$$

(c) Compare x_u and x_{umax}

$$134.43 < 220.8$$

$$x_u < x_{umax}$$

…section is under-reinforced section

(d) Calculation of moment of resistance (M_u)

$$M_u = T_u \times z = 0.87\, f_y\, A_{st}\, (d - 0.42\, x_u)$$

$$M_u = 0.87 \times 415 \times 804.25 \times$$
$$(460 - 0.42 \times 134.43)$$

$\therefore$ **M_u = 117.18 kN-m**   …(1)

(II) Calculation of safe load (w)

(a) Calculation of factored B.M. (BM_{max})

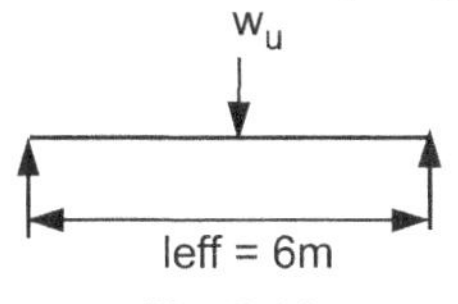

Fig. 7.10

$$BM_{max} = \frac{W_u \cdot l_{eff}}{4}$$

$$BM_{max} = \frac{W_u \cdot 6}{4}$$

$\therefore$ **BM_{max} = 1.5 W_u**   …(2)

$\therefore$ Equating equation (1) and (2)

$$BM_{max} = M_u$$

$$1.5\, W_u = 117.18$$

$\therefore$ **W_u = 78.12 kN**   …Max. concentrated load

$\therefore$ Safe concentrated load

$$W = \frac{W_u}{\gamma_f} = \frac{78.12}{1.5}$$

W = 52.08 kN

Example 7.11 : *Calculate the super imposed load the beam can carry detailed as below :*

Clear simply supported span = 5m

Overall depth of section = 520 mm

Width of support = 300 mm

Thickness of beam = 275 mm

Tension reinforcement = 4-16 mm ϕ

Clear cover = 35 mm

Take M20 and Fe415.

Solution : Given :

$b = 275$ mm, $f_{ck} = 20$ MPa, $A_{st} = 4 \times \dfrac{\pi}{4}(16)^2$

$D = 520$ mm, $f_y = 415$ MPa, $A_{st} = 804.25$ mm^2

$c' = 35$ mm,

$c = 35 + \dfrac{\phi}{2} = 35 + \dfrac{16}{2}$; $c = 43$ mm,

$d = D - C = 520 - 43$; $d = 477$ mm

Support width = 300 mm ; Clear span (L) = 5m,

$\therefore$ l_{eff} = clear span + support width

       l_{eff} = L + 0.3 = 5.3 m

(I) Calculation of actual depth of Neutral Axis (x_u)

$$C_u = T_u$$

$$x_u = \frac{0.87\, f_y\, A_{st}}{0.36\, f_{ck} \cdot b} = \frac{0.87 \times 415 \times 804.25}{0.36 \times 20 \times 275}$$

$\therefore$ **x_u = 146.65 mm**

(II) Calculation of limiting depth of Neutral axis (x_{umax})

$$x_{umax} = 0.48\, d = 0.48 \times 477$$

x_{umax} = 228.96 mm

(III) Compare x_u and x_{umax}

$\therefore$ 146.65   <   228.96

       x_u < x_{umax}

           …Section is under-reinforced section

(IV) Calculation of moment of resistance (M_u)

$$M_u = T_u \times z = 0.87\, f_y\, A_{st}(d - 0.42\, x_u)$$
$$M_u = 0.87 \times 415 \times 804.25$$
$$(4.77 - 0.42 \times 146.65)$$

$\therefore$ **M_u = 120.62 kN-m**   …(1)

(V) Calculation of factored B.M. (BM_{max})

$$BM_{max} = \frac{W_u \cdot l_e^2}{8} = \frac{W_u \times 5.3^2}{8}$$

$\therefore$ BM_{max} = 3.51 W_u …(2)

Fig. 7.11

Equating equation (1) and (2)

$$BM_{max} = M_u$$

$$3.51\, W_u = 120.61$$

$$W_u = 120.62$$

W_u = 34.35 kN/m

           …Factored / maximum udl

$\therefore$ Safe udl = w = $\dfrac{W_u}{\gamma_f} = \dfrac{34.35}{1.5}$

$\therefore$ **W = 22.90 kN/m**

$\therefore$ Safe udl (excluding self weight) / superimposed load (w') :

       w' = w – self wt. of the beam

       w' = 22.90 – (0.275 × 0.52 × 1) × 25

$\therefore$ **w' = 19.33 kN/m**   …Superimposed load

Type IV : Calculation of maximum stresses and strains

Example 7.12 : *A singly reinforced beam 230 mm wide 400 mm deep effective is reinforced with 4-16 mm ϕ. Find the maximum stresses and strain in steel and concrete. If the concrete beam is subjected to factored moment of 60 kN-m. Take M20 and Fe415 HYSD bars.*

Solution : Given :

$b = 230$ mm, $f_{ck} = 20$ MPa, $f_y = 415$ MPa

$A_{st} = 4 - 16$ mm $\phi = 4 \times \dfrac{\pi}{4} (16)^2 = 804.25$ mm^2

$d = 400$ mm,

Factored B.M. $(M_u) = 60$ kN-m

1. Calculation of actual depth of N.A. (x_u)

$$C_u = T_u$$

$$x_u = \frac{0.87\, f_y\, A_{st}}{0.36\, f_{ck} b} = \frac{0.87 \times 415 \times 804.25}{0.36 \times 20 \times 230}$$

∴   **$x_u = 175.35$ mm**

2. Calculation of limiting depth of N.A. (x_{umax})

$$x_{umax} = 0.48\, d = 0.48 \times 400$$

∴   **$x_{umax} = 192$ mm**

3. Compare x_u and x_{umax}

$$175.35 < 192$$

$$x_u < x_{umax}$$

…section is under-reinforced section

4. Calculation of maximum stress in steel $(0.87\, f_y)$

$$M_u = T_u \times z = 0.87\, f_y\, A_{st}\, (d - 0.42\, x_u)$$

$$60 \times 10^6 = 0.87\, f_y \times 804.25\, (400 - 0.42 \times 175.35)$$

∴   **$f_y = 262.76$ N/mm^2**

∴ Maximum stress in steel

$$= 0.87\, f_y$$

$$= 0.87 \times 262.76$$

∴ **Maximum stress in steel $= 228.60$ N/mm^2**

5. Calculation of maximum stress in concrete $(0.45\, f_{ck})$

$$M_u = C_u \times z$$

$$= 0.36\, f_{ck}\, x_u \cdot b\, (d - 0.42\, x_u)$$

$$60 \times 10^6 = 0.36\, f_{ck} \times 175.35 \times 230 \times$$
$$(400 - 0.42 \times 175.35)$$

∴   **$f_{ck} = 12.66$ N/mm^2**

∴ Maximum stress in concrete $= 0.45\, f_{ck}$

$$= 0.45 \times 12.66$$

∴ **Maximum stress in concrete $= 5.7$ N/mm^2**

6. Calculation of maximum strain in steel (ϵ_{su})

∴ $\epsilon_{su} = 0.002 + \dfrac{0.87\, f_y}{E_s}$

$$\epsilon_{su} = 0.002 + \frac{0.87 \times 262.75}{2 \times 10^5}$$

∴   **$\epsilon_{su} = 0.0031$**

7. Calculation of maximum strain in concrete (ϵ_{cu})

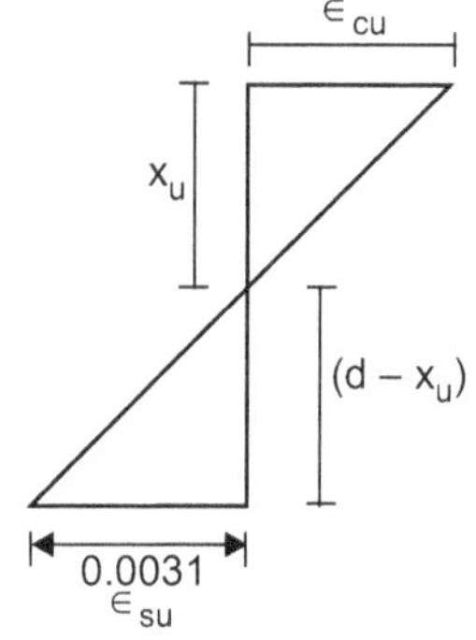

Fig. 7.12

∴ From similar triangle,

$$\frac{x_u}{d - x_u} = \frac{\epsilon_{cu}}{\epsilon_{su}}$$

$$\frac{175.35}{(400 - 175.35)} = \frac{\epsilon_{cu}}{0.0031}$$

∴   **$\epsilon_{cu} = 0.0024$**

Example 7.13 : *A singly reinforced beam 250 mm wide with overall depth 550 mm is reinforced with 4-16 mm ϕ and 4-12 mmϕ. Find the maximum stresses and strain in concrete and steel. If the concrete beam is subjected to service point load of 66 kN on effective span of 6m. Take M20 and Fe 500.*

Solution : Given

$b = 250$ mm, $f_{ck} = 20$ MPa,

$A_{st} = 4 - 16$ mm ϕ and $4 - 12$ mm ϕ

$A_{st} = 4 \times \dfrac{\pi}{4} (16^2 + 12^2) = 1256.64$ mm^2

$D = 550$ mm, $f_y = 500$ MPa

Assume cover, $c = 50$ mm, $l_{eff} = 6$ m

∴ $d = D - C = 550 - 50$, $W = 66$ kN

∴ $d = 500$ mm,

∴   Service point load $= W = 66$ kN

∴   Maximum point load $= W_u = W \times \gamma_f = 66 \times 1.5$

$$W_u = 99 \text{ kN}$$

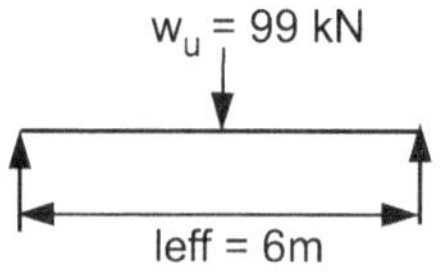

Fig. 7.13

∴ Factored B.M. (M_u)

$$M_u = \frac{W_u \cdot l_{eff}}{4}$$

$$M_u = \frac{99 \times 6}{4}$$

$M_u = 148.50$ kN-m

1. Calculation of actual depth of N.A. (x_u)

$$C_u = T_u$$

$$x_u = \frac{0.87\, f_y \cdot A_{st}}{0.36\, f_{ck} \cdot b} = \frac{0.87 \times 500 \times 1256.64}{0.36 \times 20 \times 250}$$

$$\therefore \quad x_u = 303.69 \text{ mm}$$

2. Calculation of critical depth of N.A. (x_{umax})

$$x_{umax} = 0.46\, d = 0.46 \times 500$$

$$\therefore \quad x_{umax} = 230 \text{ mm}$$

3. Compare x_u and x_{umax}

$$303.69 > 230$$

$$x_u > x_{umax}$$

…Section is over reinforced section

But, over-reinforced section is not allowed in R.C.C. design. Hence assume balance section.

4. Calculation of maximum stress in steel ($0.87\, f_y$)

$$M_u = T_u \times z = 0.87\, f_y\, A_{st}\, (d - 0.42\, x_u)$$

$$\ldots x_u = x_{umax}$$

$$148.5 \times 10^6 = 0.87 \times f_y \times 1256.64\, (500 - 0.42 \times 230)$$

$$\therefore \quad f_y = 336.71 \text{ N/mm}^2$$

$\therefore$ Maximum stress in steel $= 0.87\, f_y = 0.87 \times 336.71$

$\therefore$ **Maximum stress in steel = 292.94 N/mm^2**

5. Calculation of maximum stress in concrete ($0.45\, f_{ck}$)

$$M_u = C_u \times z$$

$$= 0.36\, f_{ck} \cdot x_{umax} \cdot b\, (d - 0.42\, x_{umax})$$

$$148.5 \times 10^6 = 0.36\, f_{ck} \times 230 \times 250\, (500 - 0.42 \times 230)$$

$$\therefore \quad f_{ck} = 17.78 \text{ N/mm}^2$$

$\therefore$ Maximum stress in concrete $= 0.45\, f_{ck} = 0.45 \times 17.78$

Maximum stress in concrete = 8 N/mm^2

6. Calculation of maximum strain in steel (ϵ_{su})

$$\therefore \quad \epsilon_{su} = 0.002 + \frac{0.87\, f_y}{E_s}$$

$$\therefore \quad \epsilon_{su} = 0.002 + \frac{0.87 \times 336.71}{2 \times 10^5}$$

$$\therefore \quad \epsilon_{su} = 0.0035$$

7. Calculation of maximum strain in concrete (ϵ_{cu})

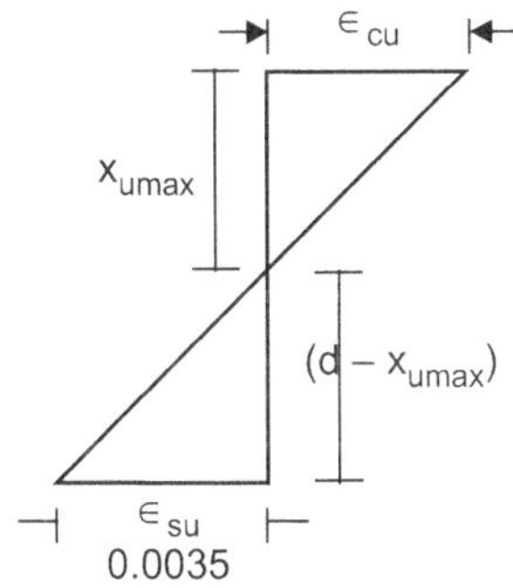

Fig. 7.14

From similar triangle,

$$\frac{x_{umax}}{d - x_{umax}} = \frac{\epsilon_{cu}}{\epsilon_{su}}$$

$$\frac{230}{500 - 230} = \frac{\epsilon_{su}}{0.0035}$$

$$\therefore \quad \epsilon_{cu} = 0.0030$$

Example 7.14 : *Determine the ultimate moment of resistance of beam section as well as corresponding stresses and strain in steel and concrete as shown in Fig. 7.15 Take M25 mix and Fe 415 steel.*

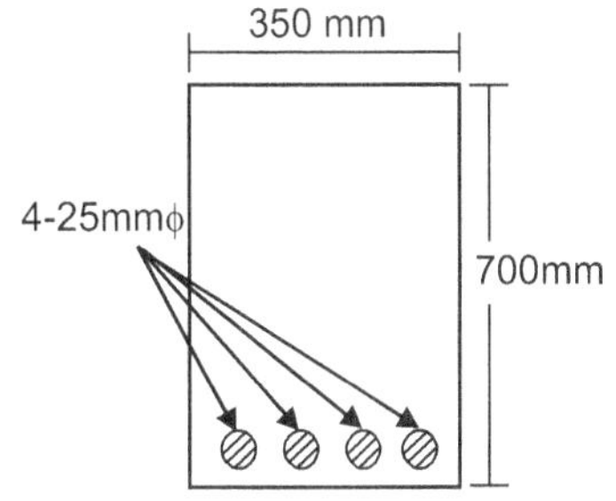

Fig. 7.15

Solution : Given:

$b = 350$ mm, $f_{ck} = 25$ MPa, $A_{st} = 4 - 25$ mm φ

$D = 700$ mm, $f_y = 415$ MPa, $A_{st} = 4 \times \dfrac{\pi}{4}(25)^2$

Assume, $c = 50$ mm $A_{st} = 1963.50$ mm^2

$\therefore$ $d = 650$ mm,

(I) Calculation of ultimate moment of resistance (M_u)

1. Calculation of actual depth of N.A. (x_u)

$$C_u = T_u$$

$$x_u = \frac{0.87\, f_y\, A_{st}}{0.36\, f_{ck} \cdot b} = \frac{0.87 \times 415 \times 1963.50}{0.36 \times 25 \times 350}$$

$$\therefore \quad x_u = 225.05 \text{ mm}$$

2. Calculation of critical depth of N.A. (x_{umax})

$$\therefore \quad x_{umax} = 0.48\, d = 0.48 \times 650$$

$$\therefore \quad x_{umax} = 312 \text{ mm}$$

3. Compare x_u and x_{umax}

$$225.05 < 312$$

$$x_u < x_{umax}$$

…Section is under-reinforced section

4. Calculation of ultimate moment of resistance (M_u)

$$M_u = T_u \times z = 0.87\, f_y\, A_{st}\, (d - 0.42\, x_u)$$

$$M_u = 0.87 \times 415 \times 1963.5$$

$$(650 - 0.42 \times 225.05)$$

$$\therefore \quad M_u = 393.80 \text{ kN-m}$$

(II) Calculation of maximum stresses and strain in steel and concrete :

5. Calculation of maximum stress in steel (0.87 f_y)

$$M_u = T_u \times z = 0.87 \times f_y \times A_{st}\,(d - 0.42\,x_u)$$

$$393.80 \times 10^6 = 0.87 \times f_y \times 1963.5\,(650 - 0.42 \times 225.05)$$

$$\therefore \quad f_y = 415 \text{ N/mm}^2$$

$\therefore$ Maximum stress in steel = 0.87 f_y = 0.87 × 415

$\therefore$ **Maximum stress in steel = 361.06 N/mm^2**

6. Calculation of maximum stress in concrete (0.45 f_{ck})

$$M_u = C_u \times z = 0.36 \cdot f_{ck} \cdot x_u \cdot b\,(d - 0.42\,x_u)$$

$$393.80 \times 10^6 = 0.36\,f_{ck} \times 225.05 \times 350$$
$$(650 - 0.42 \times 225.05)$$

$$\therefore \quad f_{ck} = 25 \text{ N/mm}^2$$

$\therefore$ Maximum stress in concrete = 0.45 f_{ck} = 0.45 × 25

$\therefore$ **Maximum stress in concrete = 11.25 N/mm^2**

7. Calculation of maximum strain in steel ($\in_{su}$)

$$\therefore \quad \in_{su} = 0.002 + \frac{0.87\,f_y}{E_s} = 0.002 + \frac{0.87 \times 415}{2 \times 10^5}$$

$$\therefore \quad \in_{su} = \mathbf{0.0038}$$

8. Calculation of maximum strain in concrete ($\in_{cu}$)

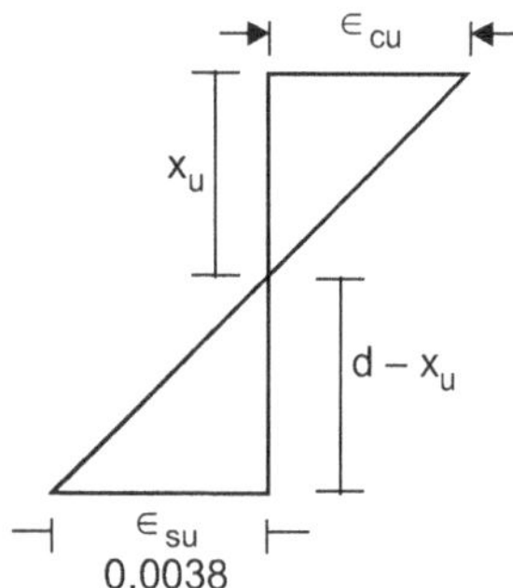

Fig. 7.16

From similar triangle

$$\frac{x_u}{d - x_u} = \frac{\in_{cu}}{\in_{su}}$$

$$\frac{225.05}{650 - 225.05} = \frac{\in_{cu}}{0.0038}$$

$$\therefore \quad \in_{cu} = \mathbf{0.002}$$

Type V : Calculation of Area of Steel (A_{st})

Example 7.15 : *Calculate the area of steel required for a R.C.C. section 200 × 450 mm effective to resist an ultimate bending moment of 150 KN.m. Assume M30 concrete and Fe415 steel.*

Solution : Given :

b = 200 mm, f_{ck} = 30 MPa, M_u = 150 kN.m

d = 450 mm, f_y = 415 MPa

1. Calculation of factored bending moment (M_u)

$$M_u = 150 \text{ kN.m} \qquad \dots\text{given}$$

2. Calculation of maximum moment of resistance (M_{umax})

$$\therefore \quad M_{umax} = 0.138\,f_{ck}\,bd^2$$
$$M_{umax} = 0.138 \times 30 \times 200 \times 450^2$$
$$\therefore \quad M_{umax} = \mathbf{167.67 \text{ kN-m}}$$

3. Compare x_u and x_{umax}

$$150 \; < \; 167.67$$
$$M_u \; < \; M_{umax}$$

$$\dots\text{Design section as under-reinforced s/c}$$

4. Calculation of area of steel (A_{st})

$$A_{st} = \frac{0.5\,f_{ck}}{f_y}\left[1 - \sqrt{1 - \frac{4.6\,M_u}{f_{ck}\,bd^2}}\right] \times b \cdot d$$

$$A_{st} = \frac{0.5 \times 30}{415}\left[1 - \sqrt{1 - \frac{4.6 \times 150 \times 10^6}{30 \times 200 \times 450^2}}\right]$$
$$\times 200 \times 450$$

$$A_{st} = \mathbf{1114.67 \text{ mm}^2}$$

Example 7.16 : *Find the limiting moment of resistance and area of tensile steel for a beam 300 × 600 mm effective. M20 and Fe 415.*

Solution : Given data ,

b = 300 mm, f_{ck} = 20 MPa

d = 600 mm, f_y = 415 MPa

1. Calculation of limiting moment of resistance (M_{umax})

$$M_{umax} = 0.138\,f_{ck} \cdot bd^2$$
$$M_{umax} = 0.138 \times 20 \times 300 \times 600^2$$
$$\therefore \quad M_{umax} = \mathbf{298.08 \text{ kN.m}}$$

2. Calculation of area of steel (A_{st})

$$A_{st} = \frac{0.5\,f_{ck}}{f_y}\left[1 - \sqrt{1 - \frac{4.6\,M_{umax}}{f_{ck} \cdot b \cdot d^2}}\right] \times b.d.$$

$$A_{st} = \frac{0.5 \times 20}{415} \times \left[1 - \sqrt{1 - \frac{4.6 \times 298.08 \times 10^6}{20 \times 300 \times 600^2}}\right]$$
$$\times 300 \times 600$$

$$A_{st} = \mathbf{1716.21 \text{ mm}^2}$$

Example 7.17 : *Compute the area of steel required for a beam section having effective span 4.8 m it carries a udl of 14.19 kN/m including self weight throughout span. Take width of beam 230mm and depth of 330 mm effective, also calculate the no. of bars required if 18 mm ϕ is used.*

Solution : Given :

b = 230 mm, w = 14.19 kN/m

d = 330 mm , Assume : f_{ck} = 20 MPa , f_y = 415 MPa

l_e = 4.8 m,

Maximum udl = W_u = w × γ_f = 14.19 × 1.5

$$W_u = \mathbf{21.29 \text{ kN/m}}$$

1. Calculation of factored bending moment (M_u)

$$M_u = \frac{W_u \cdot l_{eff}^2}{8} = \frac{21.29 \times 4.8^2}{8}$$

$$\mathbf{M_u = 61.32 \ kN/m}$$

2. Calculation of maximum moment of resistance (M_{umax})

$$M_{umax} = 0.138 \ f_{ck} \ bd^2$$
$$M_{umax} = 0.138 \times 20 \times 230 \times 330^2$$
$$\therefore \quad \mathbf{M_{umax} = 69.13 \ kN.m}$$

3. Compare M_u and M_{umax}

$$61.32 < 69.13$$
$$M_u < M_{umax}$$

…Design section as under-reinforced s/c

4. Calculation of Area of steel (A_{st})

$$A_{st} = \frac{0.5 \ f_{ck}}{f_y}\left[1 - \sqrt{1 - \frac{4.6 \ M_u}{f_{ck} \cdot bd^2}}\right] \times b.d$$

$$A_{st} = \frac{0.5 \times 20}{415}\left[1 - \sqrt{1 - \frac{4.6 \times 61.32 \times 10^6}{20 \times 230 \times 330^2}}\right]$$
$$\times 230 \times 330$$

$$\therefore \quad \mathbf{A_{st} = 620.01 \ mm^2}$$

5. Calculation of no. of bars required

$$\text{No. of bars} = \frac{\text{Area of steel } (A_{st})}{\text{Area of single bar } (a_{st})}$$

$$\text{No. of bars} = \frac{620.01}{\frac{\pi}{4}(18)^2} = 2.44 \qquad \dots a_{st} = \frac{\pi}{4}(\phi)^2$$

$$\therefore \quad \textbf{No. of bars} \cong \textbf{3 Nos.}$$

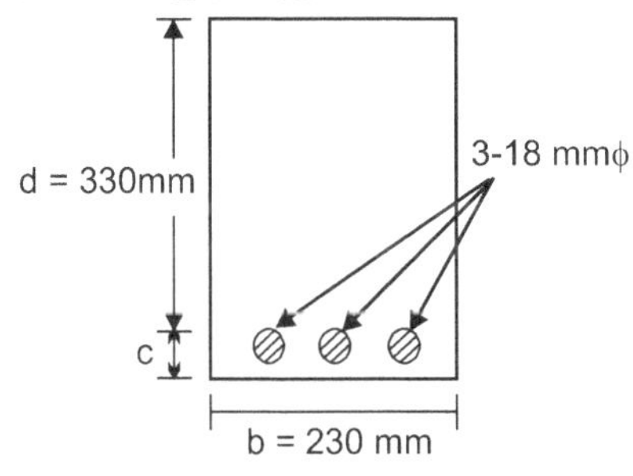

Fig. 7.17

Example 7.18 : *A simply supported beam carries a service load of 20 kN/m over an effective span of 5m. Determine the area of steel required. Use M30 and Fe500. The beam also carries a safe concentrated load of 10 kN at mid-span. For architectural reason effective dimension beam restricted to 230 mm and 500 mm, also find no. of bars required and draw neat sketch of your design.*

Solution : Given :

$$b = 230 \ mm, \ d = 500 \ mm$$
$$f_{ck} = 30 \ MPa, \ f_y = 500 \ MPa$$

$$w = 20 \ kN/m$$
$$W = 10 \ kN$$
$$l_e = 5 \ m$$

(i) ∴ Factored ud*l* (w_u)

$$= w \times \gamma_f = 20 \times 1.5$$
$$\therefore \quad w_u = 30 \ kN/m$$

(ii) Factored concentrated load (w_u) :

$$W \times \gamma_f = 10 \times 1.5$$
$$\therefore \quad W_u = 15 \ kN$$

(I) Calculation of factored bending moment (M_u) :

$$M_u = \frac{w_u \cdot l_e^2}{8} + \frac{W_u \cdot l_e}{4}$$
$$M_u = \frac{30 \times 5^2}{8} + \frac{15 \times 5}{4}$$
$$\therefore \quad M_u = 112.50 \ kN.m$$

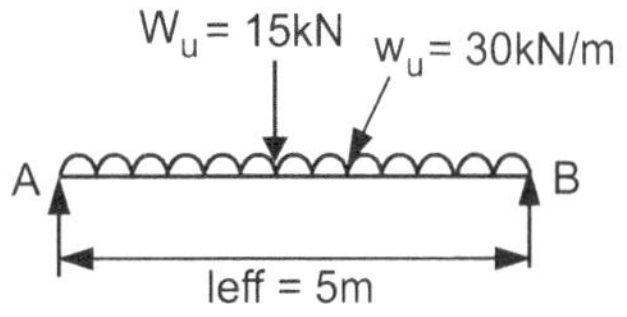

Fig. 7.18

(II) Calculation of maximum moment of resistance (M_{umax}) :

$$M_{umax} = 0.133 \ f_{ck} \ bd^2$$
$$M_{umax} = 0.133 \times 30 \times 230 \times 500^2$$
$$\therefore \quad M_{umax} = 229.43 \ kN.m$$

(III) Compare M_u and M_{umax}

$$112.50 < 229.43$$
$$M_u < M_{umax}$$

… Design section as under-reinforced s/c

(IV) Calculation of area of steel (A_{st}) :

$$A_{st} = \frac{0.5 \ f_{ck}}{f_y}\left[1 - \sqrt{1 - \frac{4.6 \ m_u}{f_{ck} \cdot b \cdot d^2}}\right] \times b \cdot d$$

$$A_{st} = \frac{0.5 \times 30}{500}\left[1 - \sqrt{1 - \frac{4.6 \times 112.50 \times 10^6}{30 \times 230 \times 500^2}}\right]$$
$$\times 230 \times 500$$

$$\therefore \quad \mathbf{A_{st} = 563.52 \ mm^2}$$

∴ Assume diameter of bar = 16 mm

$$\therefore \quad \text{No. of bars} = \frac{A_{st}}{a_{st}} = \frac{563.52}{\frac{\pi}{4}(16)^2}$$

$$\text{No. of bars} = 2.80$$

$$\therefore \quad \textbf{No. of bars} \cong \textbf{3 No's}$$

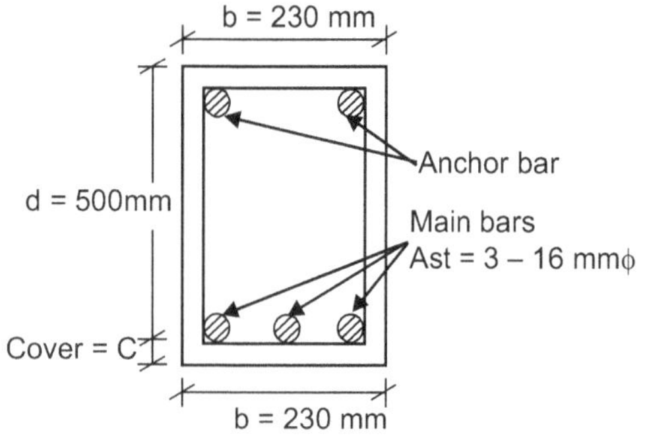

Fig. 7.19

∴ **Provide, 3 – 16 mm φ as tensile reinforcement**

Type VI : Design of Singly Reinforced Rectangular Section

Example 7.19 *: Design a rectangular R.C.C. beam section to carry an working bending moment of 70 kN.m. Use M20 and Fe415 grades material. Take width to depth ratio as 0.5.*

Solution : Given :

$$f_{ck} = 20 \text{ MPa}, f_y = 415 \text{ MPa},$$

$$M = 70 \text{ kN.m} \qquad \text{... working B.M.}$$

$$b/d = 0.5$$

∴   **b = 0.5 d**   ... (a)

1. Factored Bending moment (Mu) :

$$M_u = M \times \gamma_f = 70 \times 1.5$$

∴   **M_u = 105 kN.m**   ... (1)

2. Calculation of maximum moment of resistance (M_{umax}) :

$$M_{umax} = 0.138 \, f_{ck} \, bd^2$$

$$M_{umax} = 0.138 \times 20 \times 0.5d \times d^2$$

∴   **M_{umax} = 1.38 d^3**   ... (2)

∴ equating eq. (1) and (2)

∴ $M_{umax} = M_u$

$$1.38 \, d^3 = 105 \times 10^6$$

∴ $d^3 = \dfrac{105 \times 10^6}{1.38} = 76.09 \times 10^6$

∴   **d = 423.74 mm**

Now, from eq. (a)

$$b = 0.5 \, d = 0.5 \times 423.74$$

∴   b = 211.87 mm

Provide,   b = 211.87 mm ≅ 215 mm

d = 423.74 mm ≅ 430 mm

Assume, cover (c) = 40 mm ∴ D = 470 mm

3. Calculation of Area of steel (A_{st}) :

$$A_{st} = \frac{0.5 \, f_{ck}}{f_y}\left[1 - \sqrt{1 - \frac{4.6 \, M_u}{f_{ck} \cdot b \cdot d^2}}\right] \times b \cdot d$$

$$A_{st} = \frac{0.5 \times 20}{415}\left[1 - \sqrt{1 - \frac{4.6 \times 105 \times 10^6}{20 \times 215 \times 430^2}}\right] \times 215 \cdot 430$$

∴   **A_{st} = 832.04 mm^2**

Assume, diameter of bar = φ = 20 mm

$$\text{No. of bars} = \frac{A_{st}}{a_{st}} = \frac{832.04}{\frac{\pi}{4}(20)^2}$$

$$= 2.65 \text{ No's}$$

∴   **No. of bars ≅ 3 No's.**

∴   **Provide 3 – 20 mm φ as a tensile reinforced**

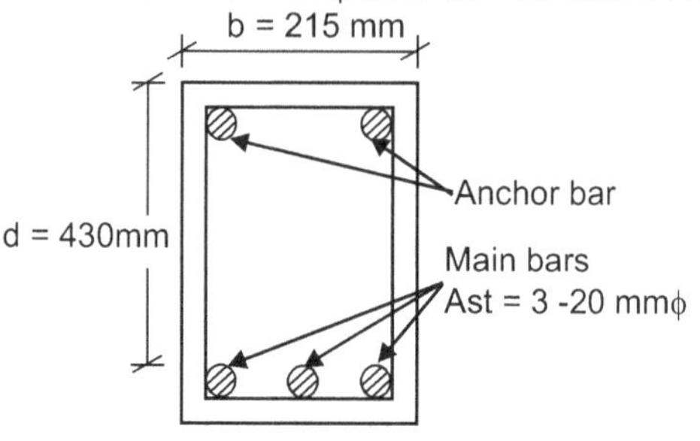

Fig. 7.20

Example 7.20 *: A simply supported rectangular beam of span 4 m carries a working load of 30 kN/m. Design a beam section using concrete mix M20 and steel Fe415. Take depth to width ratio as 1.8 and 45 mm as a effective cover.*

Solution : Given :

$$f_{ck} = 20 \text{ MPa}, \ f_y = 415 \text{ MPa}$$

$$d/b = 1.8, \ w = 30 \text{ kN/m}$$

∴   **d = 1.8 b** , l_{eff} = 4m   ... (a)

cover c = 45 mm

∴ factored load = $w_u = w \times \gamma_f$

$$= 30 \times 1.5$$

∴   **w_u = 45 kN/m**

(I) Calculation of factored bending moment (M_u) :

$$M_u = \frac{w_u \cdot l_{eff}^2}{8}$$

$$M_u = \frac{45 \times 4^2}{8}$$

∴   **M_u = 90 kN.m**   ... (1)

Fig. 7.21

(II) Calculation of maximum moment of resistance (M_{umax}) :

$$M_{umax} = 0.138 \, f_{ck} \cdot b \cdot d^2$$

$$M_{umax} = 0.138 \times 20 \times b \times (1.8\ b)^2$$

$\therefore$ **$M_{umax} = 8.94\ b^3$** ... (2)

$\therefore$ equating eq. (1) and (2) we get,

$$M_{umax} = M_u$$

$$8.94\ b^3 = 90 \times 10^6$$

$$b^3 = \frac{90 \times 10^6}{8.94} = 10.07 \times 10^6$$

$\therefore$ **b = 215.92 mm**

From eq. (a)

$$d = 1.8\ b = 1.8 \times 215.92$$

d = 388.66 mm

$\therefore$ **Provide, b = 215.92 mm $\cong$ 220 mm**

d = 388.66 mm $\cong$ 400 mm

(III) Calculation of area of steel (A_{st}) :

$$A_{st} = \frac{0.5\ f_{ck}}{f_y}\left[1 - \sqrt{1 - \frac{4.6\ M_u}{f_{ck}\ bd^2}}\right] b \cdot d$$

$$A_{st} = \frac{0.5 \times 20}{415}\left[1 - \sqrt{1 - \frac{4.6 \times 90 \times 10^6}{20 \times 220 \times 440^2}}\right]$$
$$\times 220 \times 400$$

$\therefore$ **$A_{st} = 759.52\ mm^2$**

Assume, diameter of bar = ϕ = 16 mm

$$\text{No. of bars} = \frac{A_{st}}{a_{st}} = \frac{759.52}{\frac{\pi}{4}(16)^2} = 3.78$$

No. of bars $\cong$ 4 No's

$\therefore$ **Provide, 4 – 16 mm ϕ as a tensile reinforcement**

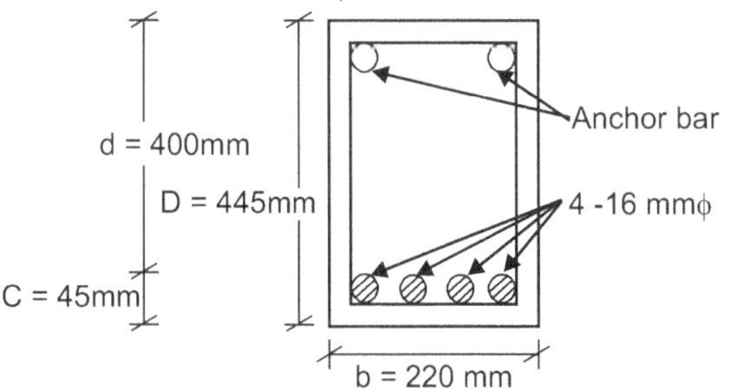

Fig. 7.22

Example 7.21 : *A RC beam has to resist an ultimate moment of 100 kN.m. Design the beam section if for some reason the depth of neutral axis is to be restricted to 0.3d.*

Assume b = 230 mm, f_{ck} = 20 MPa and f_y = 415 MPa

Solution : Given :

$$b = 230\ mm$$

$$f_{ck} = 20\ MPa$$

$$f_y = 415\ MPa$$

$$M_u = 100\ kN.m$$

$\therefore$ Assume cover C = 40 mm

(I) Calculation of factored bending moment (M_u) :

$$M_u = 100\ kN.m \qquad ...(1)\ ...\ given$$

(II) Calculation of maximum moment of resistance (M_{umax}) :

Depth of neutral axis is to be restricted to 0.3 d

$$M_u = C_u \times Z$$

$$M_{umax} = 0.36\ f_{ck} \cdot x_{umax} \cdot b \times (d - 0.42\ x_{umax})$$

$$M_{umax} = 0.36 \times 20 \times 0.3\ d \times 230$$
$$\times (d - 0.42 \times 0.3 \times d)$$

$\therefore$ **$M_{umax} = 434.20\ d^2$** ... (2)

equating (1) and (2)

$$M_{umax} = M_u$$

$$434.20\ d^2 = 100 \times 10^6$$

$\therefore$ **d = 479.90 mm**

$\therefore$ **Provide, b = 230 mm**

d = 479.90 $\cong$ 500 mm

(III) Calculation of area of steel (A_{st}) :

$$A_{st} = \frac{0.5\ f_{ck}}{f_y}\left[1 - \sqrt{1 - \frac{4.6\ M_y}{f_{ck} \cdot b \cdot d^2}}\right] \times b \cdot d$$

$$A_{st} = \frac{0.5 \times 20}{415}\left[1 - \sqrt{1 - \frac{4.6 \times 100 \times 10^6}{20 \times 230 \times 500^2}}\right]$$
$$\times 230 \times 500$$

$\therefore$ **$A_{st} = 624.61\ mm^2$**

Assume, diameter of bar = ϕ = 20 mm

$$\text{No. of bars} = \frac{A_{st}}{a_{st}} = \frac{624.61}{\frac{\pi}{4}(20)^2}$$

No. of.bars = 1.99 No's

$\therefore$ **No. of bars $\cong$ 2 No's**

$\therefore$ **Provide, 2 – 20 mm ϕ as a a tensile reinforcement**

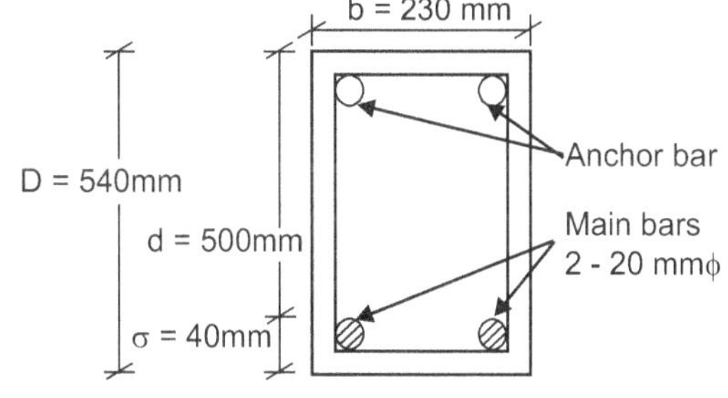

Fig. 7.23

Example 7.22 : *A singly reinforced beam of grade M20 has to resist an ultimate moment of 40 kN.m. Design the section using 0.5 %. Steel of grade Fe 250. Assume b = 230 mm*

Solution : Given:

$$b = 230 \text{ mm}$$
$$f_{ck} = 20 \text{ MPa}$$
$$f_y = 250 \text{ MPa}$$
$$M_u = 40 \text{ kN.m}$$

(I) Factored bending moment (M_u) :

$$M_u = 40 \text{ kN.m} \qquad \text{... (1)... given}$$

(II) Calculation of maximum moment of resistance (M_{umax}) :

$$\therefore \quad M_{umax} = 0.148 \, f_{ck} \cdot bd^2$$
$$M_{umax} = 0.148 \times 20 \times 230 \times d^3$$
$$\therefore \quad M_{umax} = 680.8 \, d^2 \qquad \text{... (2)}$$

$\therefore$ Equating eq. (1) and (2)

$$M_{umax} = M_u$$
$$680.8 \, d^2 = 40 \times 10^6$$
$$d^2 = \frac{40 \times 10^6}{680.8}$$
$$\therefore \quad \mathbf{d = 242.39 \text{ mm}}$$

$\therefore$ Provide, b = 230 mm

$$d = 242.39 \cong 250 \text{ mm}$$
$$c = 30 \text{ mm}$$
$$D = 280 \text{ mm}$$

(III) Calculation of area of steel (A_{st}) :

$$A_{st} = \frac{0.5 \, f_{ck}}{f_y}\left[1 - \sqrt{1 - \frac{4.6 \, M_u}{f_{ck} \cdot bd^2}}\right] b \cdot d$$

$$A_{st} = \frac{0.5 \times 20}{250}\left[1 - \sqrt{1 - \frac{4.6 \times 40 \times 10^6}{20 \times 230 \times 250^2}}\right]$$
$$\times 230 \times 250$$

$$\therefore \quad \mathbf{A_{st} = 920.0 \text{ mm}^2}$$

(IV) Assume diameter of bar = 20 mm

$$\text{No. of bars} = \frac{A_{st}}{a_{st}} = \frac{920}{\frac{\pi}{4}(20)^2}$$

No. of bars = 2.93

∴ No. of bars $\cong$ 3 No's

∴ Provide, 3 – 20 mm ϕ as a tensile reinforcement

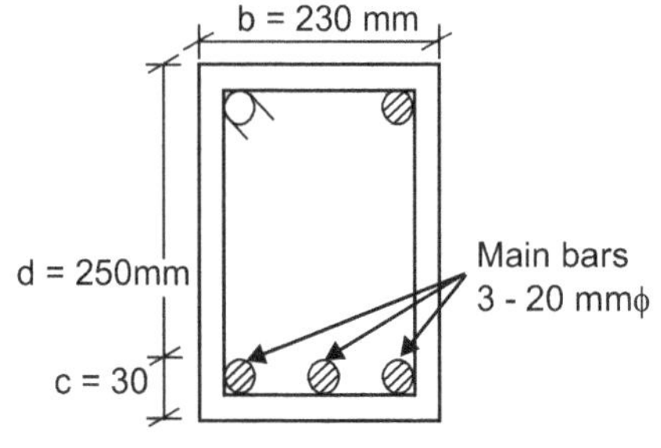

Fig. 7.24

Example 7.23 : *Design a singly reinforced beam to carry a classroom floor of clear span 7m. The beams are supported on 400 mm thick brick wall and are spaced at 2.75 m centres. Thickness of slab 100 mm, use concrete M20 and Fe415 steel.*

Solution : Given :

$$L = 7 \text{ m}$$
$$f_{ck} = 20 \text{ MPa}$$
$$f_y = 415 \text{ MPa}$$

Thickness of slab = 100 mm

support width = 400 mm

$$\therefore \quad l_{eff} = L + \frac{0.4}{2} + \frac{0.4}{2}$$
$$l_{eff} = 7.4 \text{ m}$$

c/c distance between beam = 2.75 m

(I) Calculation of factored bending moment (M_u) :

1. Calculation of loading (w_u) :

(i) Assume live load on beam = 4 × 2.75 = 11 kN/m

(ii) floor finish on slab = 1 × 2.75 = 2.75 kN/m

(iii) self wt. of slab = (2.75 × 0.1 × 1) × 25 = 6.87 kN/m

(iv) self wt. of beam = 5 = 5 kN/m

Total ud*l* (w) = 25.62 kN/m

$$\therefore \quad \text{Maximum ud}l \ (w_u) = w \times \gamma_f$$
$$= 25.62 \times 1.5$$
$$\mathbf{w_u = 38.43 \text{ kN/m}}$$

$$\therefore \quad \text{factored moment } (m_u) = \frac{w_u \cdot l_{eff}^2}{8}$$
$$= \frac{38.43 \times 7.4^2}{8}$$
$$\mathbf{M_u = 263.05 \text{ kN.m}} \qquad \text{... (1)}$$

(II) Calculation of maximum moment of resistance (M_{umax}) :

$\therefore$ Assume b = 230 mm

$$\therefore \quad M_{umax} = 0.138 \, f_{ck} \, bd^2$$
$$M_{umax} = 0.138 \times 20 \times 230 \times d^2$$
$$\therefore \quad \mathbf{M_{umax} = 634.8 \, d^2} \qquad \text{... (2)}$$

equating (1) and (2),

$$M_{umax} = M_u$$
$$634.8 d^2 = 263.05 \times 10^6$$
$$\mathbf{d = 643.73 \text{ mm}}$$

∴ Provide, b = 230 mm

$$d = 643.73 \text{ mm} \cong 650 \text{ mm}$$

(III) Calculation of Area of steel (A_{st}) :

$$A_{st} = \frac{0.5\, f_{ck}}{f_y}\left[1 - \sqrt{1 - \frac{4.6\, m_u}{f_{ck}\, bd^2}}\right] bd$$

$$A_{st} = \frac{0.5 \times 20}{415}\left[1 - \sqrt{1 - \frac{4.6 \times 263.05 \times 10^6}{20 \times 230 \times 650^2}}\right] \times 230 \times 650$$

$\therefore$ $A_{st} = 1392.48\ \text{mm}^2$

Assume, Diameter of bar = 20 mm

$$\text{No. of bars} = \frac{A_{st}}{a_{st}} = \frac{1392.48}{\frac{\pi}{4}(20)^2}$$

No. of bars = 4.43

$\therefore$ **No. of bars $\cong$ 5 No's**

$\therefore$ **Provide, 5 – 20 mm ϕ as a tensile reinforcement.**

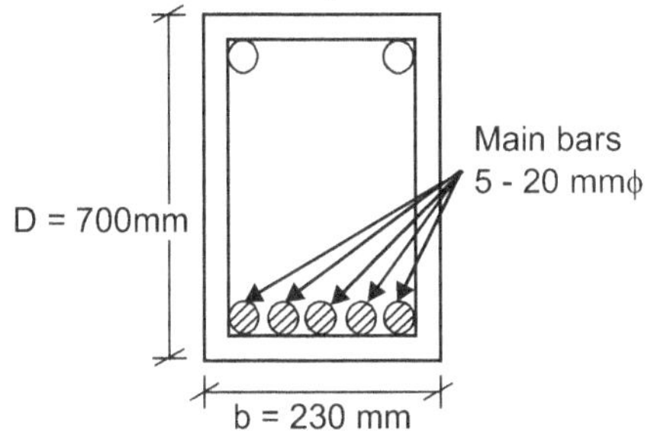

Fig. 7.25

(B) Doubly Reinforced Rectangular Section (DRRS)

7.5 STRAIN DIAGRAM AND STRESS BLOCK DIAGRAM FOR DOUBLY REINFORCED SECTION

- Tension reinforcement is designed for a stress of 0.87 f_y, but for compression reinforcement a lower stress must be used, because steel bars may buckle between shear stirrups under compressive stresses.

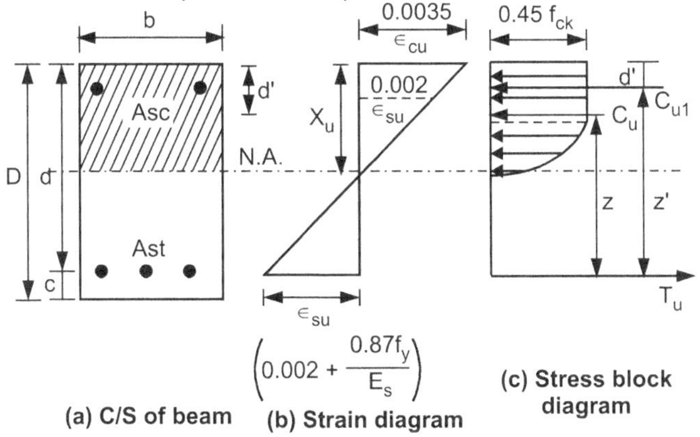

Fig. 7.26 : Stress strain diagram for DRRS

- The design stress in compression reinforcement corresponds to the strain in concrete at that level. With reference to similar triangle in strain diagram, strain at the level of compression reinforcement is equal to

$$\frac{\varepsilon_{sc}}{(x_u - d')} = \frac{0.0035}{x_u}$$

$\therefore$ $\varepsilon_{sc} = 0.0035 \times \left(\dfrac{x_u - d'}{x_u}\right)$

where, ε_{sc} = compressive strain in steel

       x_u = depth of N.A.

       d' = Effective cover to compressive reinforcement

where, b = breadth of section

       d = effective depth for tensile r/f

       d' = effective depth for compressive r/f

       A_{st} = Area of tensile r/f

       A_{sc} = Area of compressive r/f

1. Tensile force carried by tensile r/f (T_u)

   $\therefore$ $T_u = 0.87\, f_y \cdot A_{st}$

2. Compressive force carried by concrete (C_u)

   $\therefore$ $C_u = 0.36\, f_{ck} \cdot x_u \cdot b$

3. Compressive force carried by steel (C_{u1})

   $\therefore$ $C_{u1} = (f_{sc} - f_{cc}) \times A_{sc}$

where, f_{cc} = Compressive stress in concrete at the level of compressive steel

   $\therefore$ $f_{cc} = 0.45\, f_{ck}$ ... if $d' < 0.43\, x_u$

       f_{sc} = compressive stress in steel

   $\therefore$ $f_{sc} = 0.87\, f_y$ ... for mild steel (Fe-250)

Table : Stress 'f_{sc}' in Compression Reinforcement for d'/d Ratios

Grade of Steel 'f_y' N/mm²	Ratio d'/d			
	0.05	0.10	0.15	0.20
415	355	352	342	329
500	424	412	395	370

OR

Table : Salient point on the design stress-strain curve for cold worked bars

(f_{yd} = Design yield stress, f_y = characteristics stress)

Steel Stress Level	Fe-415 ; f_y = 415 N/mm²		Fe-500 ; f_y = 500 N/mm²	
	Strain	Stress (MPa)	Strain	Stress(MPa)
0.80 f_{yd}	0.00144	288.7	0.00174	347.8
0.85 f_{yd}	0.00163	306.7	0.00195	369.6
0.90 f_{yd}	0.00192	324.8	0.00226	391.3
0.95 f_{yd}	0.00241	342.8	0.00277	413.0
0.975 f_{yd}	0.00276	351.8	0.00312	423.9
1.00 f_{yd}	0.00380	360.9	0.00417	434.8

Note :

- $f_{yd} = \dfrac{f_y}{\gamma_m} = \dfrac{f_y}{1.15} \approx 0.87\, f_y$

  - Linear interpolation may be made for intermediate value.

4. Lever arm between T_u and C_u (z)

   $\therefore \qquad z = (d - 0.42\, x_u)$

5. Lever arm between T_u and C_{u1} (z′)

   $\therefore \qquad z' = (d - d')$

Note : Normally, the term f_{cc}, i.e. $0.45\, f_{ck}$ is very small, and it may be neglected without causing any appreciable error.

7.5.1 Depth of Neutral Axis

1. Actual depth of N.A. (x_u)

Equilibrium requirement in horizontal direction gives

$$C_u + C_{u1} = T_u$$

$$\therefore \qquad x_u = \frac{0.87\, f_y \cdot A_{st} - (f_{sc} - f_{cc}) \times A_{sc}}{0.36\, f_{ck} \cdot b}$$

2. Limiting depth of N.A. (x_{umax})

Grade of Steel	X_{umax}
Fe-250	0.53 d
Fe-415	0.48 d
Fe-500	0.46 d

7.5.2 Moment of Resistance

$$M_u = C_u \times z + C_{u1} \times z'$$
$$M_u = Tu_1 \times z + T_{u2} \times z'$$
$$M_u = C_u \times z + T_{u2} \times z'$$
$$M_u = T_{u1} \times z + C_{u1} \times z'$$

- Concept of doubly reinforced rectangular section.

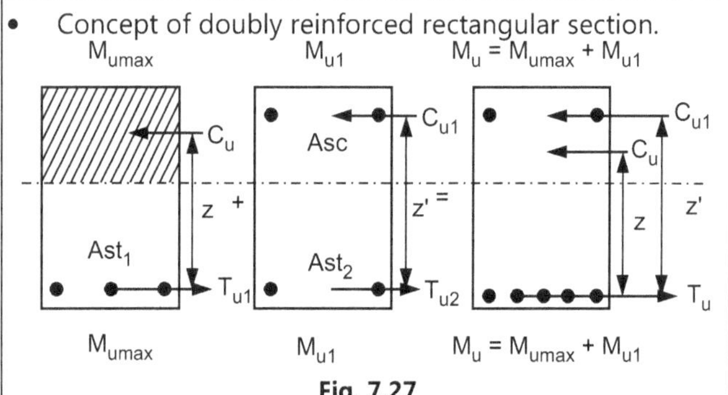

Fig. 7.27

- When applied moment exceed resisting moment doubly reinforced section provided.

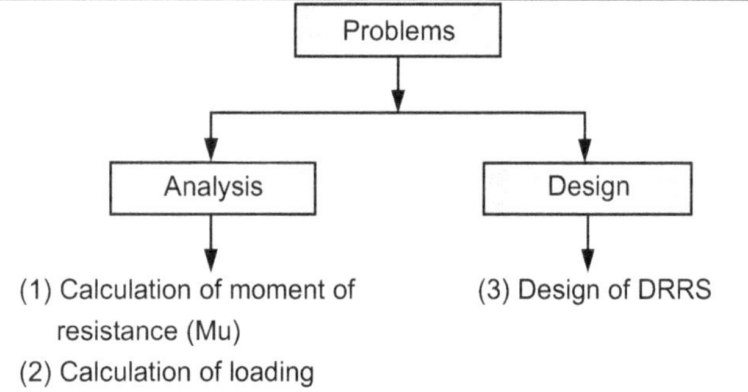

(1) Calculation of moment of resistance (Mu)

(2) Calculation of loading

(3) Design of DRRS

Type I : Calculation of Moment of Resistance

Example 7.24 : *Calculate the ultimate moment of resistance of R.C.C. beam 300 mm wide and 380 mm deep overall. The tension steel consist of 6 Nos. of 20 mm φ bars and compressive steel consist of 2 Nos. of 20 mm φ bars. Assume M20 grade concrete and Fe 250 grade steel. Effective cover to both steel is 40mm.*

Solution : Given data,

$b = 300$ mm, $f_{ck} = 20$ MPa, $\quad A_{st} = 6 - 20$ mm φ

$D = 380$ mm, $\ f_y = 250$ MPa, $A_{st} = 6 \times \dfrac{\pi}{4}(20)^2$

$c = d' = 40$ mm, $\qquad A_{st} = 1884.95$ mm^2

$\therefore \ d = D - c = 380 - 40, \quad A_{sc} = 2 - 20$ mm φ

$\therefore \ d = 340$ mm , $\qquad A_{sc} = 2 \times \dfrac{\pi}{4}(20)^2$

$\qquad\qquad\qquad\qquad\quad A_{sc} = 628.81$ mm^2

1. Calculation of actual depth of neutral axis (x_u)

$$C_u + C_{u1} = T_u$$

$$0.36\, f_{ck} \cdot x_u \cdot b + (f_{sc} - f_{cc}) \cdot A_{sc} = 0.87\, f_y\, A_{st}$$

$$x_u = \frac{0.87 \times 415 \times 1884.95 - (0.87 \times 250 - 0.45 \times 20) \times 628.31}{0.36 \times 20 \times 300}$$

$\therefore \qquad f_{sc} = 0.87\, f_y \ \dots$ for $f_y = 250$ MPa (Mild steel)

$\qquad\qquad f_{cc} = 0.45\, f_{ck} = 0.45 \times 20 = 9$ N/mm^2

$\therefore \qquad \mathbf{x_u = 129.15 \ mm}$

2. Calculation of critical depth of neutral axis ($x_{u\,max}$)

$$x_{umax} = 0.53\, d = 0.53 \times 340$$

$\therefore \qquad \mathbf{x_{umax} = 180.20 \ mm}$

3. Compare x_u and x_{umax}

$$129.15 < 180.20$$

$\therefore \qquad x_u < x_{umax}$

$\qquad\qquad\qquad$ …section is under reinforced section

4. Calculation of moment of resistance (M_u)

$$M_u = C_u \times z + C_{u1} \times z'$$

$$M_u = 0.36 \times f_{ck}\, x_u b\,(d - 0.42\, x_u) + (f_{sc} - f_{cc})$$
$$A_{sc}\,(d - d')$$

$$M_u = 0.36 \times 20 \times 129.15 \times 300$$
$$(340 - 0.42 \times 129.15) + (217.5 - 9)$$
$$\times 628.31 \times (340 - 40)$$

$\therefore \qquad M_u = 119.20$ kN.m

Example 7.25 : *A beam 230 mm × 500 mm effective is provided with 226 mm^2 steel on compression side, cover to reinforcement is 50 mm. Assume 3-16 mm φ as a tensile reinforcement. Use M20 and Fe 500. Find M_u.*

Solution : Given data

$b = 230$ mm, $f_{ck} = 20$ MPa,

$d = 500$ mm, $f_y = 500$ MPa

$d' = 50$ mm

$A_{st} = 3 - 16$ mm $\phi = 3 \times \dfrac{\pi}{4}\ (16)^2$

$A_{st} = 603.18$ mm^2

$A_{sc} = 226$ mm^2

1. Calculation of actual depth of neutral axis (x_u)

$$C_u + C_{u1} = T_u$$

$$0.36\ f_{ck} \cdot x_u \cdot b + (f_{sc} - f_{cc})\ A_{sc} = 0.87\ f_y\ A_{st}$$

$$x_u = \frac{0.87 \times 500 \times 603.18 - (412 - 0) \times 226}{0.36 \times 20 \times 230}$$

$\boxed{x_u = 102.21 \text{ mm}}$ ... for $f_{sc} - \dfrac{d'}{d} = \dfrac{50}{500} = 0.1$

$$\therefore\ f_{sc} = 412$$

$$f_{cc} = 0 \ ...\text{neglected}$$

2. Calculation of critical depth of neutral axis (x_{umax})

$$x_{umax} = 0.46\ d = 0.46 \times 500$$

$$\boxed{x_{umax} = 230 \text{ mm}}$$

3. Compare x_u and x_{umax}

$$\therefore \quad 102.21\ <\ 230$$

$$x_u\ <\ x_{umax}$$

...section is under-reinforced section

4. Calculation of moment of resistance (M_u)

$$M_u = C_u \cdot z + C_{u1} \times z'$$

$$M_u = 0.36\ f_{ck} \cdot x_u \cdot b\ (d - 0.42\ x_u) + (f_{sc} - f_{cc})$$

$$A_{sc}\ (d - d')$$

$$M_u = 0.36 \times 20 \times 102.21 \times 230$$

$$(500 - 0.42 \times 102.21) + (412 - 0)\ 226$$

$$(500 - 50)$$

$$\boxed{M_u = 119.26 \text{ kN.m}}$$

Example 7.26 : *A beam 300 × 800 mm effective is reinforced on tension and compression side with 3940 mm^2 and 795 mm^2 of steel. The compression steel is placed 40 mm below the top. If concrete grade M15 and mild steel are used, find moment of resistance (M_u).*

Solution :

Given :

$$b = 300 \text{ mm}$$
$$d = 800 \text{ mm}$$
$$d' = 40 \text{ mm}$$
$$A_{sc} = 795 \text{ mm}^2$$
$$A_{st} = 3940 \text{ mm}^2$$
$$f_{ck} = 15 \text{ N/mm}^2$$

$$f_y = 250 \text{ N/mm}^2$$

1. $x_{u,\ max}$ $x_{u,\ max} = 0.531\ d$ (for $f_y = 250$)

$$= 0.531 \times 800 = 424.8 \text{ mm}$$

2. f_{cc} $f_{cc} = 0.446\ f_{ck} = 0.446 \times 15 = 6.69 \text{ N/mm}^2$

3. f_{sc} $f_{sc} = 0.87\ f_y = 0.87 \times 250 = 217.5 \text{ N/mm}^2$

4. M_u $M_u = (0.36 f_{ck}\ x_{u,\ max} b)\ (d - 0.416\ x_{u,\ max})$

$$+ A_{sc}\ (f_{sc} - f_{cc})\ (d - d')$$

$$= (0.36 \times 15 \times 424.8 \times 300)$$

$$(800 - 0.416 \times 424.8) + 795$$

$$(217.5 - 6.69)\ (800 - 40)$$

$$= 556.3 \times 10^6 \text{ N–mm}$$

$$M_u = 556.3 \text{ kN-m}$$

Example 7.27 : *Find moment of resistance (M_u) of a beam 230 × 600 mm effective, with the following data. A_{sc} = 554 mm^2, A_{st} = 1524 mm^2, d' = 30 mm, concrete M15 and steel Fe 415.*

Solution : Given

$$b = 230 \text{ mm}$$
$$d = 600 \text{ mm}$$
$$d' = 30 \text{ mm}$$
$$A_{sc} = 554 \text{ mm}^2$$
$$A_{st} = 1524 \text{ mm}^2$$
$$f_{ck} = 15 \text{ N/mm}^2$$
$$f_y = 415 \text{ N/mm}^2$$

1. $x_{u,\ max}$ $x_{u,\ max} = 0.479\ d$ (for $f_y = 415$)

$$= 0.479 \times 600$$

$$= 287.4 \text{ mm}$$

2. f_{cc} $f_{cc} = 0.446\ f_{ck} = 0.446 \times 15 = 6.69 \text{ N/mm}^2$

3. To find f_{sc} :

$$\frac{d'}{d} = \frac{30}{600} = 0.05 \text{ from the table}$$

For $\dfrac{d'}{d} = 0.05$ and steel Fe 415, $f_{sc} = 355 \text{ N/mm}^2$

4. Find M_u $M_u = (0.36\ f_{ck}\ x_{u,\ max}\ b)\ (d - 0.416\ x_{u,\ max})$

$$+ A_{sc}\ (f_{sc} - f_{cc})\ (d - d')$$

$$= (0.36 \times 15 \times 287.4 \times 230)$$

$$(600 - 0.416 \times 287.4) + 554$$

$$(355 - 6.69)\ (600 - 30)$$

$$= 281.88 \times 10^6 \text{ N-mm}$$

$$M_u = 281.88 \text{ kN-m}$$

Example 7.28 : *Find moment of resistance (M_u), if b = 225 mm, d = 500 mm, d' = 50 mm, A_{sc} = 125 mm^2, A_{st} = 754 mm^2. Concrete M15 and steel is Fe 500.*

Given : $M_u = ?$

$d = 500$ mm

$d' = 50$ mm

$A_{sc} = 125$ mm^2

$A_{st} = 754$ mm^2

M15 and Fe 500

1. $x_{u, max}$ $x_{u, max} = 0.456\,d$ (for $f_y = 500$)

$\qquad = 0.456 \times 500$

$\qquad = 228$ mm

2. f_{cc} $f_{cc} = 0.446 \times 15 = 6.69$ N/mm^2

3. To find f_{sc} $\dfrac{d'}{d} = \dfrac{50}{500} = 0.1$

For $\qquad \dfrac{d'}{d} = 0.1$ and Fe 500 , $f_{sc} = 412$ N/mm^2

4. Find M_u $M_u = (0.36\,f_{ck}\,x_{u\,max}\,b)(d - 0.416\,x_{u, max})$

$\qquad + A_{sc}(f_{sc} - f_{cc})(d - d')$

$\qquad = (0.36 \times 15 \times 228 \times 225)$

$\qquad (500 - 0.416 \times 228) + 125\,(412 - 6.69)$

$\qquad (500 - 50)$

$\qquad = 135.038 \times 10^6$ N-mm $= 135.037$ kN-m.

Type II : Calculation of Loading

Example 7.29 : *A simply supported beam of size 230 mm × 465 mm is reinforced with 5-16 mmϕ on tension side and 4-12 mm ϕ on compression side. What superimposed load, this beam can carry. If its effective span of 6m.*

Solution : Given data :

$b = 230$ mm, Assume $f_{ck} = 20$ MPa,

$$A_{st} = 5 - 16\text{ mm } \phi = 5 \times \frac{\pi}{4}(16)^2$$

$D = 465$ mm, $f_y = 415$ MPa, $A_{st} = 1005.31$ mm^2

Assume, effective cover $C = 35$ mm

$d = D - C = 465 - 35$

$$A_{sc} = 4 - 12\text{ mm}\phi = 4 \times \frac{\pi}{4}(12)^2$$

$\therefore$ $d = 430$ mm $\qquad A_{sc} = 452.39$ mm^2

1. Calculation of actual depth of neutral axis (x_u)

$C_u + C_{u1} = T_u$

$$x_u = \frac{0.87 \times f_y \times A_{st} - (f_{sc} - f_{cc}) \cdot A_{sc}}{0.36 \times f_{ck} \times b}$$

$f_{cc} = 0.45\,f_{ck}$ $\hfill$ for f_{sc}

$f_{cc} = 0.45 \times 20$ and $\qquad \dfrac{d'}{d} = \dfrac{35}{430} = 0.08$

$\mathbf{f_{cc} = 9}$ **N/mm^2**

$\therefore$ By interpolation

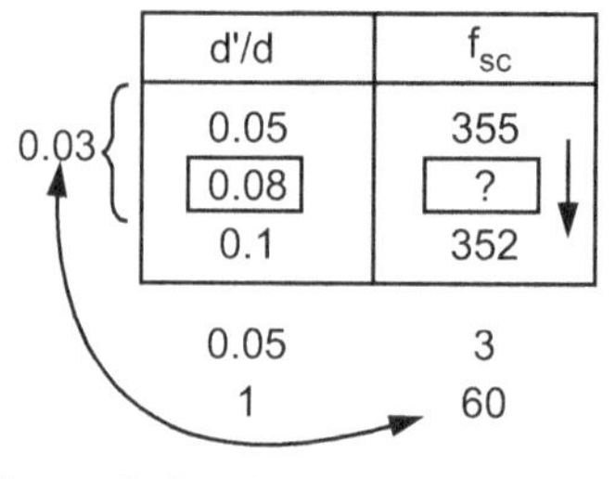

d'/d	f_{sc}
0.05	355
0.08	?
0.1	352

0.05   3

1   60

$f_{sc} = 355 - 0.03 \times 60$

$\mathbf{f_{sc} = 353.2}$ **N/mm^2**

$\therefore$ $x_u = \dfrac{0.87 \times 415 \times 1005.31 - (353.2 - 9) \times 452.39}{0.36 \times 20 \times 230}$

$\therefore$ $x_u = 125.15$ mm

2. Calculation of critical depth of Neutral Axis (x_{umax})

$\qquad x_{umax} = 0.48\,d = 0.48 \times 430$

$\therefore \qquad x_{umax} = 206.40$ mm

3. Compare x_u and x_{umax}

$\qquad 125.15 < 206.40$

$\therefore \qquad x_u < x_{umax}$

$\qquad\qquad$ …section is under-reinforced section

4. Calculation of ultimate moment of resistance (M_u)

$\qquad M_u = C_u \times z + C_{u1} \times z'$

$\qquad M_u = 0.36\,f_{ck}\,x_u \cdot b\,(d - 0.42\,x_u) + (f_{sc} - f_{cc})$

$\qquad\qquad A_{sc}(d - d')$

$\qquad M_u = 0.36 \times 20 \times 125.15$

$\qquad\qquad \times 230\,(430 - 0.42 \times 125.15)$

$\qquad\qquad + (353.2 - 9)\,452.39 \times (430 - 35)$

$\therefore \qquad \mathbf{M_u = 139.73}$ **kN.m** $\hfill …(1)$

5. Calculation of factored bending moment (B.M$_{max}$)

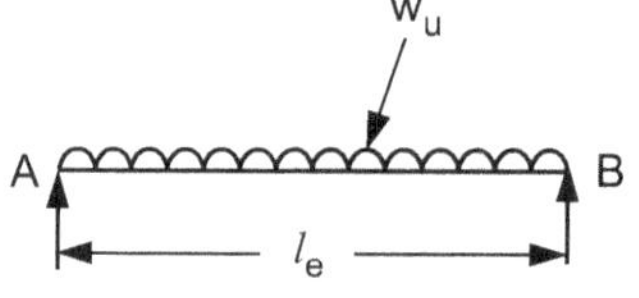

Fig. 7.28

$\therefore$ B.M$_{max} = \dfrac{W_u \cdot l_e^2}{8}$

B.M$_{max} = \dfrac{W_u \cdot 6^2}{8}$

$\therefore$ **BM$_{max}$ = 4.5 w_u** $\hfill …(2)$

$\therefore$ Equating equation (1) and (2)

$\qquad$ BM$_{max} = M_u$

$\qquad 4.5\,W_u = 139.73 \times 10^6$

$\therefore$ Maximum udl W_u = 31.05 kN/m

$\therefore$ Safe udl (W) = $\dfrac{W_u}{r_f}$ = $\dfrac{31.05}{1.5}$

        W = 20.7 kN/m

Safe udl (excluding self wt/ superimposed load) = w'

$\therefore$ w' = w – self wt. of beam

        w' = 20.7 – (0.23 × 0.465 × 1) × 25

        w' = 18.03 kN/m

Example 7.30 : *A reinforced concrete simply supported beam of 8 m span with 300 mm × 500 mm effective cross-section is reinforcement with 4-20 mm ϕ in tensile zone and 4–16 mm ϕ on compression side. Effective cover to both reinforcement is 50 mm. Determine (i) Moment of resistance (M_u), (ii) Safe udl excluding self wt of the beam can carry. Use M20 and Fe 415.*

Solution : Given data :

b = 300 mm, $A_{st} = 4 – 20$ mm $\phi = 4 \times \dfrac{\pi}{4}(20)^2$

f_{ck} = 20 MPa, $A_{st} = 1256.64$ mm^2

f_y = 415 MPa

d = 500 mm, $A_{sc} = 4 – 16$ mm $\phi = 4 \times \dfrac{\pi}{4}(16)^2$

c = d = 50 mm, $A_{sc} = 804.25$ mm^2

l_e = 8 m,

(I) For calculation of moment of resistant (M_u) :

1. Calculation of actual depth of neutral axis (x_u)

        $C_u + C_{u1}$ = T_u

$\therefore$ x_u = $\dfrac{0.87\, f_y A_{st} – (f_{sc} – f_{cc})\, A_{sc}}{0.36\, f_{ck} \cdot b}$

For f_{sc}, $\dfrac{d'}{d}$ = $\dfrac{50}{500}$ = 0.1

$\therefore$ f_{sc} = 352 N/mm^2 ...from table

Neglect f_{cc}

$\therefore$ f_{cc} = 0

        x_u = $\dfrac{0.87 \times 415 \times 1256.64 – (352 – 0) \times 804.25}{0.36 \times 20 \times 300}$

$\therefore$ x_u = 78.99 mm

2. Calculation of critical depth of Neutral axis (x_{umax})

        x_{umax} = 0.48 d = 0.48 × 500

        x_{umax} = 240 mm

3. Compare x_u and x_{umax}

        78.99 mm < 240 mm

4. Calculation of moment of resistance (M_u)

        M_u = $C_u \times z + C_{u1} \times z'$

        M_u = $0.36\, f_{ck} \cdot x_u \cdot b\, (d – 0.42\, x_u) + (f_{sc} – f_{cc})$

        $A_{sc} \times (d – d')$

        M_u = $0.36 \times 20 \times 78.99 \times 300$

        (500 – 0.42 × 78.99) + (353 – 0) 804.25

        × (500 – 50)

$\therefore$ **M_u = 207.40 kN-m**   ...(1)

(II) Calculation of safe udl excluding self wt of the beam (w') :

5. Calculation of factored bending moment (BM$_{max}$)

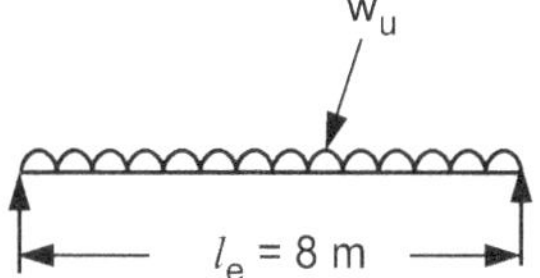

Fig. 7.29

$\therefore$ BM_{max} = $\dfrac{w_u \cdot l_e^2}{8}$ = $\dfrac{w_u \cdot 8^2}{8}$

$\therefore$ **BM_{max} = 8 w_u**   ...(2)

$\therefore$ Equating equation (1) and (2)

        BM_{max} = M_u

        $8w_u$ = 207.40

        w_u = 25.93 kN/m ...factored/maximum udl

Now, safe udl (including self wt)

$\therefore$ w = $\dfrac{w_u}{r_f}$ = $\dfrac{25.93}{1.5}$

        w = 17.29 kN/m

$\therefore$ Safe udl (excluding self wt)

i.e.   w' = w – self wt. of the beam

        w' = 17.29 – (0.3 × 0.55 × 1) × 25

        w' = 13.17 kN/m

Type III : Design of DRRS

Example 7.31 : *Design the reinforcement for reinforced concrete beam 250 mm wide and 340 mm deep to resist an ultimate moment of 90 kN.m. Use M20 and Fe415 grade. Take effective cover on both side as 40 mm.*

Solution : Given data :

        b = 250 mm, f_{ck} = 20 MPa

        D = 340 mm, f_y = 415 MPa

        c = d' = 40 mm ... assume

        $\therefore$ d = 300 mm

1. Calculation of factored bending moment (M_u)

$\therefore$ **M_u = 90 kN.m**   ...Given

2. Calculation of maximum moment of resistance (M_{umax})

        M_{umax} = 0.138 f_{ck} bd^2 ...for Fe 415

        M_{umax} = 0.138 × 20 × 250 × 300^2

$\therefore$ M_{umax} = 62.1 kN.m

3. Compare M_u and M_{umax}

$$90 > 62.1$$

$$M_u > M_{umax} \quad \text{...Design section as doubly R/f}^d \text{ rectangle section}$$

4. Calculation of tensile reinforcement (A_{st})

(i) Calculation of A_{st_1}

$$M_{umax} = T_{u1} \times z$$

$$M_{umax} = 0.87 \, f_y \, A_{st_1} \, (d - 0.42 \, x_{u\,max}) \quad \text{for Fe415}$$

$$x_{umax} = 0.48 \, d$$

$$\therefore \quad M_{umax} = 0.87 \times 415 \times A_{st_1} \, (d - 0.42 \times 0.48 \, d)$$

$$62.1 \times 10^6 = 0.87 \times 415 \times A_{st_1} \, (300 - 0.42 \times 0.48 \times 300)$$

$$A_{st_1} = 718.10 \text{ mm}^2$$

(ii) Extra moment (M_{u1})

$$M_{u1} = M_u - M_{umax} = 90 - 62.10$$

$$M_{u1} = 27.90 \text{ kN.m}$$

(iii) Calculation of A_{st_2}

$$M_{u1} = T_{u_2} \times z'$$

$$M_{u1} = 0.87 \, f_y \, A_{st_2} \, (d - d')$$

$$27.90 \times 10^6 = 0.87 \times 415 \times A_{st_2} \, (300 - 40)$$

$$\mathbf{A_{st_2} = 297.21 \text{ mm}^2}$$

(iv) Total tensile reinforcement (A_{st})

$$\mathbf{A_{st} = A_{st_1} + A_{st_2}}$$

$$A_{st} = 718.10 + 297.21$$

$$\therefore \quad \mathbf{A_{st} = 1015.31 \text{ mm}^2}$$

$\therefore$ Assume diameter of bar = 18 mmϕ

$$\text{No. of bars} = \frac{A_{st}}{\pi/4 \, (d)^2} = \frac{1015.31}{\pi/4 \, (18)^2}$$

No. of bars = 3.99

No. of bars $\cong$ 4 No.s

5. Calculation of compression reinforcement (A_{sc})

$$M_{u1} = C_{u1} \times z'$$

$$M_{u1} = (f_{sc} - f_{cc}) \cdot A_{sc} \, (d - d')$$

$\therefore$ For f_{sc} Neglect f_{cc}

$$\frac{d'}{d} = \frac{40}{300} = 0.13 \qquad \text{i.e. } f_{cc} = 0$$

$$\frac{d'}{d} = 0.13$$

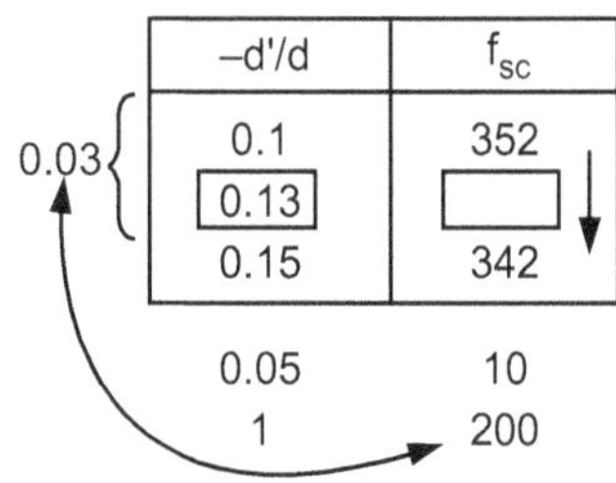

$-d'/d$	f_{sc}
0.1	352
0.13	
0.15	342
0.05	10
1	200

$$\therefore \quad f_{sc} = 352 - 0.03 \times 200$$

$$\mathbf{f_{sc} = 346 \text{ N/mm}^2}$$

$$27.9 \times 10^6 = (346 - 0) \, A_{sc} \, (300 - 40)$$

$$\therefore \quad \mathbf{A_{sc} = 310.14 \text{ mm}^2}$$

Assume ϕ = 16 mm

$$\text{No. of bars} = \frac{A_{sc}}{\pi/4 \, (d)^2}$$

$$= \frac{310.14}{\pi/4 \, (16)^2}$$

No. of bars = 1.54 $\cong$ 2 Nos.

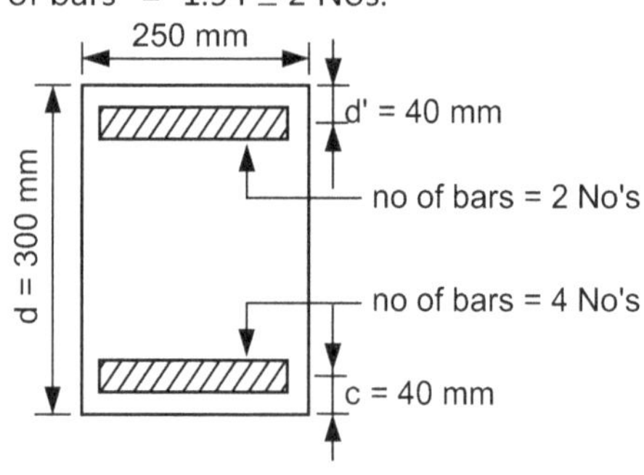

Fig. 7.30

Example 7.32 : *Calculate the moment of resistance of a doubly reinforced section. The beam size is 230 mm × 600 mm overall depth, is reinforced with 3-25 mm ϕ are tension side placed with an effective cover of 30 mm and 3-20 mm ϕ is placed on compression side with an effective cover of 30 mm. Use M20 and Fe500 steel.*

Solution : Given data ,

$$b = 230 \text{ mm}, \qquad A_{st} = 3 - 25 \text{ mm } \phi = 3 \times \frac{\pi}{4} \, (25)^2$$

$$f_{ck} = 20 \text{ MPa}, \qquad A_{st} = 1472.62 \text{ mm}^2$$

$$f_y = 500 \text{ MPa}, \qquad A_{sc} = 3 - 20 \text{ mm } \phi = 3 \times \frac{\pi}{4} \, (20)^2$$

$$D = 600 \text{ mm}, \qquad A_{sc} = 942.48 \text{ mm}^2$$

$$c = d' = 30 \text{ mm},$$

$$d = D - d' = 600 - 30 \,;$$

$$d = 570 \text{ mm}$$

1. Calculation of actual depth of neutral axis (x_u)

$$x_u = \frac{0.87 \, f_y \, A_{st} - (f_{sc} - f_{cc}) A_{sc}}{0.36 \, f_{ck} \cdot b}$$

$$x_u = \frac{0.87 \times 500 \times 1472.62 - (424 - 0) \, 942.47}{0.36 \times 20 \times 230}$$

Neglect f_{cc},   For f_{sc}

$$\therefore \quad f_{cc} = 0 \qquad\qquad \frac{d'}{d} = \frac{30}{570} = 0.05$$

$$\therefore \quad \mathbf{x_u = 145.52 \text{ mm}} \quad \therefore f_{sc} = 424 \text{ N/mm}^2 \text{... Fe 500}$$

2. Calculation of critical depth of Neutral axis (x_{umax})

$$\mathbf{x_{umax} = 262.2 \text{ mm}}$$

3.   Compare x_u and x_{umax}

$$145.52 \; < \; 262.2$$

$\therefore \qquad x_u \; < \; x_{umax}$

...Section is under reinforced section

4.   Calculation of moment of resistance (M_u)

$$M_u \; = \; C_u \times z + C_{u1} \times z'$$

$$M_u \; = \; 0.36 \times f_{ck} \times x_u \times b \times (d - 0.42 \, x_u)$$

$$+ \; (f_{sc} - f_{cc}) \, A_{sc} \, (d - d')$$

$$M_u \; = \; 0.36 \times 20 \times 145.52 \times 230 \, (570 - 0.42$$

$$\times \, 145.52) + (424 - 0) \, 942.48 \, (570 - 30)$$

$$\mathbf{M_u \; = \; 338.42 \ kN\text{-}m} \qquad \qquad ...(1)$$

Example 7.33 : *For above question, also calculate safe concentrated load the beam can carry, if it is simply supported over a span of 5m.*

Solution :

1.   Calculation of factored bending moment (BM_{max})

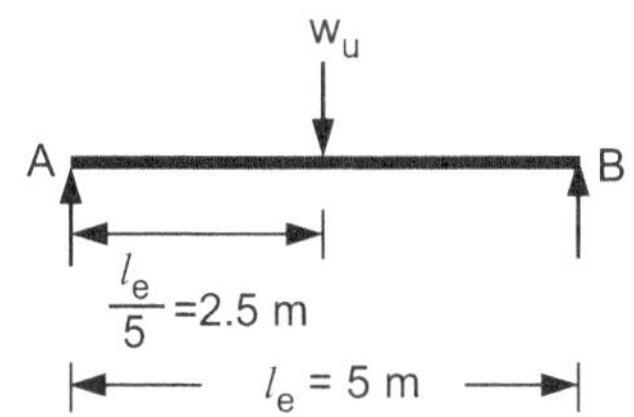

Fig. 7.31

$$BM_{max} \; = \; \frac{W_u \cdot l_{eff.}}{4} \; = \; \frac{W_u \cdot 5}{4}$$

$$BM_{max} \; = \; 1.25 \, W_u \qquad \qquad ...(2)$$

Equating equation (1) and (2)

$\therefore \qquad BM_{max} \; = \; M_u$

$$1.25 \, W_u \; - \; 338.41$$

$$\mathbf{W_u \; = \; 270.74 \ kN} \; ...\text{factored / maximum load}$$

$\therefore \quad$ Safe load $= W = \dfrac{W_u}{r_f} \; = \; \dfrac{270.74}{1.5}$

$\therefore \qquad \mathbf{W \; = \; 180.49 \ kN}$

Example 7.34 : *A beam 250 × 500 mm carries a factored moment (M_d) of 165 kN-m. Find the area of steel required, if M15 concrete and solid steel are used. Assume d' = 50 mm.*

Solution : Given

$$b \; = \; 250 \ mm$$
$$d \; = \; 500 \ mm$$
$$d' \; = \; 50 \ mm$$
$$f_{ck} \; = \; 15 \ N/mm^2$$
$$f_y \; = \; 250 \ N/mm^2$$
$$M_d \; = \; 165 \ kN\text{-}m$$
$$M_{u.\,lim} \; = \; 0.149 \, bd^2 \, f_{ck}$$

$$= \; 0.149 \times 250 \times 500^2 \times 15$$

$$= \; 139.69 \times 10^6 \ N\text{-}mm$$

For M15 and mild steel $P_{t,\,lim} = 1.32$

$\therefore \qquad A_{st\,1} \; = \; \dfrac{1.32 \times 250 \times 500}{100} \; = \; 1650 \ mm^2$

$$M_d \; > \; M_{u,\,lim}$$

Balance of $M_d, M_{d1} = M_d - M_{u\,lim}$

$$= \; 165 - 139.69$$

$$= \; 25.31 \ kN\text{-}m$$

To find A_{sc} : $M_{d1} \; = \; A_{sc} \, (f_{sc} - f_{cc}) \, (d - d')$

where, $\qquad f_{cc} \; = \; 0.446 \times 15 = 6.69 \ N/mm^2$

$$f_{sc} \; = \; 0.87 \times 250 = 217.5 \ N/mm^2$$

$\therefore \quad 25.31 \times 10^6 \; = \; A_{sc} \, (217.5 - 6.69) \, (500 - 50)$

$\therefore \qquad A_{sc} \; = \; 266.8 \ mm^2$

To find A_{st2}:

$$A_{sc} \, (f_{sc} - f_{cc}) \; = \; A_{st\,2} \, (0.87 \, f_y)$$

$$266.8 \, (217.5 - 6.69) \; = \; A_{st\,2} \, (0.87 \times 250)$$

$$A_{st\,2} \; = \; 258.59 \ mm^2$$

$\therefore \qquad A_{st} \; = \; A_{st\,1} + A_{st\,2} = 1650 + 258.59$

$$= \; 1908.59 \ mm^2$$

Example 7.35 : *An RC beam 200 × 300 mm effective, is constructed with concrete M20 and steel Fe 415. Find the steel required, stored moment on beam is 74 kN-m. Assume d' = 30 mm.*

Solution : Given

$$b \; = \; 200 \ mm$$
$$d \; = \; 300 \ mm$$
$$d' \; = \; 30 \ mm$$
$$M_d \; = \; 74 \ kN\text{-}m$$
$$f_{ck} \; = \; 20 \ N/mm^2$$
$$f_y \; = \; 415 \ N/mm^2$$
$$M_{u,\,lim} \; = \; 0.138 \, bd^2 \, f_{ck}$$

$$= \; 0.138 \times 200 \times 300^2 \times 20$$

$$= \; 49.68 \times 10^6 \ N\text{-}mm$$

$$= \; 49.68 \ kN\text{-}m$$

To find $A_{st\,1}$:

For concrete M20 and steel Fe415, $P_{t,\,lim} = 0.96$

$\therefore \qquad A_{st\,1} \; = \; \dfrac{0.96 \times 200 \times 300}{100}$

$$= \; 576 \ mm^2$$

Balance of $M_d, M_{d1} = M_d - M_{u,\,lim}$

$$74 - 49.68 = 24.32 \ kN\text{-}m$$

To find A_{sc} : $M_{d1} \; = \; A_{sc} \, (f_{sc} - f_{cc}) \, (d - d')$

Here, $\qquad f_{cc} \; = \; 0.446 \times 20 = 8.92 \ N/mm^2$

To find f_{sc} : $\dfrac{d'}{d} = \dfrac{30}{300} = 0.1$

For steel is Fe 415 , $f_{sc} = 353$ N/mm^2

$24.32 \times 10^6 = A_{sc}\,(353 - 8.92)\,(300 - 30)$

$\qquad A_{sc} = 261.78$ mm^2

To find A_{st_2} :

$\qquad A_{sc}\,(f_{sc} - f_{cc}) = A_{st\,2}\,(0.87\,f_y)$

$\qquad 261.78\,(353 - 8.92) = A_{st\,2}\,(0.87 \times 415)$

$\qquad\qquad A_{st\,2} = 249.475$ mm^2

$\qquad\qquad A_{st} = A_{st1} + A_{st2} = 576 + 249.475$

$\qquad\qquad\qquad = 825.475$ mm^2

Example 7.36 : *A beam 200 × 200 mm effective, carries a factored cement of 32 kN-m at a section. Find steel required, if concrete M 20 and steel Fe 500 are used d' = 30 mm.*

Solution : Given :

$\qquad b = 200$ mm

$\qquad d = 200$ mm

$\qquad d' = 30$ mm

$\qquad f_{ck} = 20$ N/mm^2

$\qquad f_y = 500$ N/mm^2

$\qquad M_d = 32$ kN-m

$\qquad M_{u,\,lim} = 0.133\ bd^2\ f_{ck}$

$\qquad\qquad = 0.133 \times 200 \times 200^2 \times 20$

$\qquad\qquad = 21.28 \times 10^6$ N-mm

$\qquad\qquad = 21.28$ kN-m

For concrete M 20 and Fe 500, $P_{t,\,lim} = 0.76$

$\qquad A_{st\,1} = \dfrac{0.76 \times 200 \times 200}{100} = 304$ mm^2

Balanced of factored moment,

$\qquad M_{d1} = 32 - 21.28 = 10.72$ kN-m

To find A_{sc} : $M_{d1} = A_{sc}\,(f_{sc} - f_{cc})\,(d - d')$

Here $\qquad f_{cc} = 0.446 \times 20 = 8.92$ N/mm^2

To find f_{sc} : $\dfrac{d'}{d} = \dfrac{30}{200} = 0.15$

Since steel is Fe 500,

$\qquad f_{sc} = 395$ N/mm^2

$\therefore\quad 10.72 \times 10^6 = A_{sc}\,(395 - 8.92)\,(200 - 30)$

$\qquad\qquad A_{sc} = 163.33$ mm^2

To find $A_{st\,2}$:

$\qquad A_{sc}\,(f_{sc} - f_{cc}) = A_{st\,2}(0.87\,f_y)$

$\qquad 163.33\,(395 - 8.92) = A_{st_2}\,(0.87 \times 500)$

$\therefore\qquad A_{st\,2} = 144.96$ mm^2

$\therefore\qquad A_{st} = A_{st1} + A_{st2} = 304 + 144.96$

$\qquad\qquad = 448.96$ mm^2

EXERCISE

1. What is the difference between singly reinforced beam and doubly reinforced beam?

2. Why doubly reinforced section is used ?

3. State the concept of under-reinforced, Balanced and over-reinforced section.

4. Why Over-reinforced section not permitted in R.C.C. design ?

5. Draw stress strain diagram for singly reinforced rectangular section.

6. Draw stress strain diagram for doubly reinforced rectangular section.

7. Advantages of singly reinforced section over doubly reinforced section.

8. Why doubly reinforced beam is provided ?

9. Calculate the moment of resistance of a beam 230 mm wide 460 mm effective depth. Use M-20 and Fe-415. Also find area of tensile reinforcement.

10. Design the SRRS of simply supported beam over a clear span of 5m carrying udl of 20 kN/m exclusive of self weight. Assume bearing 250 mm.

11. Design an R.C.C. lintel over 2.70 m wide opening. The lintel carries 230 mm thick brick masonry wall. Take M-20 and Fe-415 weight of brick masonry wall 20,000 N/m^3.

12. Calculate the moment of resistance of a beam detailed as below :

    b = 250 mm, d = 525 mm, d' = 35 mm

    A_{st} = 5 – 20 mm ϕ

    A_{sc} = 3 – 20 mm ϕ

    Material = M-20 and steel Fe-415.

13. Design a R.C.C. beam to resist a bending moment of 260 kN.m. The beam size 230 mm × 650 mm (overall depth). Use M-20 and Fe-415.

◈ ◈ ◈

CHAPTER - 8

LIMIT STATE OF COLLAPSE – FLEXURE (FLANGED BEAM)

8.1 INTRODUCTION

- Beams are load bearing structural members usually of a uniform cross section. A beam of a rectangular cross section is called rectangular beam. The beam must be designed to withstand the loads that are expected. Loads creates stresses which are typically not uniform across the cross section of the beam.

- You could have compressive stresses on top and tensile stresses at the bottom of the beam. Since the stresses are more at the top and bottom, you need to have more material there. Hence you need have beams of cross section shaped like on 'T' or 'L'. Such beams are called flanged beams.

8.2 NECESSITY OF FLANGED SECTION

- In R.C.C. construction, when slabs and beams are cast together the construction becomes monolithic some portion of the slab above the beam becomes the part of the beam and helps in resisting compressive stresses and it also deflects with the deflection of beam under loads.

- Therefore, the effective section of beam becomes a flanged section having T and L shape. The slab portion of the section is called 'flange' and the portion below the slab is called 'web'.

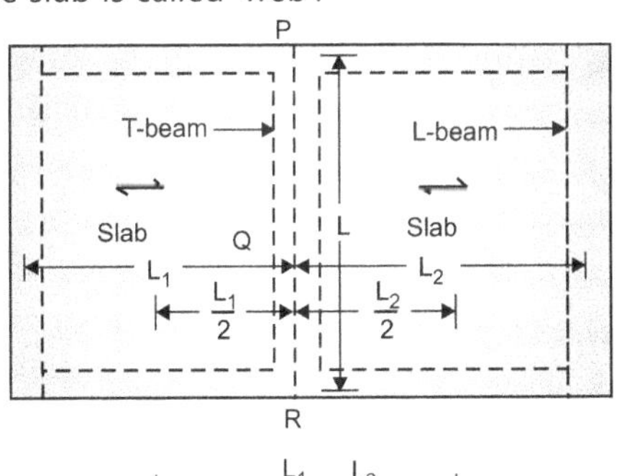

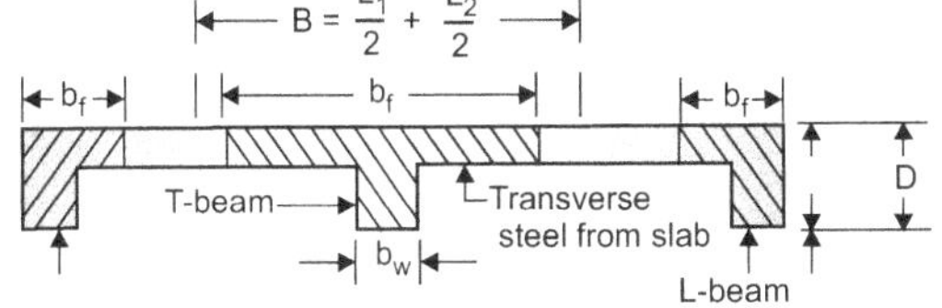

Fig. 8.1

- The intermediate beams having slab on both sides are known as T beams and the end beams having slab on one side only are known as L beams. The L-beams received loads from one side only and they are subjected to torsion also. If the torsion is less it can resisted by nominal shear reinforcement provided in the beam. If torsion is more the beam shall be designed for it.

8.3 CONDITION OF FORMATION OF FLANGED SECTION

(Refer clause 23.1.1, Pg. No. 36, IS 456 : 2000)

- The slab shall cast integrally with the web.

- If the main reinforcement of the slab is parallel to the beam, transverse reinforcement shall be provided d, such reinforcement shall not be less than 60% of the main reinforcement at mid span of the slab.

8.4 EFFECTIVE WIDTH OF FLANGE

(Refer clause 23.1.2 , Pg. No. 36, IS 456)

The effective width of flange may be taken as the following but in no case greater than the breadth of the web plus half the sum of the clear distance to the adjacent beams on either side.

(a) For T-beams,

$$b_f = \frac{l_o}{6} + bw + 6\,D_f$$

(b) For L-beams

$$b_f = \frac{l_o}{12} + bw + 3\,D_f$$

(c) For isolated beams, the effective flange width shall be obtained as below but in no case greater than the actual width

T-beam, $\quad b_f = \dfrac{l_o}{(l_o/b) + 4} + bw$

L-beam, $\quad b_f = \dfrac{0.5\,l_o}{(l_o/b) + 4} + bw$

where,

b_f = effective width of flange

l_o = distance between points of zero moments

b_w = breadth of web

D_f = Thickness of flange

b = actual width of flange

Note : For continuous beam 'l_o' may be taken as 0.7 times eff. span.

8.5 ADVANTAGES AND DISADVANTAGES OF FLANGED SECTION

Advantages :

- Since the beams casted monolithically with the slab, the flange also takes up the compressive stresses.
- Depth of beam is reduced so it gives better head room.
- For larger span, T-beams are usually preferred rather than rectangular beam as the deflection is reduced.

Disadvantages :

- Since the beam slab is monolithic (rigid), it becomes very weak in resisting lateral shear forces.
- There is a considerable increase in the shear stress at the junction of the flange and the web of the beam due to the change in cross section. So casting should be done very carefully.

8.6 STRAIN DIAGRAM AND STRESS BLOCK DIAGRAM FOR FLANGED SECTION

Case I : When ($x_u \leq DF$) N.A. lies with flange.

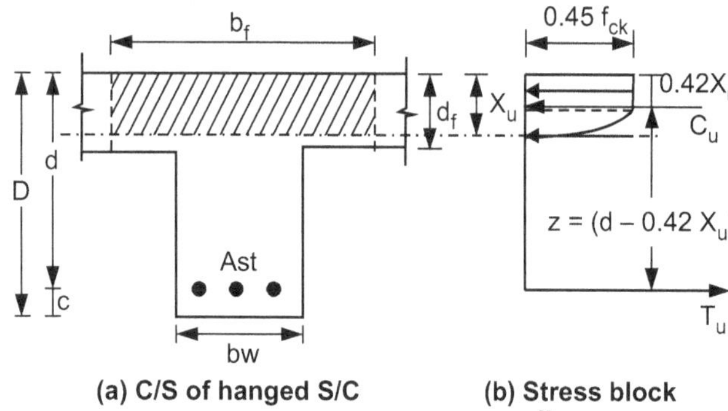

(a) C/S of hanged S/C　　　　**(b) Stress block diagram**

Fig. 8.2 : Stress-strain diagram for flanged s/c ($x_u \leq D_f$)

where,　　b_f = effective width of flange

　　　　b_w = width / breadth of web

　　　　D = overall depth

　　　　d = effective depth

　　　　D_f = thickness of flange / slab

　　　　X_u = depth of N.A.

　　　　T_u = tensile force carried by tensile r/f

　　　　C_u = compressive force carried by concrete (flange)

　　　　z = lever arm (distance between T_u and C_u)

1.　Depth of N.A. (x_u) :

Equating total tension and compression.

$$C_u = T_u$$

$$\therefore \quad x_u = \frac{0.87 \, f_y \cdot A_{st}}{0.36 \, f_{ck} \cdot b_f}$$

2.　Moment of Resistance (M_u)

(a)　For under-reinforced section

$$M_u = T_u \times z$$

$$\therefore \quad M_u = 0.87 \, f_y \cdot A_{st} \times (d - 0.42 \, x_u)$$

(b)　For balanced section / over-reinforced section

$$M_{umax} = 0.36 \, f_{ck} \cdot x_{umax} \cdot b_f \, (d - 0.42 \, x_{umax})$$

$$\therefore \quad M_{umax} = 0.148 \, f_{ck} \, bd^2 \, \dots \, Fe \, 250$$

$$\therefore \quad M_{umax} = 0.138 \, f_{ck} \, bd^2 \, \dots \, Fe \, 415$$

$$\therefore \quad M_{umax} = 0.133 \, f_{ck} \, bd^2 \, \dots \, Fe \, 500$$

(c)　Area of steel (A_{st})

$$A_{st} = \frac{0.5 \, f_{ck}}{f_y} \left[1 - \sqrt{1 - \frac{4.6 \, M_u}{f_{ck} \, b_f \cdot d^2}} \right] \times b_f d$$

Case II : When N.A. lies outside the flange ($x_u > D_f$)

(A) When $x_u > D_f$ and $D_f \leq 0.43 \, x_u$

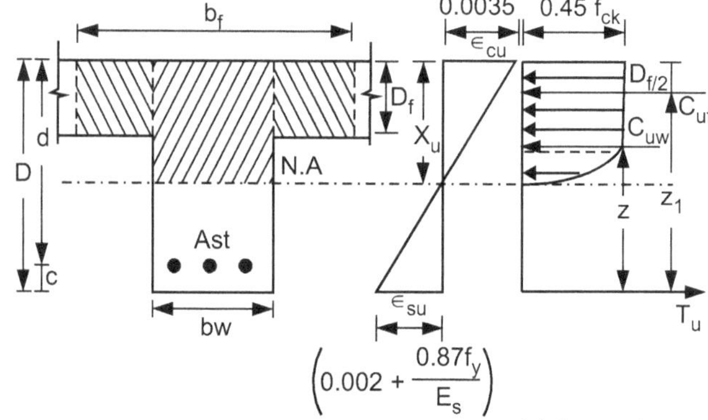

(a) C/S of beam　**(b) Strain diagram**　**(c) Stress block diagram**

Fig. 8.3 : Stress strain diagram for flanged s/c

($x_u > D_f : D_f \leq 0.43 \, x_u$)

where,　　C_{uw} = compressive force carried by web portion

　　　　C_{uf} = compressive force carried by flange portion

　　　　z = lever arm (distance bet. T_u and C_{uw})

　　　　z_1 = lever arm (distance bet. T_u and C_{uf})

1.　Depth of N.A (x_u)

Equating total tension and total compression

$$C_{uw} + C_{uf} = T_u$$

$$0.36 \, f_{ck} \cdot x_u \cdot bw + 0.45 \, f_{ck} \, (b_f - b_w) \times D_f = 0.87 \, f_y \cdot A_{st}$$

$$\therefore \quad x_u = \frac{0.87 \, f_y \cdot A_{st} - 0.45 \, f_{ck} \, (b_f - b_w) \times D_f}{0.36 \, f_{ck} \cdot b_w}$$

2.　Moment of Resistance (M_u)

Moment of resistance

$$M_u = C_{uw} \times z + C_{uf} \times z_1$$

$$\therefore \quad M_u = 0.36 \, f_{ck} \cdot x_u \cdot bw - (d - 0.42 \, x_u) + 0.45 \, f_{ck} \, (b_f - b_w) \, D_f \times (d - d')$$

3. Area of Steel (A_{st})

$$T_u = C_{uw} + C_{uf}$$

$$\therefore \quad A_{st} = \frac{0.36 f_{ck} \cdot x_u \cdot b_w + 0.45 f_{ck} (b_f - b_w) \cdot D_f}{0.87 f_y}$$

(B) When $x_u > D_f$ and $D_f > 0.43 x_u$

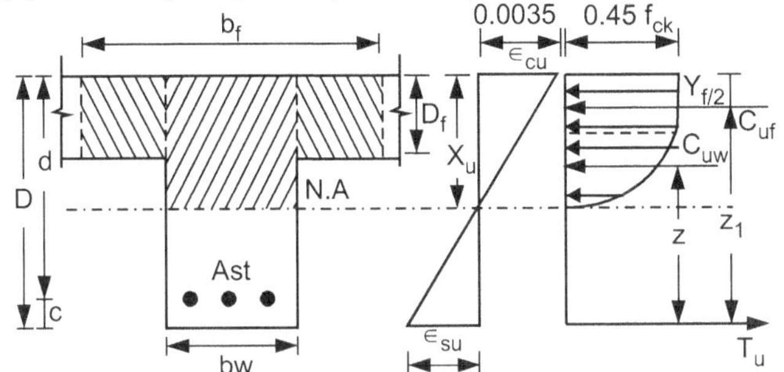

Fig. 8.4 : Stress-strain diagram for flanged section
($x_u > D_f$; $D_f > 0.43 x_u$)

where, z_2 = lever arm (distance bet. T_u and C_{uf})

$y_f = 0.15 x_u + 0.65 D_f$

1. Depth of N.A. (x_u) :

Equating total tension and total compression

$$T_u = C_{uw} + C_{uf}$$

$$\therefore \quad x_u = \frac{0.87 f_y \cdot A_{st} - 0.45 f_{ck} (b_f - b_w) \times y_f}{0.36 f_{ck} \cdot b_w}$$

where, $y_f = 0.15 x_u + 0.65 D_f$

2. Moment of Resistance (M_u)

$$M_u = C_{uw} \times z + C_{uf} \times z_2$$

$$\therefore \quad M_u = 0.36 f_{ck} \cdot x_u \cdot b_w \times (d - 0.42 x_u) + 0.45 f_{ck} (b_f - b_w) \cdot y_f$$

3. Area of steel (A_{st})

Equating T_u and C_u

$$T_u = C_{uw} + C_{uf}$$

$$\therefore \quad A_{st} = \frac{0.36 f_{ck} \cdot x_u \cdot b_w + 0.45 f_{ck} (b_f - b_w) \cdot y_f}{0.87 f_y}$$

```
                    Problems
                       |
         +-------------+-------------+
         |                           |
      Analysis                    Design
         |                           |
 (1) Calculation of moment of   (2) Design of flanged
     resistance (Mu)                section
```

SOLVED EXAMPLES

Type I : Calculation of Effective Flange Width

Example 8.1 : *Determine the flange width of simply supported beam of 4m thickness of rib is 230 mm and thickness of slab is 120 mm. The centre to centre distance between rib as 2.5 m*

Solution : Given data :

l_o = 4m, c/c distance between rib = 2.5 m

b_w = 230 mm

D_f = 120 mm

$\therefore$ Effective flange width (b_f)

$$\therefore \quad b_f = \frac{l_o}{6} + b_w + 6D_f$$

$$b_f = \frac{4000}{6} + 230 + 6 \times 120$$

$$b_f = 1616.67 \text{ mm}$$

$$\therefore \quad \mathbf{b_f = 1.616 \text{ m}} \not> 2.5 \text{ m}$$

Example 8.2 : *Calculate the effective flange width of a simply supported isolated T-beam of span 4 m. The actual width of flange is 1.6 m and width of web is 230 mm.*

Solution : Given data :

$$l_o = 4\text{m}$$
$$b = 1.6 \text{ m}$$
$$b_w = 230 \text{ mm}$$

$\therefore$ Effective flange width for isolated T-beam (b_f)

$$b_f = \frac{l_o}{(l_o/b) + 4} + b_w$$

$$= \frac{4000}{(4000/1600) + 4} + 230$$

$$\therefore \quad b_f = 845.38 \text{ mm}$$

$$\therefore \quad \mathbf{b_f = 0.845 \text{ m}} \not> 1.6 \text{ m}$$

Example 8.3 : *Determine the effective flange width for T-beam and L-beam for the following details :*

Hall size – 5m × 18 m

Spacing of web – 3m

Width of web – 225 mm

Thickness of flange – 100 mm

Solution : Given data,

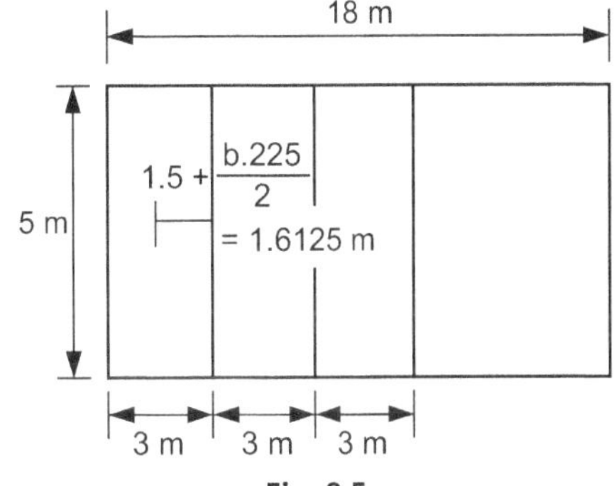

Fig. 8.5

hall size = 5m × 18 m

Spacing of web = 3m

Width of web = b_w = 225 mm

Thickness of flange = D_f = 100 mm

1. Effective flange width for T-beam (b_f)

$$b_f = \frac{l_o}{6} + b_w + 6\,D_f$$

$$b_f = \frac{3000}{6} + 225 + 6 \times 100$$

$$\mathbf{b_f = 1325\ mm}$$

2. Effective flange width for L-beam (b_f)

$$b_f = \frac{l_o}{12} + b_w + 3\,D_f$$

$$b_f = \frac{3000}{12} + 225 + 3 \times 100$$

$$b_f = 775\ mm$$

Type II : Calculation of Moment of Resistance

Example 8.4 : *A T-beam has a permissible effective flange width 1500 mm. Thickness of floor slab 120 mm and width of rib 300 mm, effective depth of section 500 mm is reinforced with 4-25 mm ϕ. Calculate the factored moment of resistance. Use M25 concrete and Fe415 steel grade.*

Solution : Given data :

b_f = 1500 mm, f_{ck} = 25 MPa , A_{st} = 4 – 25 mm ϕ

D_f = 120 mm, f_y = 415 MPa, $A_{st} = 4 \times \dfrac{\pi}{4}\,(25)^2$

b_w = 300 mm ,   d = 500 mm $\therefore$ A_{st} = 1963.50 mm^2

1. Calculation of actual depth of Neutral Axis (x_u)

(a) Assuming neutral axis lies within flange ($x_u < D_f$)

$$C_u = T_u$$

$$x_u = \frac{0.87\ f_y\ A_{st}}{0.36\ f_{ck}\ b_f} = \frac{0.87 \times 415 \times 1963.50}{0.36 \times 25 \times 1500}$$

$$\mathbf{x_u = 52.51\ mm} < D_f = 120\ mm$$

$$\text{…Assumption is correct}$$

2. Calculation of critical depth of Neutral axis (x_{umax})

$$x_{umax} = 0.48\ d = 0.48 \times 500$$

$$\mathbf{x_{umax} = 240\ mm}$$

3. Compare x_u and x_{umax}

$$52.51 < 240$$

$$x_u < x_{umax}$$

$$\text{…section is under-reinforced section}$$

4. Calculation of moment of resistance (M_u)

$$M_u = T_u \times z = 0.87\ f_y\ A_{st}\ (d - 0.42\ x_u)$$

$$M_u = 0.87 \times 415 \times 1963.50$$

$$(500 - 0.42 \times 52.51)$$

$$M_u = 338.83\ kN.m$$

Example 8.5 : *A singly reinforced T-beam has following sectional properties, estimate the flexural strength of the section using IS 456.*

Width of flange = 1250 mm

Thickness of flange = 160 mm

Thickness of rib = 250 mm

Effective depth = 750 mm

Area of tensile steel = 5000 mm^2

Material M20 and Fe 415

Solution : Given data ,

b_f = 1250 mm, f_{ck} = 20 MPa , A_{st} = 5000 mm^2

D_f = 160 mm, f_y = 415 MPa

d = 750 mm

b_w = 250 mm

1. Calculation of actual depth of netural axis (x_u)

(a) Assuming Neutral axis lies within flange ($x_u < D_f$)

$$C_u = T_u$$

$$x_u = \frac{0.87\ f_y\ A_{st}}{0.36\ f_{ck}\ b_f} = \frac{0.87 \times 415 \times 5000}{0.36 \times 20 \times 1250}$$

$$x_u = 200.58\ mm \not< D_f = 160\ mm$$

$$\text{…Assumption is wrong.}$$

(b) Assuming neutral axis lies outside the flange ($x_u > D_f$) and $D_f \leq 0.43\ x_u$

$$C_{uw} + C_{uf} = T_u$$

$$0.36\ f_{ck}\ x_u \cdot b_w + 0.45\ f_{ck}\ (b_f - b_w)\ D_f = 0.87\ f_y\ A_{st}$$

$$0.36 \times 20 \times x_u \times 250 + 0.45 \times 20\ (1250 - 250)\ 160$$

$$= 0.87 \times 415 \times 5000$$

$$\mathbf{x_u = 202.92\ mm} > D_f = 160\ mm$$

and　　　$D_f \leq 0.43\ x_u$

$$160 \not< 0.43 \times 202.92$$

$$160 \not< 87.25 \qquad \text{…Assumption is wrong.}$$

(c) Assuming ($x_u > D_f$ and $D_f > 0.43\ x_u$)

$$C_{uw} + C_{uf} = T_u$$

$$0.36\ f_{ck} \cdot x_u \cdot b_w + 0.45\ f_{ck}\ (b_f - b_w) \cdot (0.15\ x_u + 0.65\ D_f)$$

$$= 0.87\ f_y\ A_{st}$$

$$0.36 \times 20 \times x_u \times 250 + 0.45 \times 20\ (1250 - 250) \cdot$$

$$(0.15 \times x_u + 0.65 \times 160)$$

$$= 0.87 \times 415 \times 5000$$

$$\mathbf{x_u = 275.95\ mm} > D_f = 160\ mm$$

and　　　$D_f > 0.43\ x_u$

$$160 > 0.43 \times 275.95$$

$$160 > 118.66\ mm \qquad \text{…Assumption is correct}$$

2. Calculation of critical depth of neutral axis (x_{umax}) :

$$x_{umax} = 0.48\, d = 0.48 \times 750$$

$$\mathbf{x_{umax} = 360\ mm}$$

3. Compare x_u and x_{umax}

$$275.95 < 360$$

$$x_u < x_{umax} \qquad \text{...section is under-reinforced}$$

4. Calculation of ultimate flexural strength (M_u) :

$$M_u = C_{uw} \times z + C_{uf} \times z_2$$

$$M_u = 0.36\, f_{ck} \cdot x_u \cdot b_w\,(d - 0.42\, x_u)$$

$$\qquad + 0.45\, f_{ck}\,(b_f - b_w)\, y_f\,(d - y_{f/2})$$

$$\therefore \qquad y_f = 0.15\, x_u + 0.65\, D_f$$

$$y_f = 0.15 \times 275.95 + 0.65 \times 160$$

$$\mathbf{y_f = 145.39\ mm}$$

$$M_u = 0.36 \times 20 \times 275.95 \times 250\,(750 - 0.42$$

$$\times 275.95) + 0.45 \times 20\,(1250 - 250)$$

$$\times 145.39 \left(750 - \frac{145.39}{2}\right)$$

$$\therefore \qquad \mathbf{M_u = 1201.22\ kN.m}$$

Example 8.6 : *Calculate the moment of resistance of beam shown in Fig. 8.6 use M20 and Fe415 grade of concrete and steel.*

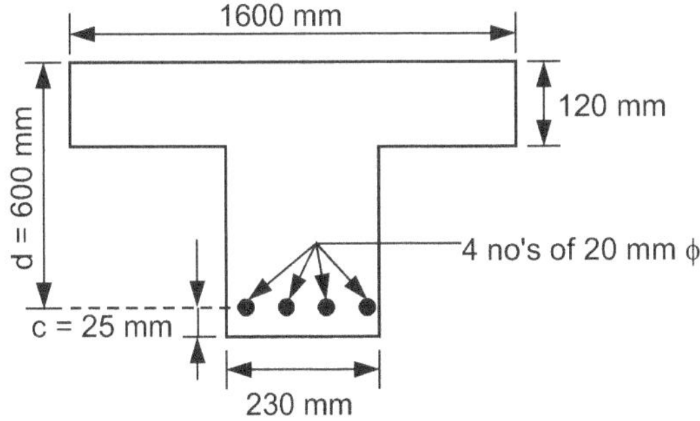

Fig. 8.6

Solution : Given data :

$$b_f = 1600\ mm,\ f_{ck} - 20\ MPa,\ A_{st} - 4 - 20\ mm\phi$$

$$D_f = 120\ mm,\ \ f_y = 415\ MPa,\ A_{st} = 4 \times \frac{\pi}{4}\,(20)^2$$

$$d = 600\ mm$$

$$c = d' = 25\ mm\,,\ A_{st} = 1256.64\ mm^2$$

$$b_w = 230\ mm$$

1. Calculation of actual depth of neutral axis (x_u)

(a) Assuming neutral axis lies within flange ($x_u < D_f$)

$$C_u = T_u$$

$$x_u = \frac{0.87\, f_y\, A_{st}}{0.36\, f_{ck}\, b_f} = \frac{0.87 \times 415 \times 1256.64}{0.36 \times 20 \times 1600}$$

$$\mathbf{x_u = 39.38}\ mm < D_f = 120\ mm$$

$$\text{...Assumption is correct.}$$

2. Calculation of critical depth of neutral axis (x_{umax})

$$x_{umax} = 0.48\, d = 0.48 \times 600$$

$$\therefore \qquad \mathbf{x_{umax} = 288\ mm}$$

3. Compare x_u and x_{umax}

$$39.38 < 288$$

$$x_u < x_{umax}$$

4. Calculation of ultimate moment of resistance (M_u) :

$$M_u = T_u \times z$$

$$M_u = 0.87\, f_y\, A_{st} \times (d - 0.42\, x_u)$$

$$M_u = 0.87 \times 415 \times 1256.64$$

$$(600 - 0.42 \times 39.38)$$

$$\therefore \qquad \mathbf{M_u = 264.72\ kN.m}$$

Example 8.7 : *A T-beam of flange width 1200 mm flange thickness 100 mm rib width 275 mm has an effective depth of 550 mm and is reinforced with 6-25 mm ϕ and 4-16 mm ϕ. Find the ultimate moment of resistance. Use M20 concrete and Fe 415 steel.*

Solution : Given data

$$b_f = 1200mm,\ f_{ck} = 20\ MPa,$$

$$A_{st} = 6 \times \frac{\pi}{4}\,(25)^2 + 4 \times \frac{\pi}{4}\,(16)^2$$

$$D_f = 100\ mm,\ \ f_y = 415\ MPa,\ b_w = 275\ mm,$$

$$A_{st} = 3749.44\ mm^2,\ d = 550\ mm$$

1. Calculation of actual depth of neutral axis (x_u)

(a) **Assuming neutral axis lies within flange ($x_u < D_f$)**

$$x_u = \frac{0.87\, f_y\, A_{st}}{0.36\, f_{ck} \cdot b_f} = \frac{0.87 \times 415 \times 3749.49}{0.36 \times 20 \times 1200}$$

$$x_u = 156.68\ mm \nless D_f = 100\ mm$$

$$\text{...Assumption is wrong.}$$

(b) **Assuming neutral axis lies outside the flange**

$$(x_u > D_f\ \text{and}\ D_f \le 0.43\, x_4)$$

$$C_{uw} + C_{uf} = T_u$$

$$0.36\, f_{ck}\, x_u\, b_w + 0.45\, f_{ck}\,(b_f - b_w)\, D_f = 0.87\, f_y\, A_{st}$$

$$0.36 \times 20 \times x_u \times 275 + 0.45 \times 20\,(1200 - 275) \times 100$$

$$= 0.87 \times 415 \times 3749.49$$

$$\therefore \qquad \mathbf{x_u = 263.26\ mm} > D_f = 100\ mm$$

$$\text{and} \qquad D_f \le 0.43\, x_u$$

$$100 \le 0.43 \times 263.25$$

$$100 \le 113.20 \qquad \text{...assumption is correct}$$

2. Calculation of critical depth of neutral axis (x_{umax})

$$x_{umax} = 0.48\, d = 0.48 \times 550$$

$$\mathbf{x_{umax} = 264\ mm}$$

3. Compare x_u and x_{umax}

$$263.25 \; < \; 264$$

$$x_u \; < \; x_{umax}$$

…section is under-reinforced section

4. Calculation of ultimate flexural strength (M_u)

$$M_u \; = \; C_{uw} \times z + C_{uf} \times z_1$$

$$M_u \; = \; 0.36 \, f_{ck} \, x_u \, b_w \cdot (d - 0.42 \, x_u)$$
$$+ \; 0.45 \, f_{ck} \, (b_f - b_w) \, D_f \, (d - D_f/2)$$

$$M_u \; = \; 0.36 \times 20 \times 263.25 \times 275 \, (550 - 0.42$$
$$\times \, 263.25) + 0.45 \times 20 \, (1200 - 275)$$
$$\times \, 100 \times (550 - 100/2)$$

$$\therefore \qquad M_u \; = \; 645.30 \text{ kN.m}$$

Example 8.8 : *Find M_u of a T beam with the following data, $b_f = 1500$ mm, $b_w = 300$ mm, $d = 600$ mm, $D_f = 100$ mm, $A_{st} = 4500$ mm^2. Concrete M20 and steel Fe 415.*

Solution :

$$A_{st} \; = \; \frac{0.36 \, f_{ck} b_f D_f}{0.87 \, f_y}$$

$$x_u \; = \; ?,$$

$$0.36 \, f_{ck} \, x_u \, b_w + 0.446 \, f_{ck} \, D_f \, (b_f - b_w) \; = \; 0.87 \, f_y A_{st} \quad …A_{st} > A_{st}$$

$$M_u \; = \; M_{u,\,web} + M_u \text{ flange}$$

$$…A_{st} > A_{st} \text{ and } x_u > D_f, \; D_f \leq 0.43 \, x_u$$

$$= \; 0.36 \, f_{ct} \, x_u \, b_w \, (d - 0.416 \, x_u) + 0.446 \, f_{ck} \, D_f$$
$$(b_f - b_w) \, (d - D_f/2)$$

To find approximate area of steel required for flange :

$$A_{sf} \; = \; \frac{0.36 \, f_{ck} D_f b_f}{0.87 \, f_y}$$

$$= \; \frac{0.36 \times 20 \times 100 \times 1500}{0.87 \times 415}$$

$$= \; 2991.275 \text{ mm}$$

$$\therefore \; A_{st} > A_{sf} \; ; \text{ as such } x_u > D_f$$

To find x_u :

$$C_{uw} + C_{uf} \; = \; T_u$$

$$0.36 \, f_{ck} \, x_u \, b_w + 0.446 \, f_{ck} \, D_f \, (b_f - b_w) \; = \; 0.87 \, f_y \, A_{st}$$

$$0.36 \times 20 \times x_u \times 300 + 0.446 \times 20 \times 100 \, (1500 - 300)$$
$$= \; 0.87 \times 415 \times 4500$$

$$\therefore \qquad x_u \; = \; 256.63 \text{ mm}$$

$$x_{u,\,max} \; = \; 0.479 \, d$$

$$= \; 0.479 \times 600 = 316.96 \text{ mm}$$

$$\therefore \qquad x_{u,\,max} \; > \; x_u$$

This is all right.

$$0.43 \, x_u \; = \; 0.43 \times 256.63$$

$$= \; 110.35 > 100 \text{ mm}$$

$$\therefore \qquad D_f \; < \; 0.43 \, x_u$$

Hence, this is a case II problem. Stress in the flange is uniform.

To find M_u :

$$M_u \; = \; M_{u,\,web} + M_{u,\,flange}$$

$$= \; 0.36 \, f_{ck} \, x_u \, b_w \, (d - 0.416 \, x_u) + 0.446 \, f_{ck}$$
$$D_f \, (b_f - b_w) \, (d - D_f/2)$$

$$= \; 0.36 \times 20 \times 256.63 \times 300 \, (600 - 0.416$$
$$\times \, 256.63) + 0.446 \times 20 \times 100$$
$$(1500 - 300) \, (600 - 100/2)$$

$$= \; 273.41 \times 10^6 + 588.72 \times 10^6$$

$$= \; 862.12 \times 10^6 \text{ N–mm} = 862.13 \text{ kN-m}$$

Example 8.9 : *Find M_u of a T beam with the following data, $b_f = 1500$ mm, $b_w = 300$ mm, $d = 700$ mm, $D_f = 100$ mm, $A_{st} = 4510$ mm^2. Concrete M15 and $f_y = 250$ N/mm^2.*

Solution :

$$A_{sf} \; = \; \frac{0.36 \, f_{ck} b_f D_f}{0.87 f_y}$$

$$x_u \; = \; ?,$$

$$0.36 \, f_{ck} \, x_u \, b_w + 0.446 \, f_{ck} \, D_f \, (b_f - b_w) = 0.87 \, f_y A_{st}$$

$$…A_{st} > A_{st}$$

$$M_u \; = \; 0.36 \, f_{ck} x_u b_w \, (d - 0.416 \, x_u)$$
$$+ \; 446 \, f_{ck} Y_f \, (b_f - b_w) \left(d - \frac{Y_f}{2} \right)$$

$$…A_{st} > A_{sf} \text{ and } x_u > D_f \text{ and } D_f > 0.43 \, x_u$$

$$Y_f \; = \; (0.15 \, x_u + 0.65 \, D_f)$$

To find approximate area of steel required for flange :

$$A_{sf} \; = \; \frac{0.36 \, f_{ck} D_f b_f}{0.87 \, f_y}$$

$$= \; \frac{0.36 \times 15 \times 100 \times 1500}{0.87 \times 250} = 3724.13 \text{ mm}^2$$

$$\therefore \quad A_{st} > A_{sf} \text{ and } x_u > D_f$$

To find x_u :

$$0.36 \, f_{ck} x_u \, b_w + 0.446 \, f_{ck} D_f \, (b_f - b_w) = 0.87 \, f_y \, A_{st}$$

$$0.36 \times 15 \times x_u \times 300 + 0.446 \times 15 \times 100 \, (1500 - 300)$$
$$= \; 0.87 \times 250 \times 4510$$

$$\therefore \qquad x_u \; = \; 109.98 \text{ or } 110 \text{ mm}$$

$$\therefore \qquad x_{u,\,max} \; = \; 0.531 \, d$$

$$= \; 0.531 \times 700 = 371.7 \text{ mm}$$

$$x_{u,\,max} \; > \; x_u \quad \text{OK}$$

$$0.43 \, x_u \; = \; 0.43 \times 110 = 47.3 < 100 \text{ mm}$$

$$\therefore \qquad D_f \; < \; 0.43 \, x_u$$

and stress in the flange is not uniform.

$$Y \; = \; (0.15 \, x_u + 0.65 \, D_f)$$

$$= \; (0.15 \times 110 + 0.65 \times 100) = 81.5 \text{ mm}$$

To find M_u

$$\begin{aligned}
M_u &= M_{u,\,web} + M_{u,\,flange} \\
&= 0.36\, f_{ck}\, x_u\, b_w\, (d - 0.416\, x_u) + 0.446\, f_{ck}\, Y_f \\
&\quad (b_f - b_w)\, (d - Y_f/2) \\
&= 0.36 \times 15 \times 110 \times 300\, (700 - 0.416 \\
&\quad \times 110) + 0.446 \times 15 \times 81.5\, (1500 - 300) \\
&\quad (700 - 81.5/2) \\
&= 116.59 \times 10^6 + 431.34 \times 10^6 \\
&= 547.93 \times 10^6 \text{ N-mm} \\
&= 547.93 \text{ kN-m}
\end{aligned}$$

Example 8.10 : *Find M_u of a T beam with the following data, b_f = 1250 mm, D_f = 100 mm, b_w = 250 mm, d = 650 mm, A_{st} = 2800 mm^2, f_y = 410 N/mm^2, f_{ck} = 20 N/mm^2.*

Solution :

$$A_{sf} = \frac{0.36\, f_{ck} b_f D_f}{0.87\, f_y}$$

$$x_u = ?,$$

$$0.36\, f_{ck}\, x_u\, b_w + 0.446\, f_{ck}\, D_f\, (b_f - b_w) = 0.87\, f_y A_{st}$$

$$\dots A_{st} > A_{st}$$

$$M_u = 0.36\, f_{ck}\, x_u\, b_w\, (d - 0.416\, x_u) + 0.446\, f_{ck}\, Y_f$$

$$(b_f - b_w)\left(d - \frac{Y_f}{2}\right)$$

$$Y_f = (0.15\, x_u + 0.65\, D_f)$$

To find approximate area of steel required for flange :

$$A_{sf} = \frac{0.36\, f_{ck} D_f b_f}{0.87 f_y}$$

$$= \frac{0.36 \times 20 \times 1250 \times 100}{0.87 \times 415}$$

$$= 2492.73 \text{ mm}^2 < 2800 \text{ mm}^2$$

$\therefore \qquad A_{st} > A_{sf}\,;$

$\therefore \qquad x_u > D_f$

To find x_u :

$$0.36\, f_{ck}\, x_u\, b_w + 0.446\, f_{ck}\, D_f\, (b_f - b_w) = 0.87\, f_y\, A_{st}$$

$$0.36 \times 20 \times x_u \times 250 + 0.446 \times 20 \times 100\, (1250 - 250)$$

$$= 0.87 \times 415 \times 2800$$

$\therefore \qquad x_u = 66.07$ mm

$\qquad\qquad x_u = \; < D_f$

This shows that x_u is slightly less than D_f, the stress in the flange is not uniform, as such D_f is to be replaced by Y_f.

$$Y_f = 0.15\, x_u + 0.65\, D_f$$

$$= 0.15\, x_u + (0.65 \times 100)$$

$\therefore \quad 0.36 \times 20 \times x_u \times 250 + 0.446 \times 20 \times (0.15\, x_u + 0.65$

$\qquad \times 100)\, (1250 - 250)$

$$= 0.87 \times 415 \times 2800$$

$\therefore \qquad x_u = 104.19$ mm

$$x_u > D_f$$

$$x_{u,\,max} = 0.479\, d$$

$$= 0.479 \times 650 = 311.35 \text{ mm} > x_u$$

$$0.43\, x_u = 0.43 \times 104.19 = 44.80 < 100$$

$\therefore \qquad D_f > 0.43\, x_u$

Hence $\qquad Y_f = 0.15\, x_u + 0.65\, D_f$

$$= 0.15 \times 104.19 + 0.65 \times 100$$

$$= 80.63 \text{ mm}$$

To find M_u,

$$\begin{aligned}
M_u &= M_{u,\,web} + M_{u,\,flange} \\
&= 0.36\, f_{ck}\, x_u\, b_w\, (d - 0.416\, x_u) + 0.446\, f_{ck}\, Y_f \\
&\quad (b_f - b_w)\, (d - Y_f/2) \\
&= 0.36 \times 20 \times 104.19 \times 250 \\
&\quad (650 - 0.416 \times 104.19) \\
&\quad + 0.446 \times 20 \times 80.63\, (1250 - 250) \\
&\quad \left(650 - \frac{80.63}{2}\right) \\
&= 113.77 \times 10^6 + 438.5 \times 10^6 \\
&= 552.27 \times 10^6 \text{ N-mm} \\
&= 552.27 \text{ kN-m}
\end{aligned}$$

Example 8.11 : *Calculate the moment of resistance by LSM for T-beam section detailed as below :*

   (i)   *Width of rib = 300 mm*

   (ii) *Effective flange width = 1200 mm*

   (iii) *Thickness of flange = 120 mm*

   (iv) *Effective depth = 565 mm*

   (v) *Tension steel = 4 no 25 mm diameter*

   (vi) *M25 grade of concrete and Fe 500 grade of steel.*

Solution :

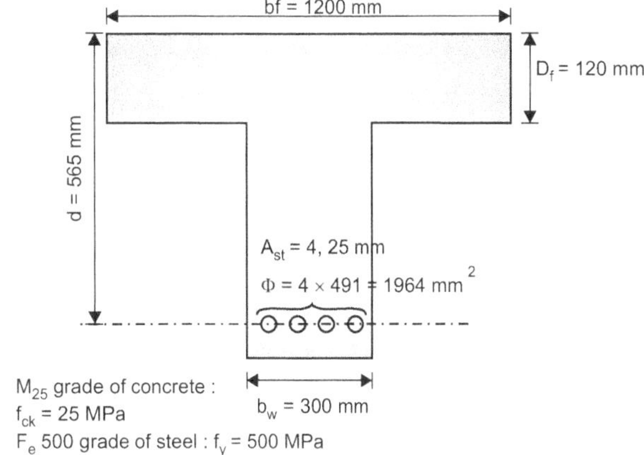

Fig. 8.7

Fe 500 grade of steel $\therefore f_y$ = 500 MPa

1.   Calculation of actual depth of N.A.

Let us assume N.A. lies in flange : $x_{u(act)} < D_f$

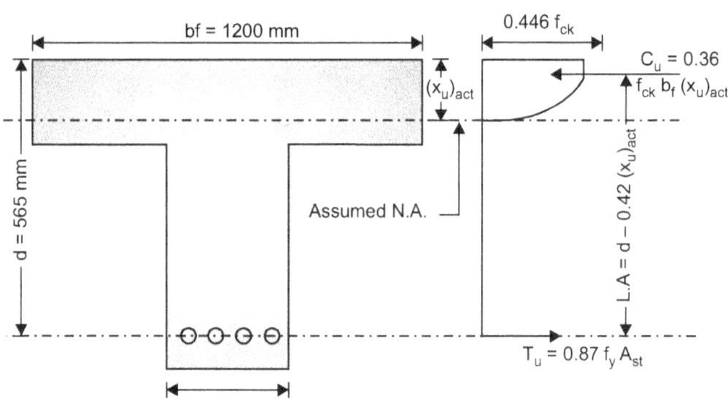

Fig. 8.8

2. From stress diagram;

For equilibrium

$$C_u = T_u$$

$$0.36\, f_{ck}\, b_f\, (x_u)_{act} = 0.87\, f_y\, A_{st}$$

$$(x_u)_{act} = \frac{0.87\, f_y\, A_{st}}{0.36\, f_{ck}\, b_f} = \frac{0.87 \times 500 \times 14 \times 491}{0.36 \times 25 \times 1200}$$

$$= 79.10 \text{ mm}$$

3. Validating assumption,

As $(x_u)_{act} < D_f$ ∴ Assumption is correct.

4. Calculation of $(x_u)_{bal}$

$$(x_u)_{bal} = (k_u)_{bal} \cdot d$$

where,

$$(k_u)_{bal} = \text{Design constant for N.A.}$$

$$= \frac{700}{1100 + 0.87 \times 500}$$

$$= 0.46$$

$$\therefore \quad (x_u)_{bal} = 0.46 \times 565 = 257.65 \text{ mm}$$

5. Classification of section :

As $(x_u)_{act} < (x_u)_{bal}$ the flanged section is under reinforced.

6. Moment of Resistance :

$$M_u = C_u \times \text{L.A. or } T_u \cdot L_A$$

$$M_u = 0.36 \cdot f_{ck} \cdot b_f\, (x_u)_{act} \times [d - 0.42\, (x_u)_{act}] \text{ or}$$

$$\quad 0.87 \cdot f_y \cdot A_{st}\, [d - 0.42\, (x_u)_{act}]$$

$$= 0.36 \times 25 \times 1200 \times 79.10 \times [565 - 0.42 \times 79.10] \text{ or } 0.87 \times 500 \times 1964$$

$$\times [565 - 0.42 \times 79.10]$$

$$= 454.32 \times 10^6 \text{ N.mm}$$

$$= 454.32 \text{ kNm}$$

Example 8.12 : Calculate the moment of resistance by LSM for flanged section as below details :

(a) *Width of rib = 230 mm,*

(b) *Effective flange width = 1400 mm*

(c) *Thickness of flange = 125 mm*

(d) *Effective depth = 565 mm*

(e) *Tension steel = 2 – # 20 through plus 2 - # 16 curtail at midspan.*

(f) *Use M20 grade of concrete and Fe 415 grade of steel.*   **[Nov. 2016, 8M]**

Solution : Given

$$b_w = 230 \text{ mm}$$
$$b_f = 1400 \text{ mm}$$
$$D_f = 125 \text{ mm}$$
$$d = 565 \text{ mm}$$
$$A_{st} = 2 \times \pi/4 \times 20^2 = 628 \text{ mm}^2$$

M20 and Fe 415

Approximate area of steel required for flange

$$A_{sf} = \frac{0.36\, f_{ck}\, D_f \cdot b_f}{0.87\, f_y}$$

$$= \frac{0.36 \times 20 \times 125 \times 1400}{0.87 \times 415}$$

$$= 3489.82 \text{ mm}^2$$

$$\therefore \qquad A_{st} < A_{sf}$$

$$x_u = \frac{0.87\, f_y\, A_{st}}{0.36\, f_{ck}\, b_f}$$

$$= \frac{0.87 \times 415 \times 628}{0.36 \times 20 \times 1400} = 22.49 \text{ mm}$$

$$M_u = 0.87\, f_y\, A_{st}\, (d - 0.416\, x_u)$$

$$= 0.87 \times 415 \times 628\, (565 - 0.416 \times 22.49)$$

$$= 125986613.9 \text{ N.mm}$$

$$= 125.966 \text{ kN.m}$$

Type III : Design of Flanged Section

Example 8.13 : *Design the reinforcement for the T-beam for the following data :*

Width of flange = 900 mm

Breadth of beam = 230 mm

Thickness of flange = 100 mm

Effective depth = 500 mm

Applied factored moment = 250 kN-m

Use M20 and Fe 415.

Solution : Given data ;

$b_f = 900$ mm, $M_u = 250$ kN-m

$D_f = 100$ mm, $f_{ck} = 20$ MPa

$b_w = 230$ mm , $f_y = 415$ MPa

$d = 500$ mm

1. Calculation of actual depth of neutral axis (x_u)

(a) Assuming neutral axis lies within flange ($x_u < D_f$)

$$M_u = C_u \times z$$
$$M_u = 0.36\, f_{ck}\, x_u \cdot b_f\, (d - 0.42\, x_u)$$
$$250 \times 10^6 = 0.36 \times 20 \times x_u \times 900 \times (500 - 0.42 \times x_u)$$
$$\therefore \quad x_u = 82.94 \text{ mm} < D_f = 100 \text{ mm}$$
$$\dots \text{Assumption is correct}$$

2.　Calculation of critical depth of neutral axis (x_{umax})

$$x_{umax} = 0.48\, d = 0.48 \times 500$$
$$\mathbf{x_{umax} = 240 \text{ mm}}$$

3.　Compare x_u and x_{umax}

$$82.94 < 240$$
$$x_u < x_{umax}$$
$$\dots \text{section is under-reinforced section}$$

4.　Calculation of area of steel required (A_{st})

$$A_{st} = \frac{0.5\, f_{ck}}{f_y}\left[1 - \sqrt{1 - \frac{4.6\, M_u}{f_{ck} \cdot b_w d^2}}\right] b_w d$$

$$A_{st} = \frac{0.5 \times 20}{415}\left[1 - \sqrt{1 - \frac{4.6 \times 250 \times 10^6}{20 \times 230 \times 500^2}}\right]$$
$$\times 230 \times 500$$

$$A_{st} = 2771.08 \text{ mm}^2$$

Assume, diameter of bar = 20 mmϕ

$$\text{No. of bar} = \frac{A_{st}}{\pi/4\,(d)^2}$$
$$= \frac{2771.08}{\pi/4\,(20)^2} = 8.82$$

No. of bars $\cong$ 9 Nos.

∴　Provide 9-20 mm ϕ on tensile reinforcement.

Example 8.14 : *A T-beam has a flange width of 1200 mm, flange thickness of 100 mm. A rib width 250 mm and effective depth of 400 mm. Find the reinforcement required. For the balanced section limiting moment of resistance. Use M20 concrete and Fe 416 steel.*

Solution : Given data ,

b_f = 1200 mm,　f_{ck} = 20 MPa

D_f = 100 mm,　　f_y = 415 MPa

b_w = 250 mm,　　d = 400 mm

1.　Calculation of critical depth of neutral axis for balanced section (x_{umax})

$$x_{umax} = 0.48\, d = 0.48 \times 400$$
$$\mathbf{x_{umax} = 192 \text{ mm}}$$

(a)　Assuming N.A. lies within flange ($x_{umax} < D_f$)

$$x_{umax} = 192 \text{ mm} \nless D_f \quad \dots \text{assumption is wrong}$$

(b)　Assuming N.A. lies outside the flange

($x_{umax} > D_f$ and $D_f \leq 0.43\, x_{umax}$)

$$\therefore \quad x_{umax} = 192 > D_f$$

and
$$D_f \leq 0.43\, x_{umax}$$
$$100 \leq 0.43 \times 192$$
$$100 \leq 82.56 \qquad \dots \text{assumption is wrong}$$

(c)　Assuming ($x_{umax} > D_f$ and $D_f > 0.43\, x_{umax}$)

$$x_{umax} = 192 \text{ mm} > D_f$$

and
$$D_f > 0.43\, x_{umax}$$
$$100 > 0.43 \times 192$$
$$100 > 82.56 \qquad \dots \text{assumption is correct}$$

2.　Calculation of area of steel (A_{st})

$$C_{uw} + C_{uf} = T_u$$
$$0.36\, f_{ck} \cdot x_{umax} \cdot b_w + 0.45\, f_{ck}\,(b_f - b_w)\, y_f = 0.87\, f_y A_{st}$$
$$\therefore \quad 0.36 \times 20 \times 192 \times 250 + 0.45 \times 20\,(1200 - 250)$$
$$(0.15\, x u_{max} + 0.65\, D_f) = 0.87 \times 415 \times A_{st}$$
$$\therefore \quad 0.36 \times 20 \times 192 \times 250 + 0.45 \times 20\,(1200 - 250)$$
$$(0.15 \times 192 + 0.65 \times 100) = 0.87 \times 415 \times A_{st}$$
$$\therefore \quad \mathbf{A_{st} = 3178.48 \text{ mm}^2}$$

Assume diameter of bar = 20 mmϕ

$$\text{No. of bar} = \frac{A_{st}}{\pi/4\,(d)^2}$$
$$\text{No. of bar} = \frac{3178.48}{\pi/4\,(20)^2} = 10.11$$

No. of bars $\cong$ 11 Nos.

∴　Provide 11-20 mm ϕ as a tensile reinforcement.

3.　Calculation of ultimate flexural strength (M_u)

$$M_u = C_{uw} \times z + C_{uf} \times z_2$$
$$M_{umax} = 0.36\, f_{ck} \cdot x_{umax} \cdot b_w\,(d - 0.42\, x_{u\,max})$$
$$+ 0.45\, f_{ck}\,(b_f - b_w)\, y_f\,(d - y_f/2)$$
$$y_f = 0.15\, x_u + 0.65\, D_f$$
$$= 0.15 \times 192 + 0.65 \times 100$$
$$y_f = 93.8 \text{ mm}$$
$$M_{umax} = 0.36 \times 20 \times 192 \times 250\,(400 - 0.42 \times 192) + 0.45 \times 20\,(1200 - 250)\,93.8$$
$$(400 - 93.8/2)$$
$$\therefore \quad M_{umax} = 393.55 \text{ kN.m}$$

Example 8.15 : *Calculate the area of reinforcement for a T-beam of flange width 1000 mm, flange thickness 100 mm, width of rib 230 mm, effective depth 550 mm to resist an ultimate bending moment of 310 kNm. Use M25 and Fe 415.*

Solution : Given data

b_f = 1000 mm, f_{ck} = 25 MPa, M_u = 310 kN-m

D_f = 100 mm,　f_y = 415 MPa

b_w = 230 mm,　d = 550 mm

1.　Calculation of actual depth of neutral axis (x_u)

(a)　Assuming neutral axis lies within flange (i.e. $x_u < D_f$)

$$M_u = C_u \times z$$
$$310 \times 10^6 = 0.36\, f_{ck}\, x_u\, b_f \cdot (d - 0.42\, x_4)$$
$$310 \times 10_6 = 0.36 \times 25 \times x_4 \times 1000\, (550 - 0.42\, x_4)$$
$$\therefore \quad \mathbf{x_u = 65.94\ mm} < D_f = 100\ mm$$

…Assumption is correct.

2. **Calculation of critical depth of neutral axis (x_{umax})**

$$x_{umax} = 0.48\, d = 0.48 \times 550$$
$$\therefore \quad x_{umax} = 264\ mm$$

3. **Compare x_u and x_{umax}**

$$65.94 < 264$$
$$x_u < x_{umax}$$

…section is under-reinforced section

4. **Calculation area of steel (A_{st})**

$$A_{st} = \frac{0.5\, f_{ck}}{f_y} \left[1 - \sqrt{\frac{4.6\, M_u}{f_{ck} \cdot b_f \cdot d^2}} \right] b_f \cdot d$$

$$A_{st} = \frac{0.5 \times 25}{415} \left[1 - \sqrt{1 - \frac{4.6 \times 310 \times 10^6}{25 \times 1000 \times 550^2}} \right]$$
$$\times 1000 \times 550$$

$$\therefore \quad \mathbf{A_{st} = 1643.40\ mm^2}$$

Assume diameter of bar = ϕ = 18 mm

$$\text{No. of bars} = \frac{A_{st}}{a_{st}} = \frac{1643.40}{\pi/4\,(18)^2}$$

$$\text{No. of bars} = 6.46$$

$\therefore$ No. of bars $\cong$ 7 No's

$\therefore$ **Provide 7-18 mm ϕ as a tensile reinforcement.**

Example 8.16 : *A T-beam of flange width 1200 mm, flange thickness 100 mm width of rib 275 mm as an effective depth of 600 mm. Use M20 and Fe 500 grades of material. Calculate the safe moment of resistance if the area of steel is 6-25 mm ϕ.*

Solution : Given data ,

$$b_f = 1200\ mm, \quad f_{ck} = 20\ MPa, \quad A_{st} = 6 - 25\ mm\phi$$

$$D_f = 100\ mm, \quad f_y = 500\ MPa, \quad A_{st} = 6 \times \frac{\pi}{4}\,(25)^2$$

$$b_w = 275\ mm, \quad d = 600\ mm, \quad A_{st} = 2945.24\ mm^2$$

1. **Calculation of actual depth of neutral axis (x_u)**

(a) **Assuming N.A. lies within flange ($x_u < D_f$)**

$$C_u = T_u$$
$$x_u = \frac{0.87\, f_y\, A_{st}}{0.36\, f_{ck}\, b_f} = \frac{0.87 \times 500 \times 2945.24}{0.36 \times 20 \times 1200}$$
$$x_u = 148.28\ mm \not< D_f = 100\ mm$$

…Assumption is wrong

(b) **Assuming N.A. lies outside the flange ($x_u > D_f$ and $D_f \leq 0.43\, x_u$)**

$$C_{uw} + C_{uf} = T_u$$
$$0.36\, f_{ck}\, x_u \cdot b_w + 0.45\, f_{ck}\, (b_f - b_w)\, D_f = 0.87\, f_y\, A_{st}$$

$$0.36 \times 20 \times x_u \times 275 + 0.45 \times 20\,(1200 - 275) \times 100$$
$$= 0.87 \times 500 \times 2945.24$$
$$\therefore \quad x_u = 226.60\ mm > D_f = 100\ mm$$
and $\qquad D_f \leq 0.43\, x_u$
$$100 \leq 0.43 \times 226.60$$
$$100 \leq 97.44 \qquad \text{…assumption is wrong}$$

(c) **Assuming ($x_u > D_f$ and $D_f > 0.43\, x_u$)**

$$C_{uw} + C_{uf} = T_u$$
$$0.36\, f_{ck} \cdot x_u \cdot b_w + 0.45\, f_{ck}\, (b_f - b_w)\, y_f = 0.87\, f_y \cdot A_{st}$$
$$\therefore \quad y_f = 0.15\, x_u + 0.65\, D_f$$
$$\therefore \quad 0.36 \times 20 \times x_u \times 275 + 0.45 \times 20\,(1200 - 275)$$
$$(0.15\, x_u + 0.65 \times 100) = 0.87 \times 500 \times 2945.24$$
$$\therefore \quad \mathbf{x_u = 229.21\ mm} > D_f = 100\ mm$$
and $\qquad D_f > 0.43 \times x_u$
$$100 > 0.43 \times 229.21$$
$$100 > 98.56 \qquad \text{…assumption is correct}$$

2. **Calculation of critical depth of neutral axis ($x_{u\,max}$)**

$$x_{u\,max} = 0.46 \times d = 0.46 \times 600$$
$$\mathbf{x_{u\,max} = 276\ mm}$$

3. **Compare x_u and $x_{u\,max}$**

$$229.21 < 276$$
$$x_u < x_{u\,max}$$

…section is under-reinforced section

4. **Calculation of safe moment of resistance (M)**

$$M_u = C_{uw} \cdot z + C_{uf} + z_2$$
$$M_u = 0.36\, f_{ck}\, x_u\, b_w\, (d - 0.42\, x_u) + 0.45\, f_{ck}$$
$$(b_f - b_w)\, y_f\, (d - y_{f/2})$$
$$\therefore \quad y_f = 0.15 \times 229.21 + 0.65 \times 100$$
$$\mathbf{y_f = 99.38\ mm}$$
$$M_u = 0.36 \times 20 \times 229.21 \times 275\,(600 - 0.42$$
$$\times 229.21) + 0.45 \times 20\,(1200 - 275)$$
$$\times 99.38 \times \left(600 - \frac{99.38}{2}\right)$$
$$M_u = 683.90\ kN/m$$

$\therefore$ Safe moment of resistance $M = \dfrac{M_u}{r_f}$

$$\therefore \quad M = \frac{683.90}{1.5}$$

$$\therefore \quad \mathbf{M = 455.94\ kN.m}$$

Example 8.17 : *A hall of building has an internal dimension as 5m $\times$ 24 m a floor slab of 125 mm thick is cast monolithically with beam the floor slab is subjected to floor finish of 1 kN/m^2 and live load of 3 kN/m^2. The beams are spaced at 4m centre to centre having clear span 5m. Assume* *beam is simply supported on 300 mm thick brick wall. Design the intermediate beam and also design the beam for shear reinforcement and take all necessary check. Use M20 and Fe 415 grade. Show reinforcement details.*

Solution : Given data ,

Support width = 300 mm,　f_{ck} = 20 MPa

Live load = 3 kN/m^2,　f_y = 415 MPa

Floor finish = 1 kN/m^2

∴　Effective span = l_o = l + b = 5 + 0.3

l_o = 5.3 m

1.　Calculation of dimensions

(i)　Overall depth (D)

$$D = \frac{l_o}{12} = \frac{5300}{12}$$

D = 441.67 mm

∴　**D ≅ 450 mm**

∴　Assume effective cover c = 35 mm

d = D – C = 450 – 35

∴　**d = 415 mm**

(ii)　Flange thickness (D_f)

∴　**D_f = 125 mm**　　　　...given

(iii)　Assume width of web (b_w)

b_w = 230 mm

(iv)　Effective width of flange (b_f)

$$b_f = \frac{l_o}{6} + b_w + 6\,D_f$$

$$= \frac{5300}{6} + 230 + 6 \times 125$$

∴　**b_f = 1863.33 mm**

2.　Calculation of factored bending moment (B.M.) (M_u)

Calculation of loading

(i)　Live load on slab = 3×4 = 12 kN/m

(ii)　Floor finish on slab = 1×4 = 4 kN/m

(iii)　Self weight of slab = $(4 \times D_f)\,25 = (4 \times 0.125) \times 25$
= 12.5 kN/m

(iv)　Self weight of rib = $[0.23 \times (0.45 - 0.125)] \times 25$
= 1.87 kN/m

∴　Total udl = w = 30.37 kN/,

∴　Maximum udl = W_u = w $\times$ r_f = 30.37 $\times$ 1.5

∴　W_u = 45.56 kN/m

∴　Factored B.M. = $M_u = \dfrac{w_u \cdot l_o^2}{8} = \dfrac{45.56 \times 53^2}{8}$

M_u = 159.97 kN/m

3.　Calculation of actual depth of N.A. (x_u)

(a)　Assuming N.A. lies within flange ($x_u < D_f$)

$$M_u = C_u \times z = 0.36\,f_{ck}\,x_u\,b_f \cdot (d - 0.42\,x_u)$$

$$159.97 \times 10^6 = 0.36 \times 20 \times x_u \times 1863.33\,(415 - 0.42\,x_u)$$

∴　**x_u = 29.62 mm** < D_f = 125 mm

　　　　　...Assumption is correct

4.　Calculation of critical depth of N.A. (x_{umax})

x_{umax} = 0.48 d = 0.48 $\times$ 415

x_{umax} = 199.2 mm

5.　Compare x_u and x_{umax}

29.62 < 199.2

x_u = x_{umax}

　　　...section is under-reinforced section

6.　Calculation of area of steel (A_{st})

$$A_{st} = \frac{0.5\,f_{ck}}{f_y}\left[1 - \sqrt{1 - \frac{4.6\,M_u}{f_{ck} \cdot b_f \cdot d^2}}\right] \cdot b_f \cdot d$$

$$A_{st} = \frac{0.5 \times 20}{415}\left[1 - \sqrt{1 - \frac{4.6 \times 159.97 \times 10^6}{20 \times 1863.33 \times 415^2}}\right]$$
$$\times 1863.33 \times 415$$

∴　**A_{st} = 1100.68 mm^2**

Assume diameter of bar ϕ = 20 mm.

∴　No. of bars = $\dfrac{A_{st}}{a_{st}} = \dfrac{1100.68}{\pi/4\,(20)^2}$

No. of bars = 3.50

∴　No. of bars ≅ 4 No's.

∴　**Provide 4-20 mm ϕ as a tensile reinforcement.**

Example 8.18 : *Design the reinforcement for a T-beam for the following data : Effective span = 6m ends are simply supported, spacing of beams = 3.5 m c/c. Thickness of slab = 110 mm. Width of web = 300 mm, Total depth = 480 mm. Live load on floor 8 kN/m^2. Floor finish load 0.75 kN/m^2. The beam also supports a partition wall which transmits a load of 15 kN/m. Use M25 and Fe 425.*

Solution: Given data :

l_o = 6 m,　Spacing of beams = 3.5 m

D_f = 110 mm, live load = 8 kN/m^2

b_w = 300 mm, floor finish load = 0.75 kN/m^2

D = 480 mm,　partition wall = 15 kN/m

Assume effective cover = c = 40 mm

∴　　d = 480 – 40 = 440 mm

f_{ck} = 25 MPa

f_y = 415 MPa

∴　Effective flange width (b_f)

$$b_f = \frac{l_o}{6} + b_w + 6D_f$$

$$= \frac{6000}{6} + 300 + 6 \times 110$$

∴　**b_f = 1960 mm**

1. Calculation of loading (w_u)

(i) Live load on floor = $8 \times 3.5 = 28$ kN/m

(ii) Floor finish on slab = $0.75 \times 3.5 = 2.63$ kN/m

(iii) Self weight of slab = $(3.5 \times 0.11 \times 1)\,25$

　　 = 9.63 kN/m

(iv) Self weight of rib = $[0.3\,(0.48 - 0.11)]\,25$

　　 = 2.78 kN/m

(v) Partition wall = 15 kN/m

∴　Total udl = $w = 58.04$ kN/m

∴　Maximum udl = $w_u = w \times r_f = 58.04 \times 1.5$

　　　$w_u = 87.06$ kN/m

2. Calculation of factored bending moment (M_u)

$$M_u = \frac{w_u \cdot l_e^2}{8} = \frac{87.06 \times 6^2}{8}$$

∴　　　$M_u = 391.77$ kN/m

3. Calculation of actual depth of neutral axis (x_u)

(a) Assuming N.A. lies within flange (i.e. $x_u < D_f$)

　　　$M_u = c_u \times z = 0.36\, f_{ck}\, x_u\, b_f\, (d - 0.42\, x_u)$

$391.77 \times 10^6 = 0.36 \times 25 \times x_u \times 1960\,(440 - 0.42\, x_u)$

∴　　　$x_u = 53.17$ mm $< D_f = 110$ mm

　　　　　　　　…Assumption is correct

4. Calculation of critical depth of N.A. (x_{umax})

　　　$x_{umax} = 0.48d = 0.48 \times 440$

　　$x_{umax} = 211.2$ mm

5. Compare x_u and x_{umax}

　　　$53.17 < 211.2$

　　　　$x_u < x_{umax}$

　　　　…Section is under-reinforced section

6. Calculation of area of steel (A_{st})

$$A_{st} = \frac{0.5\, f_{ck}}{f_y}\left[1 - \sqrt{1 - \frac{4.6\, M_u}{f_{ck} \cdot b_f d^2}}\right] \times b_f \cdot d$$

$$A_{st} = \frac{0.5 \times 25}{415}\left[\sqrt{1 - \frac{4.6 \times 391.77 \times 10^6}{25 \times 1960 \times 440^2}}\right]$$
$$\times 1960 \times 440$$

∴　　　$A_{st} = 2597.17$ mm^2

Example 8.19 : *A hall of a building has an internal dimension as 7m × 15 mm a floor slab of 115 mm thickness with a floor finish of 1 kN/m^2 and live load of 3 kN/m^2 cost monolithically with beams. The beams are spaced at 8 m*

c/c. Assuming beam as simply supported of 300 mm thick brick wall. Design an intermediate beam, design for shear reinforcement details and all necessary check. Use M20 and Fe 415.

Solution : Given data

　　　$D_f = 115$ mm, $f_{ck} = 20$ MPa , clear span = 7m

　　　$b_w = 300$ mm, $f_y = 415$ MPa

1. Calculation of dimension

(i) Overall depth (D)

$$D = \frac{l_o}{12} = \frac{7300}{12}$$

　　　$D = 608.33$ mm

　　　$D \cong 610$ mm

Assume　cover = $c = 40$ mm

　　∴　Effective depth = $d = D - C = 610 - 40$

　　　　$d = 570$ mm

(ii) Effective width of flange (b_f)

$$b_f = \frac{l_o}{6} + b_w + 6D_f = \frac{7300}{6} + 300 + 6 \times 115$$

∴　　　$b_f = 2206.67$ mm

2. Calculation of loading (W_u)

(i) Live load = $3 \times 3 = 9$ kN/m

(ii) Floor finish = $1 \times 3 = 3$ kN/m

(iii) Self wt. of slab = $(3 \times 0.115 \times 1)\,25 = 8.63$ kN/m

(iv) Self wt. of rib = $[0.3 \times (0.61 - 0.115)\,1]\,25 = 3.71$ kN/m

　∴　Total udl = $w = 24.34$ kN/m

　∴　Maximum udl = $w_u = w \times r_f = 24.34 \times 1.5$

3. Factored bending moment (M_u)

$$M_u = \frac{w_u \cdot l^2}{8} = \frac{36.51 \times 7^2}{8}$$

　　　$M_u = 223.62$ kN.m

4. Calculation of actual depth of neutral axis (x_u)

(a) Assuming N.A. lies within flange ($x_y < D_f$)

　　　$M_u = C_u \times z = 0.36\, f_{ck} \cdot x_u \cdot b_f\,(d - 0.42\, x_u)$

$223.62 \times 10^6 = 0.36 \times 20 \times x_u \times 2206.67\,(570 - 0.42\, x_u)$

∴　　　$x_u = 25.16$ mm $< D_f = 115$ mm

　　　　　　　…assumption is correct

5. Calculation of critical depth of N.A. (x_{umax})

　　　$x_{umax} = 0.48\, d = 0.48 \times 570$

　　　$x_{umax} = 273.6$ mm

6. Compare x_u and x_{umax}

$$25.16 \; < \; 273.6$$

$$x_u \; < \; x_{umax}$$

…section is under-reinforced section

7. Calculation of area of steel (A_{st})

$$A_{st} \; = \; \frac{0.5\,f_{ck}}{f_y}\left[1-\sqrt{1-\frac{4.6\,M_y}{f_{ck}\,b_f\cdot d^2}}\right] \times b_f \cdot d$$

$$A_{st} \; = \; \frac{0.5 \times 20}{415}\left[1-\sqrt{1-\frac{4.6 \times 223.62 \times 10^6}{20 \times 2206.67 \times 570^2}}\right]$$

$$\times\, 2206.67 \times 570$$

$$\mathbf{A_{st} \; = \; 1107.37 \; mm^2}$$

Example 8.20 : *A T beam slab floor has 125 mm thick slab forming part of T-beams which are of 8 m clear span. The end bearing are 450 mm thick. Spacing of T-beam is 3.5 m. The live load on the floor is 3 kN/m². Design one of the intermediate beams. Use M20 concrete and Fe 415 steel.*

Solution : $D_f \; = \; 125$ mm,

$$\text{Spacing} \; = \; 3.5 \text{ m}$$
$$LL \; = \; 3 \text{ kN/m}^3$$
$$f_{ck} \; = \; \text{N/mm}^2$$
$$f_y \; = \; 415 \text{ N/mm}^2$$
$$\text{Clear span} \; = \; 8 \text{ m}$$

End bearing width = 450 mm

$$d \; = \; \left(\frac{\text{clear span}}{12}\right) \text{ to } \left(\frac{\text{clear span}}{15}\right)$$

Simply supported beam

1. Dimensions of Beam

$$\text{Depth} \; = \; \frac{1}{12}^{th} \text{ to } \frac{1}{15}^{th} \text{ span}$$

$$= \; \frac{1}{12} \times 8000 \text{ to } \frac{1}{15} \times 8000$$

Let, $d \; = \; 600$ mm

and $D \; = \; 650$ mm

$$b_w \; = \; \frac{1}{2} \text{ to } \frac{1}{3} d$$

$$b = (d/3) \text{ to } (d/2) > 200 \text{ mm}$$

Let, $b \; = \; 250$ mm

2. Effective Span

$$L_{eef} \; = \; lx + d_{ava.} \;\; OR \;\; l_x + \text{ width of support}(t)$$

adopt whichever is less

Clear span + d = 8 + 0.6 = 8.6 m

c/c of bearings = 8 + 0.45 = 8.45 m

∴   Effective span = 8.45 m

3. Flange Width

l_o - distance betweens points of zero moments

(a) $b_f \; = \; \dfrac{l_o}{\sigma} + b_w + 6D_f$

Since it is simply supported,

$$l_o \; = \; L = 8.45 \text{ m} = 8450 \text{ mm}$$

$$b_f \; = \; \frac{8450}{6} + 250 + 6 \times 125 \; = \; 2408 \text{ mm}$$

(b) $b_f = 0.5\,(L_1 + L_2) + b_w$ i.e. c/c of adjacent slabs

Max. = 3.5 m = 3500 mm

∴ $b_f \; = \; 2408$ mm

The cross section of the beam to be designed is as shown in Fig. 8.9.

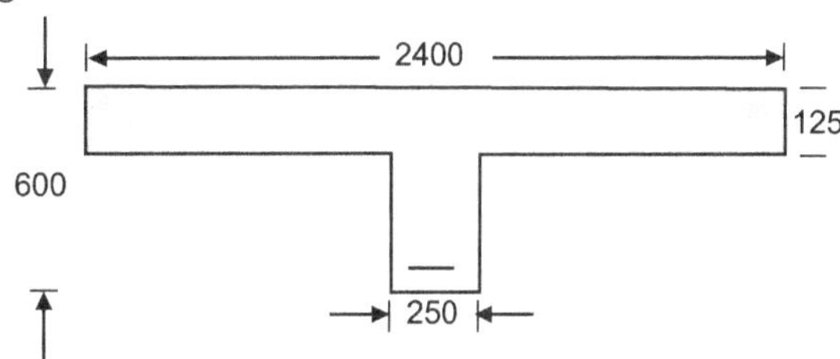

Fig. 8.9

4. Design Moment (M_u) and Shear Force (V_u) :

Load per meter run of beam is calculated as shown below :

Load from slab

Self weight of slab = 0.125 × 1 × 1 × 25 = 3.125 kN/m²

Weight of floor finish = 0.6 kN/m² (assumed)

Live load = 3 kN/m²

Total = 6.725 kN/m²

Since beam is taking case of 3.5 m width of slab.

Load on beam from slab = 6.725 × 1 × 3.5 = 23.577 kN/m

Self weight of rib :

Width of rib = 250 mm = 0.25 m

Depth of rib = 650 – 125 = 525 mm = 0.525 m

∴ Self weight of rib = 0.25 × 0.525 × 1 × 25 = 3.28 kN/m

Weight of plaster to rib = 0.5 kN/m (assumed)

∴ Total load on beam = 23.577 + 3.28 + 0.5

$$= 27.3 \text{ kN/m}$$

Factored load $w_u = 1.5 \times 27.3 = 40.95$ kN/m

$$M_u = \frac{w_u \, l^2}{8}$$

$$= 40.95 \times \frac{8.45^2}{8} = 365.5 \text{ kN-m}$$

$$V_u = \frac{1}{2} \times w_u l = \frac{1}{2} \times 40.95 \times 8.46 = 173 \text{ kN}$$

5. Design of Longitudinal Bars : Assuming neutral axis coincides with flange

$$x_u = D_f = 125 \text{ mm} \qquad (x_u \le D_i \dots \text{Case I})$$

N-A is within the flange. Let depth of N-A from extreme compression flange be x_u. Then equating moment to movement of resistance, we get,

$$M_u = 0.87 \, f_y \, A_{st} \, (d - 0.416 \, x_u)$$

$$365.5 \times 10^6 = 0.87 \times 415 \times A_{st} \, (600 - 0.416 \times 125)$$

$$\therefore \qquad A_{st} = 1847.30 \text{ mm}^2$$

Provide 6 bars of 20 mm diameter

$$A_{st \, provided} = 6 \times \frac{\pi}{4} \times 20^2 = 1885 \text{ mm}^2$$

6. Design of Shear Reinforcement

$$P_t = \frac{A_{st}}{b.d}$$

$$P_t = \frac{1885}{250 \times 600} \times 100 = 1.257$$

From table 19 in Is 456

$$\tau_c = 0.67 \text{ N/mm}^2$$

$$\tau_v = \frac{V_u}{bd} = \frac{173 \times 1000}{250 \times 600} = 1.153 \text{ N/mm}^2$$

From table 20 in IS 456

$$\tau_{c \, max} = 2.8 \text{ N/mm}^2$$

$$\therefore \qquad \tau_c < \tau_v < \tau_{c \, max}$$

Shear reinforcement is to be designed.

$$V_{us} = V_u - \tau_c \, bd = 173000 - 0.67 \times 250 \times 600$$

$$= 72500 \text{ N}$$

Using 2 legged 8 mm diameter Fe 415 steel as strirrups spacing S_v can be found from the relations,

$$V_{us} = \frac{0.87 \, f_y \, A_{sv} \, d}{S_v}$$

$$72500 = \frac{0.87 \times 415 \times 2 \times \frac{\pi}{4} \times 8^2 \times 600}{S_v}$$

$$S_v = 300.39 \text{ mm}$$

Maximum spacing allowed is $0.75 \times 600 = 450$ mm or 300 mm, whichever is less. Hence provide 2 legged 8 mm dia Fe 415 stirrups throughout at 300 mm c/c.

7. Check for Deflection Control : As it is simply supported beam,

$$\text{Basic} \left(\frac{L}{d}\right) = 20$$

$$\frac{L}{d} \text{ provided} = \frac{8450}{600} = 14 < \frac{L}{d} \text{ max}$$

Hence, deflection control is satisfactory.

As 6 bars of 20 mm diameter cannot be accommodated in 250 mm width with sufficient side cover and gap between the bars, provide bars in two rows as shown in Fig. 8.10.

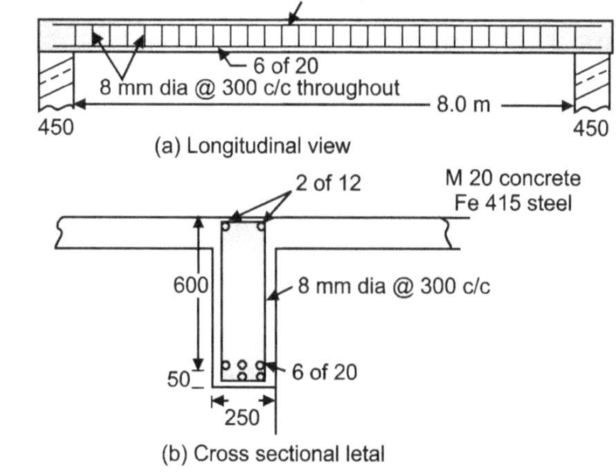

Fig. 8.10

Example 8.21 : *A floor beam 230 mm wide is simply supported at its ends and carries a uniformly distributed load of 30 kN/m. The span of the beam is 5 m and centre to centre spacing of beams is 3 m. The thickness of the slab is 120 mm. Design the central T-beam including shear reinforcement use concrete M20 and steel Fe415.*

Given : $f_{ck} = 20$ N/mm^2, $f_y = 415$ N/mm^2, $b_w = 230$ mm, L = 5 m, $D_f = 120$ mm, c/c spacing of beams = 3 m

Solution :

1. Loads : The depth of the beam may be assumed between L/12 to L/16. Small depths for light loads and small spans and more depths for heavy loads and large spans.

$$d = (\text{clear span}/12) \text{ to } (\text{clear span}/15)$$

$$\dots \text{Simply supported beam}$$

$$\text{Depth of beam D} = \frac{L}{15} = \frac{5000}{15} = 333 \text{ mm}$$

Assume size of beam 230 × 400 mmb = (d/3) to (d/2) > 200 mm

$\therefore$ Assume effective cover = 40 mm

$\therefore$ Effective depth d = 400 – 40 = 360 mm

Self weight of beam = $25 \times (0.23 \times 0.4)$ = 2.3 kN/m

$\therefore$ Load on beam w = 30 + 2.3 = 32.3 kN/m

Ultimate load w_u = 1.5×32.3 = 48.45 kN/m

2. Design Moment

$$M_u = w_u \times \frac{L^2}{8}$$

$$= 48.45 \times \frac{5^2}{8} = 151.4 \text{ kN/m}$$

L_o = Ll_o - distance betweens points of zero moments

= 5000 mm for simply supporter beam

$$b_f = \frac{L_o}{6} + 6 D_f + b_w$$

$$= \frac{5000}{6} + 6 \times 120 + 230$$

b_f = 1783 mm

For $x_u = D_f = 120$ mm

Mu (in term of Ast) = ? , Mu = 0.87 fy Ast (d – 0.416 xu)

Ast = ?, Equating moment to moment of resistance

3. Main Steel

$$A_{st} = \frac{0.5 f_{ck}}{f_y}\left[1 - \sqrt{1 - \frac{4.6 M_u}{f_{ck} b_f d^2}}\right] b_f \, d$$

$$A_{st} = \frac{0.5 \times 20}{415}\left[1 - \sqrt{1 - \frac{4.6 \times 151.4 \times 10^6}{20 \times 1783 \times 360^2}}\right]$$

$$\times 1783 \times 360$$

$$= 1214 \text{ mm}^2$$

Provide 4-20 mm bars

Area provided A_{st} = 1256 mm^2

4. Design for Shear

No bent up bars

Maximum shear at support

$$V_u = w_u \times \frac{L}{2}$$

$$= 48.45 \times \frac{5}{2}$$

$$= 121.12 \text{ kN}$$

$$P_t = \frac{100 A_{st}}{b \times d} = \frac{100 \times 1256}{230 \times 360}$$

$$P_t = 1.5 \%$$

For P_t = 1.5 % τ_{uc} = 0.72   (Ref. table 19,P.N.73,IS 456)

Shear to be taken by steel reinforcement $(V_{us}) = V_u - \tau_c.b.d$

$$V_{uc} = 0.72 \times 230 \times \frac{360}{1000}$$

$$= 59.6 \text{ kN}$$

Shear to be resisted by stirrups,

$$V_{us} = V_u - V_{uc}$$

$$= 121.12 - 59.6 = 61.12 \text{ kN}$$

Using 8 mm 2-legged stirrups

Area A_{sv} = 2×50 = 100 mm^2

Spacing of design stirrups

$$s = \frac{0.87 \times f_y \times A_{sv} \, xd}{V_{us}}$$

$$s = \frac{0.87 \times 415 \times 100 \times 360}{61.12 \times 1000}$$

$$= 212 \text{ mm}$$

$$= \text{say } 210 \text{ mm}$$

$$< (0.75 \times 360 \text{ or } 300 \text{ mm})$$

$\therefore$ Provide 8 mm 2-legged stirrups at 210 mm c/c.

EXERCISE

1. What is flanged beam ?

2. What are the conditions of formation of flanged section ?

3. What is effective flange width ? How do you calculate the effective flange width ?

4. State advantages and disadvantages of flanged section.

5. Draw stress strain diagram for following condition with their moment of resistance expression.

    (i)   N.A. lies within flange ($x_u \leq D_f$)

    (ii) N.A. lies outside the flange ($x_u > D_f$)

        (a) $x_4 > D_f$ and $D_f \leq 0.43 \, x_u$

        (b) $x_4 > D_f$ and $D_f > 0.43 \, x_u$

6. Determine the moment of resistance for a T-beam of flange width 1200 mm, flange thickness 100 mm, width of rib 275 mm, flange thickness 100 mm, width of rib 275 mm and effective depth 600 mm. Use M-20, Fe-500 and tensile r/f as 6 No. of 25 mm ϕ.

7. Calculate the moment of resistance of flanged beam for the following data :

    Width of flange = 1200 mm

Depth of slab = 110 mm

Effective depth = 600 mm

Width of web = 300 mm, concrete grade M20 and Steel grade Fe415. Area of tension steel provided is 4-25 mm bars.

8. T beam consists of flange 1100 mm wide and 120 mm deep. The depth of the beam is 600 mm up to the centre of steel and width of web is 275 mm. Find the area of steel required for an ultimate moment of 380 kN.m.

9. Calculate effective flange width for a T beam for the following details

Width of web = 230 mm

slab thickness = 100 mm

size of hall = 12 m × 6 m

Width of support for beam = 230 mm

c/c distance of beams = 3 m.

9.1 INTRODUCTION

- Reinforced concrete members are normally designed for the limit state of collapse in flexure rather than in shear. Shear failure, which in reality, occurs under the combined action of shearing forces and bending moments, is characterized by very small deflection and lack of ductility.

- This failure is, many times, sudden and without any warning. For this reason, the shear failure is considered very undesirable and is usually avoided. The code provisions for shear are, therefore, more conservative as compared to bending.

9.2 SHEAR STRESSES IN BEAMS

- A beam, when loaded, is subjected to a varying bending moment from section to section and has different stresses at any two adjacent sections in the same fibres. This inequality of stresses produces a tendency in each fibre to slide over the lower one in a horizontal plane i.e. horizontal shear stress is created and it is also accompanied by a complementary shear between two adjacent sections in the vertical direction. The horizontal and vertical shear stresses are to be accounted for in the designs of beams.

9.3 SHEAR STRESSES IN HOMOGENEOUS BEAMS

- Fig. 9.1 shows the distribution of bending and shear stresses in a homogeneous section. The total difference of bending stresses acting on two adjacent sections of a beam above any plane AB, which causes shear on plane AB, goes on increasing as the plane AB moves towards the neutral axis. As the plane AB moves below the neutral axis, the unbalanced force, causing shearing, goes on reducing till it becomes zero at bottom. So the shear stress is zero at the top and bottom of the beam and increases to a maximum value at the neutral plane. The variation of shear stress is parabolic.

- For a beam of homogeneous section, the shear stress at any section is given by

$$\tau = \frac{V \cdot A\bar{y}}{b \cdot I} \qquad \text{... (9.1)}$$

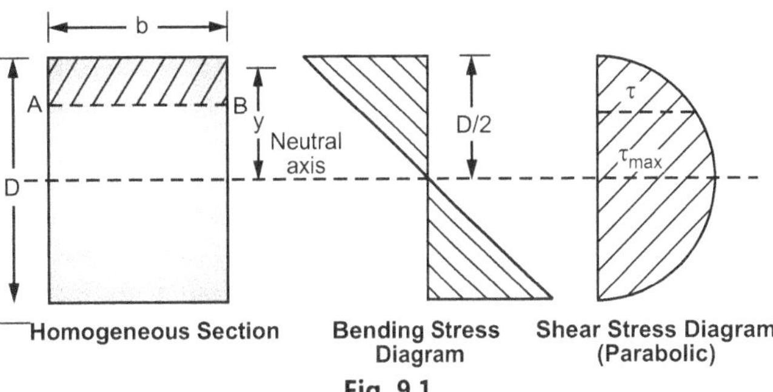

Fig. 9.1

where, V = shear force at any section

$A\bar{y}$ = moment of area above the section, about the neutral axis.

   b = breadth of the section

   I = moment of inertia of beam section, about the neutral axis.

9.4 SHEAR STRESSES IN R.C. SECTION

- In the case of a R.C. section, shown in Fig. 9.2, the distribution of bending stress above the neutral axis being similar to that in a homogeneous section, the shear stress distribution will also be parabolic upto the neutral axis.

- Below the neutral axis, concrete is not supposed to have any tension and so whatever difference in forces on the sections existed at neutral plane between two adjacent sections of the beam, the same will exist below the neutral plane. Hence, the value of shear stress below the neutral axis will remain constant as shown by the rectangular portion of the diagram in Fig. 9.2.

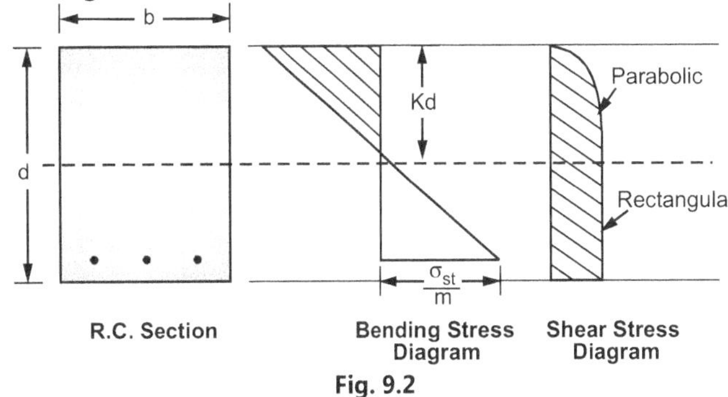

Fig. 9.2

- As total tension on any section is equal to the total compression, the difference in the tensile forces in the steel bars between two adjacent sections will be equal to the difference in compressive forces between those sections.

- Thus, at a plane passing through the C.G. of steel bars, the total difference in compressive force, which was causing the shear stress, will be neutralised by an equal difference in tensile forces. Hence, the shear stress suddenly falls to zero at the level of steel bars shown by the horizontal line in the shear stress diagram of Fig. 9.2. If the shear force on a vertical section is V, then

$$V = \text{(Area of the shear stress diagram)} \times \text{(Breadth of the section)}$$

$$= \left[\frac{2}{3} \cdot \tau \cdot Kd + \tau (d - Kd)\right] \times b$$

where, τ is maximum shear stress.

$$\therefore \quad V = \tau \cdot b \cdot d \left[\frac{2}{3} K + 1 - K\right]$$

$$= \tau \cdot b \cdot d \left(1 - \frac{K}{3}\right)$$

$$= \tau \cdot b \cdot j \cdot d$$

$$\therefore \quad \tau = \frac{V}{b \cdot j \cdot d} \qquad \dots (9.2)$$

Equation (9.2) gives the value of maximum shear stress developed in the section.

- Actually, the behaviour of concrete is very complicated and hence it has not been possible to determine the exact shear stress distribution across the section. In fact, the designer is interested in knowing the shear resisting capacity of the section rather than the actual stress condition. Therefore, a simple concept of nominal shear stress has been introduced and is expressed as

Nominal shear stress,

$$\tau_v = \frac{V}{(bd)} \qquad \dots (9.3)$$

Using limit state method, the above formula can be written as (clause 40.1) for uniform depth of beam,

$$\tau_v = \frac{V_u}{(bd)} \qquad \dots (9.4)$$

where, V_v = ultimate shear force due to design loads

b = breadth of the member, which for flanged section shall be taken as the breadth of web, b_w; and

d = effective depth.

Beams of Varying Depth (Clause 40.1.1) :

In the case of beams of varying depth, the equation shall be modified as,

$$\tau_v = \frac{V_u \pm \dfrac{M_u}{d} \tan \beta}{(bd)} \qquad \dots (9.5)$$

where, M_u = bending moment at the section

β = angle between the top and the bottom edges of the beam

The negative sign in the formula applies when bending moment M_u increases numerically in the same direction as the effective depth d increases (e.g. tapered cantilever beam) and the positive sign when the moment decreases numerically in this direction (e.g. at the haunch of a simply supported end of the beam).

9.5 DIAGONAL TENSION

- Consider a small element ABCD along the length of the beam which is taken between the neutral axis and extreme fibres (at top or bottom). This is subjected to the shear stress (τ) parallel to four sides and the tensile bending stress (σ) along the length of the beam as shown in Fig. 9.3.

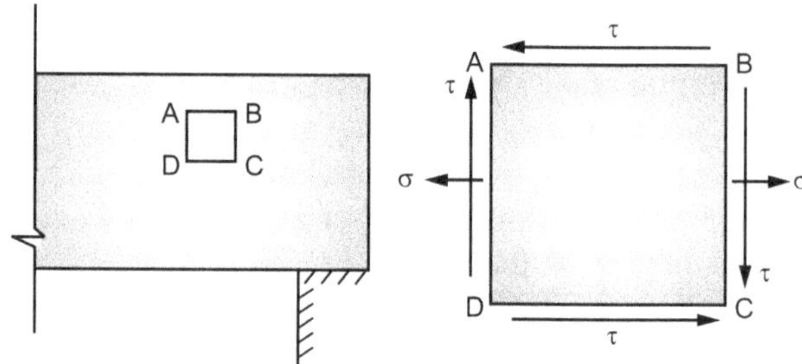

(a) Elevation of Beam Indicating the Element ABCD

(b) Small Element along the Length of Beam

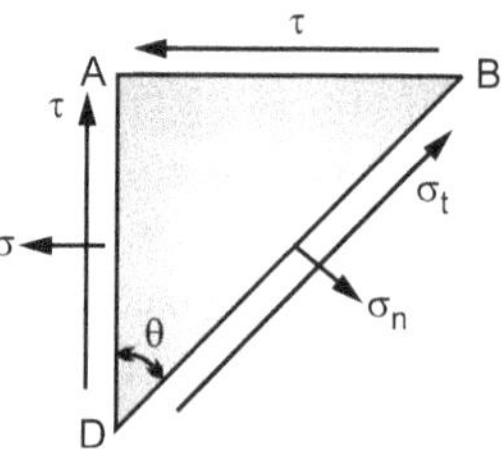

(c) Principal Stresses

Fig. 9.3 : Forces on small element along the length of beam

The principal stresses on this element are given by

$$\sigma_n = \frac{\sigma}{2} + \sqrt{\left(\frac{\sigma}{2}\right)^2 + \tau^2} \qquad \dots (9.6)$$

This stress σ_n is tensile. There will be another principal stress (σ_t) at right angle to σ_n, which will be compressive.

$$\sigma_t = \frac{\sigma}{2} - \sqrt{\left(\frac{\sigma}{2}\right)^2 + \tau^2} \qquad \text{... (9.7)}$$

and the inclination of principal plane is given by

$$\tan 2\theta = \frac{2\tau}{\sigma}$$

Two important cases are discussed below :

Case I : If   B.M. $= 0$ i.e. $\sigma = 0$

then $\sigma_n = \tau$ and $\sigma_t = -\tau$

$\tan 2\theta = \infty = \tan 90°$

i.e. $\theta = 45°$ or $135°$

This means that near the support for a simply supported beam, where bending moment is zero, the principal tension (σ_n) is equal to shear stress (τ) and is inclined at 45°. This is known as **Diagonal Tension** and the other principal stress (σ_t) will be at right angle to σ_n, which will be compressive. This is known as **Diagonal Compression** and is of the same value as the shear stress. The principal tensile stresses and compressive stresses act along the diagonal BD and AC respectively. This is the case of pure shear because $\sigma = 0$, as shown in Fig. 9.4.

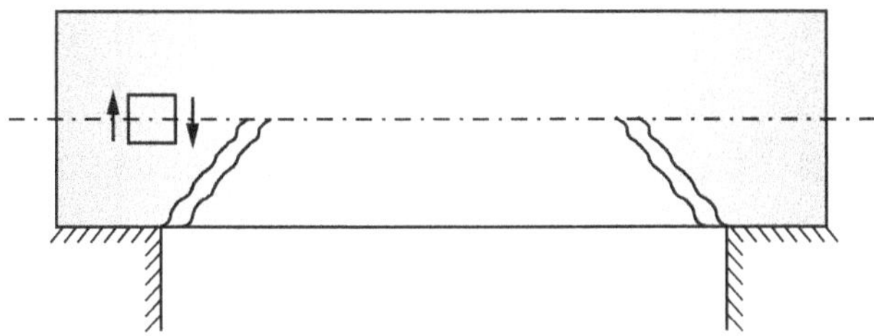

(a) Shear cracks in R.C.C. beam

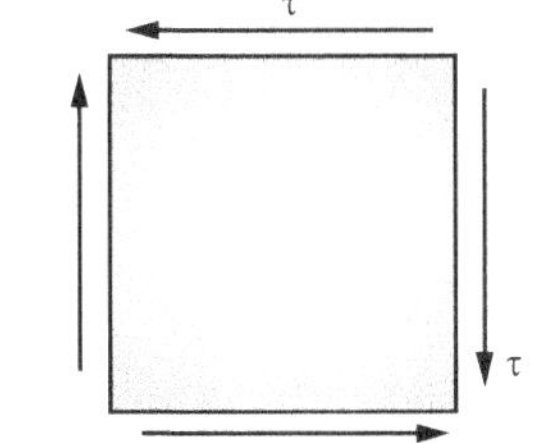

(b) Pure shear

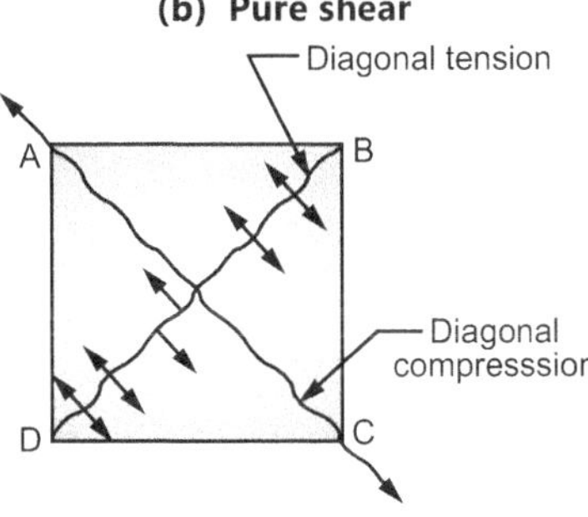

(c) Cracking under pure shear

Fig. 9.4

As the concrete is weak in tension, the concrete near the support cracks at 45° (i.e. perpendicular to the diagonal BD) with horizontal. These are known as web-shear cracks or diagonal tension cracks. To avoid the shear cracks, the beam should be reinforced across the cracks.

Case II : When bending moment is maximum at mid-span of a simply supported i.e. consider an element at the bottom fibre at the mid-section of the beam. The bending stress (σ) is maximum, while shear stress is zero (i.e. $\tau = 0$), we get,

$$\sigma_n = \sigma, \quad \sigma_t = 0$$

and $\theta = 90°$ i.e. principal plane is perpendicular to the beam axis.

This means that principal tensile stress acts in horizontal direction and shear cracks will be vertical as shown in Fig. 9.5.

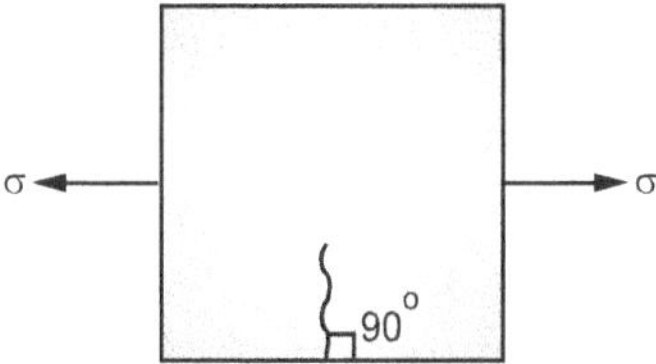

Fig. 9.5 : Pure flexure

Thus, at mid-section, where bending stresses are predominant, the cracks will start developing vertically. These cracks are called

Flexural Cracks.

From above discussion, it is clear that between the two limits, if concrete cracks due to diagonal tension, the cracks will change from a vertical direction at a point of zero shear to a direction inclined at an angle of 45° at a point where bending stress is zero. This is illustrated in Fig. 9.6, which shows diagonal tension failure of a beam.

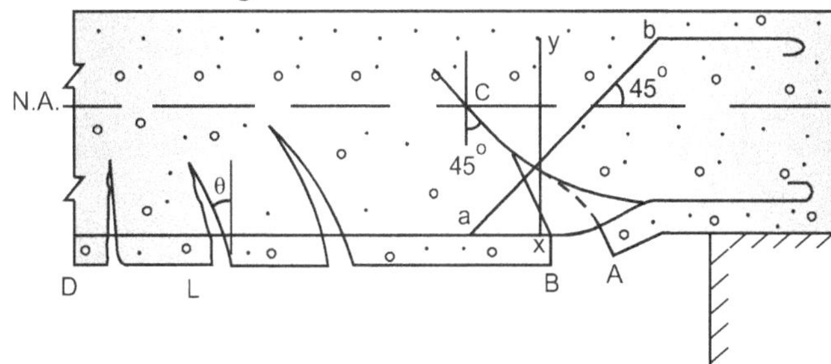

Fig. 9.6 : Diagonal tension cracks in a beam

At the neutral axis in any section, the diagonal tension is equal to the maximum shear intensity as the bending stress is zero and is inclined at 45° to the vertical. Below the neutral axis, the diagonal tension is more in magnitude, as the concrete shares the bending stress before cracking and 'σ' is a positive quantity. The direction of the principal tensile stress makes an angle larger than 45° with the vertical. Thus, the cracks at the bottom of the beam start vertically and gradually take an inclined shape, becoming diagonal at the neutral axis. Above the neutral axis, where the bending stress is compressive, diagonal tension goes on decreasing rapidly as the value of 'σ' is negative.

9.6 DESIGN SHEAR STRENGTH OF CONCRETE

The design shear strength of beam shall be taken is as follows :

1. Without shear reinforcement.

2. With shear reinforcement.

1. **Without Shear Reinforcement :** The mechanics of shear transfer when concrete cracks due to shear without providing the shear reinforcement is illustrated in Fig. 9.7. The shear is resisted by beam consisting of the following shear carrying components :

(a) **Shear Carried by Concrete in Compression :** The uncracked concrete in compression has the resisting effect to prevent the propagation of cracks in the compression zone (i.e. above neutral axis). The shear force carried by concrete in compression is between 20% to 40%.

(b) **Shear Carried by Aggregate-Concrete Interlock :** Along the crack, shear resistance is provided by the vertical component of force due to the interlocking of aggregates. The contribution of aggregate interlock is 35% to 50%.

(c) **Shear Carried by Dowel Action :** At the tensile reinforcement, shear is resisted by the dowel action of the longitudinal bar. The shear force carried by dowel action varies between 15% to 30%.

The capacity of all the three shear resisting components increases with the concrete strength. The design shear strength of concrete in beams without shear reinforcement is given in Table 9.1 which is reproduced from table 19 of IS : 456-2000.

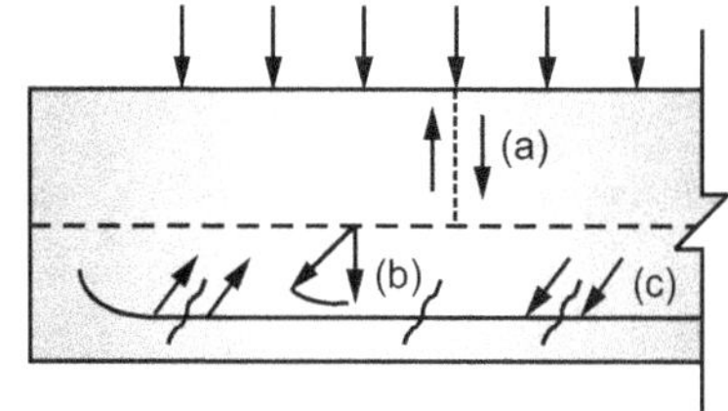

(a) Uncracked concrete   (b) Aggregate-interlock   (c) Dowel action

Fig. 9.7 : Shear resistance mechanism of a cracked beam without shear reinforcement

The values given in the Table 9.1 are based on the following equation :

$$\tau_c = 0.85 \frac{\sqrt{0.8\, f_{ck}}\,(\sqrt{1 + 5\beta} - 1)}{6\beta} \qquad ...\,(9.8)$$

where, $\beta = \dfrac{0.8\, f_{ck}}{6.89\, p_t} \not< 1.0$

$$p_t = \frac{100\, A_{st}}{b \cdot d}$$

for flanged section b shall be taken as b_w, the width of rib.

Table 9.1 : Design Shear Strength of Concrete, τ_c, N/mm^2

(Clauses 40.2.1, 40.2.2, 40.3, 40.4, 40.5.3, 41.3.2, 41.3.3 and 41.4.3)

$100\dfrac{A_s}{b \cdot d}$	Concrete Grade					
	M 15	M 20	M 25	M 30	M 35	M 40 and above
(1)	(2)	(3)	(4)	(5)	(6)	(7)
≤ 0.15	0.28	0.28	0.29	0.29	0.29	0.30
0.25	0.35	0.36	0.36	0.37	0.37	0.38
0.50	0.46	0.48	0.49	0.50	0.50	0.51
0.75	0.54	0.56	0.57	0.59	0.59	0.60
1.00	0.60	0.62	0.64	0.66	0.67	0.68
1.25	0.64	0.67	0.70	0.71	0.73	0.74
1.50	0.68	0.72	0.74	0.76	0.78	0.79
1.75	0.71	0.75	0.78	0.80	0.82	0.84
2.00	0.71	0.79	0.82	0.84	0.86	0.88
2.25	0.71	0.81	0.85	0.88	0.90	0.92
2.50	0.71	0.82	0.88	0.91	0.93	0.95
2.75	0.71	0.82	0.90	0.94	0.96	0.98
3.00 and above	0.71	0.82	0.92	0.96	0.99	1.01

Note : The term A_s is the area of longitudinal tension reinforcement which continues atleast one effective depth beyond the section being considered except at support where the full area of tension reinforcement may be used, provided the detailing conforms to the code requirements.

• Close observation of Table 9.1 shows that the design shear strength τ_{uc} in concrete depends on the percentage of tension steel.

This is because :

(a) When the amount of tension steel increases, the depth of neutral axis increases and thus, the depth of uncracked concrete increases. This increases the capacity of concrete in shear.

(b) When the amount of tension steel increases, the cracks formed are smaller, which improves the aggregate inter-lock. Also because of larger steel area, the dowel action is improved. This further improves the capacity of section in shear.

- The design shear strength of concrete also depends on the grade of concrete.

$$V_{uc} = \text{Ultimate shear resisted by concrete}$$
$$\text{without shear reinforcement}$$
$$V_{uc} = \tau_c \cdot b \cdot d \qquad \qquad \dots (9.9)$$

If the nominal shear force ($V_u = \tau_v \cdot b \cdot d$) does not exceed the value of ultimate shear force (V_{uc}), the section is safe for shear and shear reinforcements are not required theoretically. However, some minimum shear reinforcement shall be provided.

Table 9.2

Maximum Shear Stress $\tau_{uc,\,(max)}$ N/mm^2					
Concrete Grade	M 15	M 20	M 25	M 30	M 35	M 40 and above
τ_c (max) (N/mm^2)	2.5	2.8	3.1	3.5	3.7	4.0

For solid slab, the design shear strength for concrete shall be $\tau_c \cdot k$, where 'k' has the values given below

Overall depth of slab, mm	300 or more	275	250	225	200	175	150 or less
k	1.00	1.05	1.10	1.151	1.20	1.25	1.30

2. With Shear Reinforcement :

- Shear reinforcements are provided against diagonal tension. As the concrete is strong in compression, generally a beam is safe against diagonal compression. However, for any beam, suitably reinforcement against diagonal tension, the nominal shear shall not exceed the maximum shear stress values given in table 20 of IS : 456 and are reproduced in Table 9.2. By this provision the failure of the beam by diagonal compression is prevented.

- For any section, if nominal shear stress exceeds the maximum shear stress value, the section shall be redesigned either by using richer concrete mix or by increasing the size of section.

- The empirical design of shear as described in code is based on the assumption that the shear failure of the section occurs on a vertical plane when the nominal shear stress as defined by equation $\left(\tau_u = \dfrac{V_u}{b \cdot d} \right)$ exceeds the vertical shear strength of concrete (i.e. τ_{uc}). By limiting the shear stress on a section and design the shear reinforcement when it exceeds the shear strength of concrete, code ensures that the possibility of brittle failure caused by the diagonal tension is eliminated. The shear reinforcement is designed either by using vertical stirrups or by a combination of stirrups and bent bars crossing the cracks.

- Now, concrete alone can resist a shear force V_{uc} (or shear capacity of concrete V_{uc}), which is given by

$$V_{uc} = \tau_c \cdot b \cdot d$$

- According to IS : 456, the shear reinforcement shall be designed for a shear force V_{us}, which is given by

$$V_{us} = V_u - V_c \qquad \qquad \dots (9.10)$$
$$V_{us} = V_u - \tau_c \cdot b \cdot d$$

where, V_{us} = strength of shear reinforcement

V_u = shear force due to external loads

V_{uc} = shear capacity of concrete

9.7 TYPE OF SHEAR REINFORCEMENT

When τ_u exceeds τ_{uc}, shear reinforcement shall be provided in any of the following forms :

1. Vertical stirrups.

2. Inclined stirrups, and

3. Bent-up bars along with stirrups.

These are discussed briefly.

1. **Vertical Stirrups :** The strength of shear reinforcement for vertical stirrups is calculated as,

$$V_{usv} = \frac{0.87\, f_y \cdot A_{sv} \cdot d}{S_v} \qquad \qquad \dots (9.11)$$

These stirrups are usually two-legged. However, four-legged, six-legged etc. stirrups can also be formed as shown in Fig. 9.8. More than two-legged stirrups are used for heavy shear. Unless otherwise, specified stirrups are always two-legged.

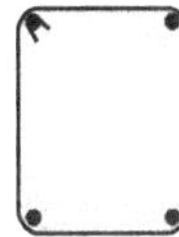
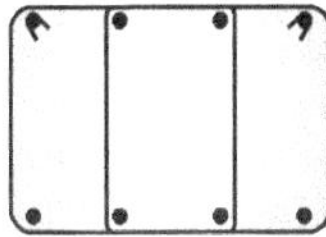

(a) Two-legged stirrup   (b) Four-legged stirrup

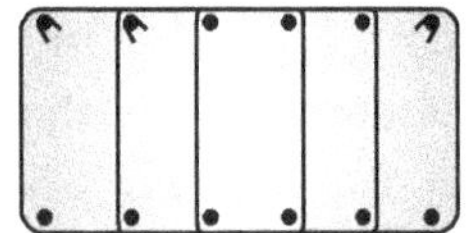

(c) Six-legged stirrup

Fig. 9.8 : Types of stirrups

2. Inclined Stirrups or a Series of Bars Bent up at Different Cross-Section : The strength of shear reinforcement for inclined stirrups is calculated as,

$$V_{usi} = \frac{0.87\,f_y \cdot A_{sv} \cdot d}{S_v}\,(\sin \alpha + \cos \alpha) \qquad \dots (9.12)$$

A series of inclined stirrups is shown in Fig. 9.9.

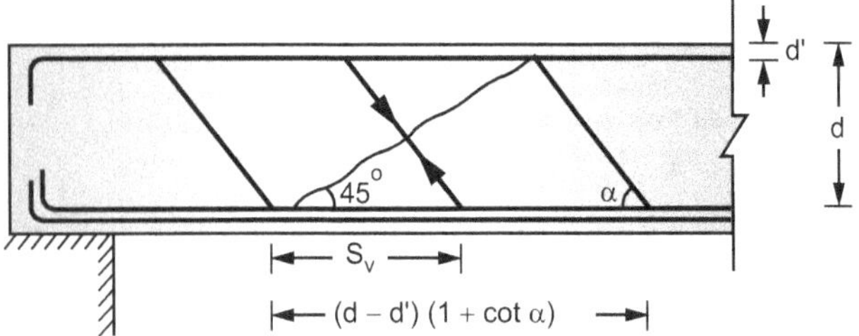

Fig. 9.9 : Inclined stirrups as shear reinforcement

3. Bent-up Bars : For a single bar or a group of bars, all bent up at the same cross-section, the strength of shear reinforcement is calculated as :

$$V_{usb} = 0.87\,f_y \cdot A_{sv} \cdot \sin \alpha \qquad \dots (9.13)$$

The bent bars alone are not satisfactory for the shear reinforcements. This is because the exact behaviour of bent bars in resisting shear is not clearly understood. When bent bars are provided, their contribution towards shear resistance shall not be more than half of the total shear reinforcement. The remaining shear shall be taken by stirrups.

where, A_{sv} = Total cross-sectional area of stirrup legs or bent-up bars within a distance of S_v

     S_v = Spacing of the stirrups or bent-up bars along the length of the member.

     b = Breadth of the member which for flanged beams shall be taken as the breadth of the web b_w.

     f_y = Characteristic strength of the stirrup or bent-up reinforcement which shall not be taken greater than 415 N/mm^2

     α = Angle between the inclined stirrup or bent-up bar and the axis of the member, not less than 45°, and

     d = Effective depth.

9.8 MINIMUM SHEAR REINFORCEMENT (CLAUSE 26.5.1.6)

- The study of mode of failure has revealed that if shear reinforcement is not provided, failure due to shear occurs suddenly without giving any warning. Therefore, the code specifies the provision of minimum shear reinforcement in the form of stirrups.

- The minimum shear reinforcement in the form of stirrups shall be provided such that,

$$\frac{A_{sv}}{bS_v} \geq \frac{0.4}{0.87\,f_y} \qquad \dots (9.14)$$

where, A_{sv} = Total cross-sectional area of stirrup legs effective in shear

     S_v = Stirrup spacing along the length of the member

     b = Breadth of the beam or breadth of the web of flanged beam, and

     f_y = Characteristic strength of the stirrup reinforcement in N/mm^2, which shall not be taken greater than 415 N/mm^2.

- However, in the members of minor structural importance such as lintels or where the maximum shear stress is less than half the design shear strength, this provision need not be complied with.

9.9 MAXIMUM SPACING OF SHEAR REINFORCEMENT (CLAUSE 26.5.1.5)

- The maximum spacing of shear reinforcement measured along the axis of the member shall not exceed 0.75 d for vertical stirrups and d for inclined stirrups at 45°, where d is the effective depth of the section. In no case shall the spacing exceed 300 mm.

Important Notes :

- The code does not specify the minimum spacing of stirrups. But it is considered that the minimum spacing of the stirrups should be preferably not less than 100 mm to effect better concreting.

  The first stirrup can be provided at a distance of half the stirrup spacing from the face of support.

- **Distance of First Bent Bar from Support :** Usually, the bars are bent at 45° to the longitudinal axis of the beam. The minimum and maximum limit of the horizontal distance of a bent bar from support d and 2d respectively. If a bar is bent at a distance from support less than d, the bent bar will enter into the support which is not practical. Therefore, first bent bar should start a distance from support between d and 2d, as shown in Fig. 9.10.

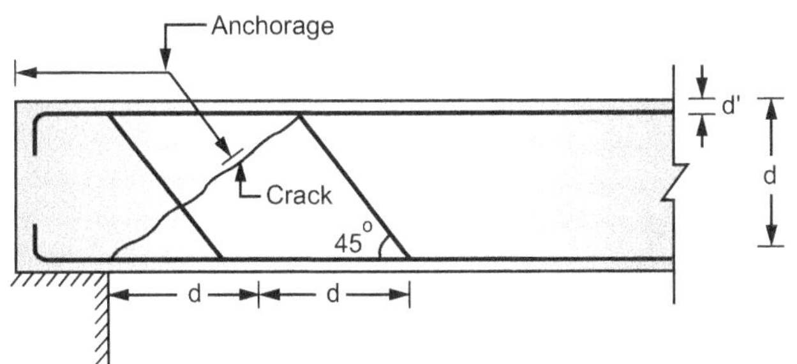

Fig. 9.10 : Limits of distance of first bent from support

9.10 CRITICAL SECTION FOR SHEAR (CLAUSE 22.6.2)

- When the reaction in the direction of the applied shear introduces compressive reaction in the end region, the diagonal crack gets displaced away from the face of support, therefore sections located at a distance less than 'd' from the face of support may be designed for the same shear as that computed at distance 'd'. While the support reaction introduces tension in the end region, a diagonal crack is likely to develop from the face and hence the shear is computed at the face of support. [Fig. 9.11 (a) and (b)]

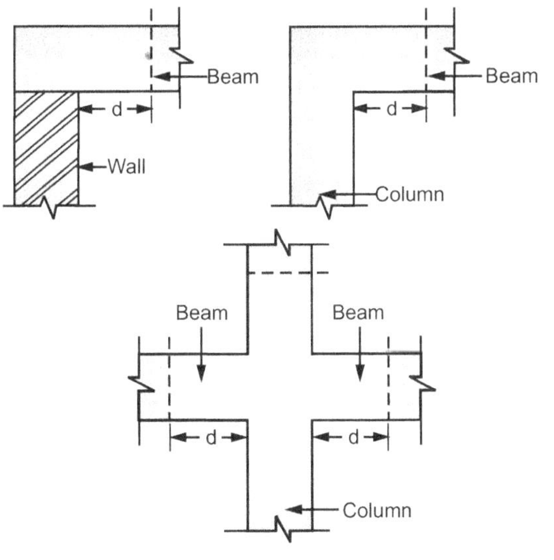

(a) Critical section at a distance 'd' from face to support

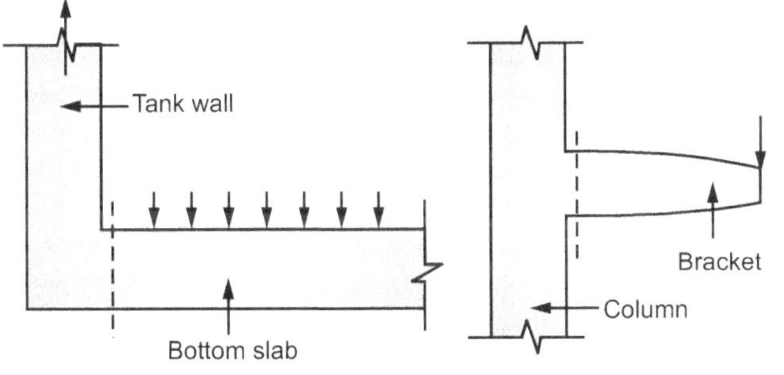

(b) Critical section at face of support

Fig. 9.11 : Critical sections for shear

SOLVED EXAMPLES

Example 9.1 : *Design the shear reinforcement for the beam of size 250 mm × 440 mm with effective cover 40 mm of span 4.25 m carries a udl of 17.5 kN/m including self weight. The beam is reinforced with 3-16mm ϕ. Use M20 and Fe 415.*

Solution : b = 250 mm, w = 17.5 kN/m, A_{st} = 3-16 mm ϕ

$$D = 440 \text{ mm}, \quad \therefore \text{ Max. udl} = w_u, \qquad \therefore A_{st} = 3 \times \frac{\pi}{4}(16)^2$$

$$d = 400 \text{ mm}, \quad w_u = w \times r_f = 17.5 \times 1.5$$

$$l_e = 4.25 \text{ m}, \quad w_u = 26.25 \text{ kN/m}, \qquad A_{st} = 603.19 \text{ mm}^2$$

Factored shear force $(V_u) = \dfrac{w_u \cdot l_e}{2}$

$$V_u = \frac{26.25 \times 4.25}{2}$$

$$\mathbf{V_u = 55.78 \ kN}$$

1. **Calculation of nominal shear stress (τ_v)**

$$\tau_v = \frac{V_u}{b.d} = \frac{55.78 \times 10^3}{250 \times 400}$$

$$\therefore \qquad \mathbf{\tau_v = 0.56 \ N/mm^2}$$

Check for τ_{cmax}

$$0.56 < 2.8$$

$$\tau_v = \tau_{cmax} \qquad \qquad \text{...safe okay}$$

2. **Calculation of design shear strength of concrete (τ_c) :** τ_c depends upon $P_{t \ lim}$

$$\therefore \qquad P_{t \ lim} = \frac{A_{st}}{bd} \times 100 = \frac{603.19}{250 \times 400} \times 100$$

$$\therefore \qquad P_{t \ lim} = 0.6 \ \%$$

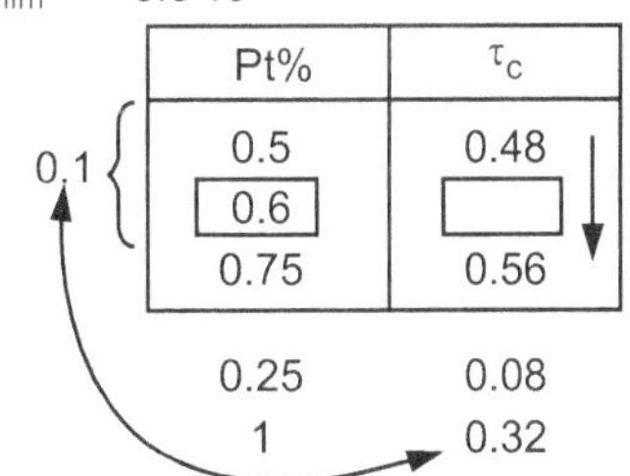

Pt%	τ_c
0.5	0.48
0.6	
0.75	0.56
0.25	0.08
1	0.32

By interpolation

$$\therefore \qquad \tau_c = 0.48 + 0.1 \times 0.32$$

$$\tau_c = 0.51 \ N/mm^2$$

3. **Compare τ_v and τ_c**

$$0.56 > 0.51$$

$\therefore$ Shear reinforcement is required.

4. **Shear force resisted by shear reinforcement (V_{us})**

$$V_{us} = V_u - V_{uc}$$

$$V_{us} = 55.78 - (0.51 \times 250 \times 400) \times 10^{-3}$$

$$V_{us} = 55.78 - 51$$

$$V_{us} = 4.78 \ kN$$

5. **Assume, 8mmϕ – 2 legged vertical stirrups as a shear reinforcement**

6. **Spacing of shear reinforcement (S_r)**

(i) $$S_v = \frac{0.87\, f_y \cdot A_{sv} \cdot d}{V_{us}}$$

$$= \frac{0.87 \times 415 \left(2 \times \dfrac{\pi}{4}\,(8)^2\right) \times 400}{4.78 \times 10^3}$$

$\therefore$ $S_v = 3037.38$ mm

(ii) $$S_v = \frac{0.87\, f_y A_{sv}}{0.4\, b} = \frac{0.87 \times 415 \left(2 \times \dfrac{\pi}{4}\right)(8)^2}{0.4 \times 250}$$

$\therefore$ $S_v = 362.97$ mm

(iii) $S_v = 0.75\, d = 0.75 \times 400$

$S_v = 300$ mm

(iv) $S_v = 300$ mm

$\therefore$ $S_v = 300$ mm

$\therefore$ Provide, 8mm ϕ – 2 legged vertical stirrups @ 300 mm c/c spacing.

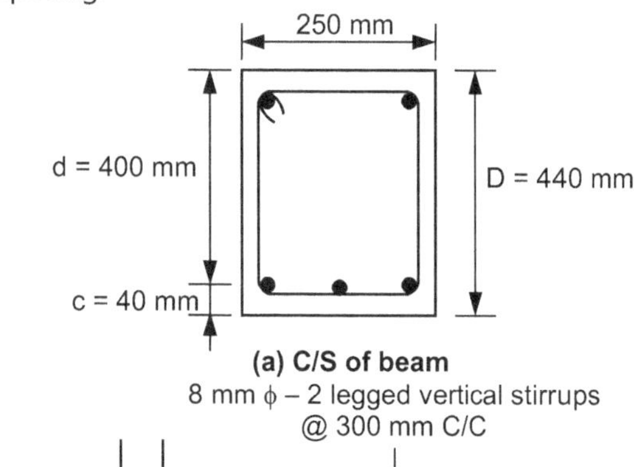

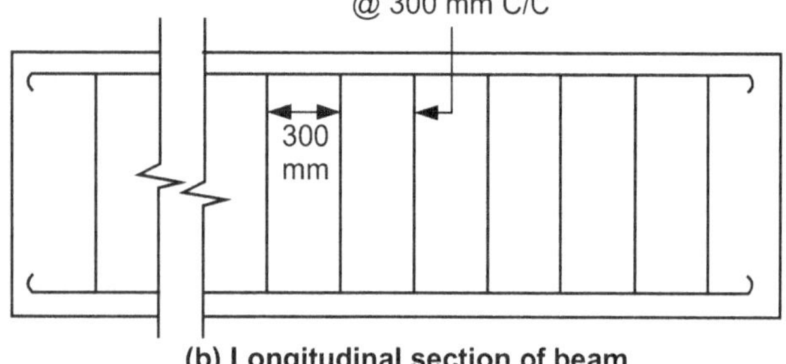

(a) C/S of beam

8 mm ϕ – 2 legged vertical stirrups @ 300 mm C/C

(b) Longitudinal section of beam

Fig. 9.12

Example 9.2 : *A beam 230 mm × 450 mm effective subjected to a factored shear force of 30 kN. Design the shear reinforcement. The beam is reinforced with 2-20 mm ϕ. Use M15 and Fe 415.*

Solution : Given data :

$b = 230$ mm, $f_{ck} = 15$ MPa, $A_{st} = 2 - 20$ mm ϕ

$d = 450$ mm, $f_y = 415$ MPa, $A_{st} = 2 \times \dfrac{\pi}{4}\,(20)^2$

$v_u = 30$ kN, $A_{st} = 628.32$ mm^2

---

1. **Calculation of nominal shear stress (τ_v)**

$$\tau_v = \frac{V_u}{b.d} = \frac{30 \times 10^3}{230 \times 450}$$

$\therefore$ $\tau_v = 0.29$ N/mm^2

Check for τ_{cmax}

$0.29 < 2.1$

$\tau_v < \tau_{c\,max}$ ...safe okay

2. **Calculate of design shear strength of concrete (τ_c)**

τ_c depends upon P_{tlim}

$$P_{tlim} = \frac{A_{st}}{b.d} \times 100 = \frac{628.32}{230 \times 450} \times 100$$

$P_{tlim} = 0.6$ %

By interpolation

$\tau_c = 0.46 + 0.1 \times 0.32$

$\therefore$ $\tau_c = 0.5$ N/mm^2

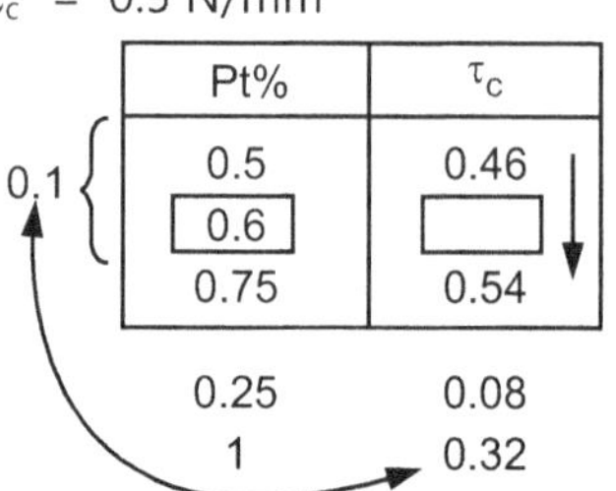

Pt%	τ_c
0.5	0.46
0.6	
0.75	0.54
0.25	0.08
1	0.32

3. **Compare τ_v and τ_c**

$0.29 < 0.5$

$\tau_v < \tau_c$

...Shear reinforcement is not required

But provide minimum shear reinforcement

4. **Assume, 8mm ϕ – 2 legged vertical stirrups**

5. **Spacing of shear reinforcement (S_v)**

(i) $$S_v = \frac{0.87\, f_y\, A_{sv}}{0.4\, b} = \frac{0.87 \times 415 \times [2 \times \pi/4\,(8)^2]}{0.4 \times 230}$$

$S_v = 394.53$ mm

(ii) $S_v = 0.75\, d = 0.75 \times 450$

$S_v = 337.5$ mm

(iii) $S_v = 300$ mm

$\therefore$ $S_v = 300$ mm

Provide 8mm ϕ – 2 legged vertical stirrups @ 300 mm c/c.

(a) C/S of beam

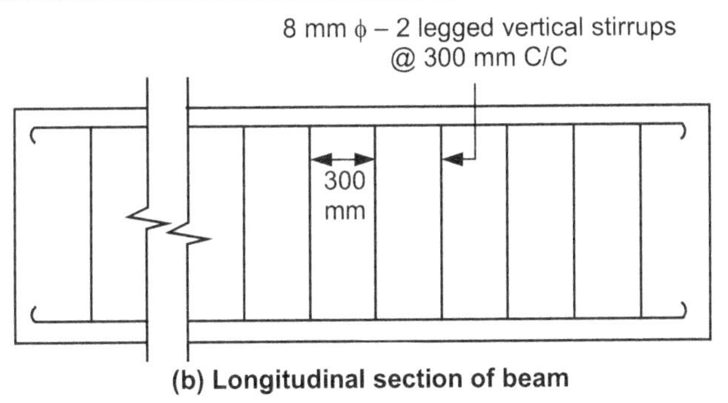

(b) Longitudinal section of beam

Fig. 9.13

Example 9.3 : *A R. C. beam 230 mm wide and 450 mm effective depth is reinforced with 3-16 mm φ of grade Fe 415 on tension side. If the beam is subjected to factored shear force of 85 kN. Design the shear reinforced for only two bars on tension side are available and third bar of beam curtailed. Assume concrete grade M20.*

Solution :

$b = 230$ mm, $f_{ck} = 20$ MPa, $A_{st} = 3 - 16$ mm φ

$d = 450$ mm , $f_y = 415$ MPa, $A_{st} = 3 \times \dfrac{\pi}{4}(16)^2$

$v_u = 85$ kN, $A_{st} = 603.19$ mm^2

1. Calculation of nominal shear stress (τ_v)

$$\tau_v = \frac{V_u}{b.d} = \frac{8.5 \times 10^3}{230 \times 450}$$

$$\tau_v = 0.82 \text{ N/mm}^2$$

Check for $\tau_{c\,max}$

$$0.82 < 2.8$$

$$\tau_c < \tau_{c\,max} \qquad \text{...Safe okay}$$

2. Calculation of design shear strength of concrete (τ_c)

τ_c depends upon $P_{t\,lim}$

$$P_{t\,lim} = \frac{A_{st}}{b.d} \times 100 = \frac{603.19}{230 \times 450} \times 100$$

$$P_{t\,lim} = 0.58\%$$

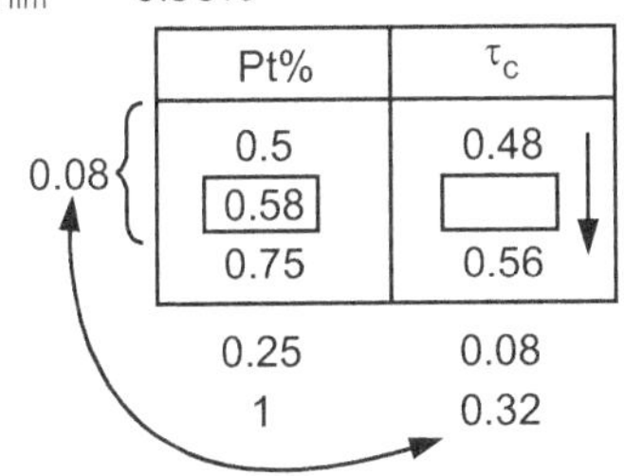

Pt%	τ_c
0.5	0.48
0.58	
0.75	0.56
0.25	0.08
1	0.32

By interpolation

$$\tau_c = 0.48 + 0.08 \times 0.32$$

$$\tau_c = 0.51 \text{ N/mm}^2$$

3. Compare τ_v and τ_c

$$0.82 > 0.51$$

$$\tau_v > \tau_c \quad \text{...Shear reinforcement is required}$$

4. Shear force resisted by shear reinforcement (V_{us})

$$V_{us} = V_u - V_{uc} = 85 \times 10^3 - (0.5 \times 230 \times 450)$$

$$V_{us} = 85 \times 10^3 - 51.75 \times 10^3$$

$$\mathbf{V_{us} = 33.25 \text{ kN}}$$

5. Assume, 8mm φ – 2 legged vertical stirrups

6. Spacing of shear reinforcement (S_v)

(i)
$$S_v = \frac{0.87\, f_y \cdot A_{sv} \cdot d}{V_{us}}$$

$$= \frac{0.87 \times 415 \times [2 \times \pi/4 \times (8)^2] \times 450}{33.25 \times 10^3}$$

$$S_v = 491.23 \text{ mm}$$

(ii)
$$S_v = \frac{0.87\, f_y\, A_{sv}}{0.4b} = \frac{0.87 \times 415 \times [2 \times \pi/4\,(8)^2]}{0.4 \times 230}$$

$$S_v = 394.53 \text{ mm}$$

(iii)
$$S_v = 0.75\, d = 0.75 \times 450$$

$$S_v = 337.5 \text{ mm}$$

(iv)
$$S_v = 300 \text{ mm}$$

$$\therefore \qquad \mathbf{S_v = 300 \text{ mm}}$$

$\therefore$ **Provide 8 mm φ – 2 legged vertical stirrups @ 300 mm c/c.**

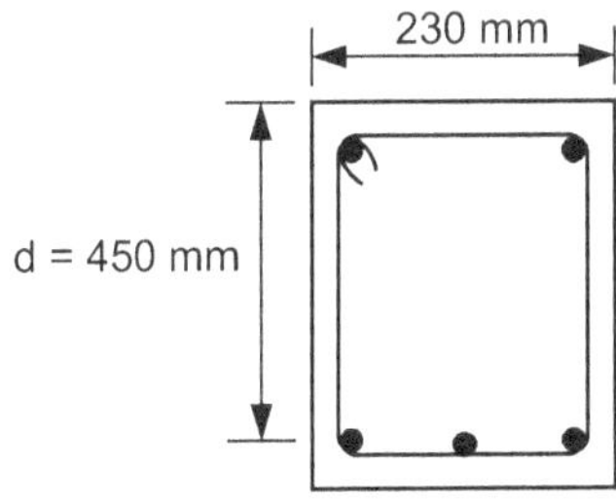

(a) C/S of beam

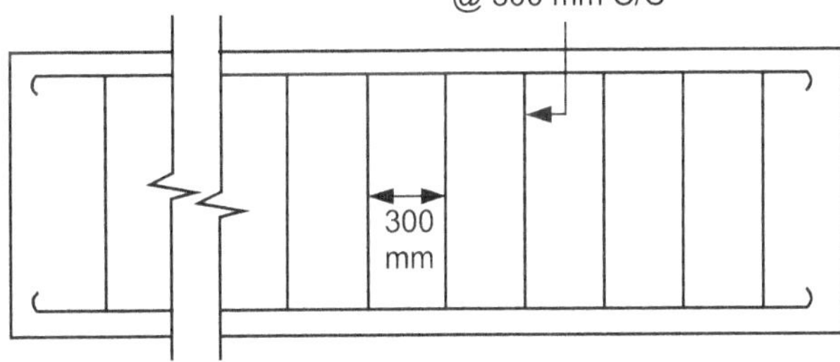

(b) Longitudinal section of beam

Fig. 9.14

Example 9.4 : *A simply supported R.C. beam 230 mm wide and 700 mm deep effective is reinforcement with 4-20 mm φ of centre, out of which 2 bars are bent-up at the top near the support and remaining bars are continued in addition to this vertical stirrups 8 mm φ at spacing of 110 mm c/c are provided. Determine the shear strength resisted by beam, at a section crossing the bent-up bors.*

Solution :

$b = 230$ mm, $f_{ck} = 20$ MPa,

$A_{st} = 4 - 20$ mm $\phi = 4 \times \dfrac{\pi}{4}(20)^2$

$d = 700$ mm, $f_y = 451$ MPa, $A_{st} = 1256.64$ mm^2

$S_v = 110$ mm, $A_{sv} = 8$ mm $\phi - 2$ legged vertical stirrups

Shear strength resisted by beam at s/c crossing the bent-up bars = [Shear st. resisted by concrete + shear st. resisted by vertical stirrups + Shear st. resisted by bent-up bars]

$$V_u = V_{uc} + V_{usv} + V_{usb} \qquad \ldots(1)$$

For τ_c $\quad P_{t\,lim} = \dfrac{A_{st}}{b.d} \times 100 = \dfrac{1256.64}{230 \times 700} \times 100$

$\quad P_{t\,lim} = 0.78\ \%$

$0.03\Big\{$

Pt%	τ_c
0.75	0.56
0.78	
1.0	0.62

| 0.25 | 0.06 |
| 1 | 0.24 |

By interpolation

$\quad \tau_c = 0.56 + 0.03 \times 0.24$

$\quad \boldsymbol{\tau_c = 0.57\ N/mm^2}$

$V_u = [0.57 \times 230 \times 700]$

$\qquad + \left[\dfrac{0.87\ f_y\ A_{sv}\ d}{V_{us}}\right] + [0.87\ f_y\ A_{sv} \cdot \sin \alpha]$

$V_u = [0.57 \times 230 \times 700]$

$\qquad + \left[\dfrac{0.87 \times 415 \times (2 \times \pi/4\,(8)^2) \times 700}{110 \times 10^3}\right]$

$\qquad + \left[0.87 \times 415 \times (2 \times \pi/4\,(8)^2) \cdot \sin 45\right]$

$\therefore \qquad V_u = 483.15$ kN

Example 9.5 : *An R.C. beam 350 mm wide and 500 mm effective depth is reinforced with 4-25 mm ϕ. It is having a udl of 75 kN/m including self weight over a simply supported span of 7m. Design the shear reinforcement using vertical stirrups and bent-up bars.*

Solution :

$b = 350$ mm, $f_{ck} = 20$ MPa,

$A_{st} = 4 - 25$ mm $\phi = 4 \times \dfrac{\pi}{4}(25)^2$

$d = 500$ mm, $f_y = 415$ MPa, $A_{st} = 1963.5$ mm^2

$l_e = 7$m, $w = 75$ kN/m

$\therefore$ Max. udl $= w_u = 75 \times 1.5 = 112.5$ kN/m

$\therefore$ Factored S.F. $= V_u = \dfrac{w_u \cdot l_e}{2} = \dfrac{112.5 \times 7}{2}$

$\quad \boldsymbol{V_u = 393.75\ kN}$

1. Calculation of nominal shear stress (τ_v)

$$\tau_v = \frac{V_u}{b.d} = \frac{393.75 \times 10^3}{350 \times 500}$$

$\quad \boldsymbol{\tau_v = 2.25\ N/mm^2}$

Check for $\tau_{c\,max}$

$\quad 2.25 < 2.8$

$\quad \tau_v < \tau_{c\,max} \qquad\qquad \ldots$safe okay

2. Design shear strength of concrete (τ_c)

τ_c depends upon $P_{t\,lim}$

$\therefore \quad P_{t\,lim} = \dfrac{A_{st}}{b.d} \times 100 = \dfrac{1963.5}{350 \times 500} \times 100$

$\quad P_{t\,lim} = 1.12\ \%$

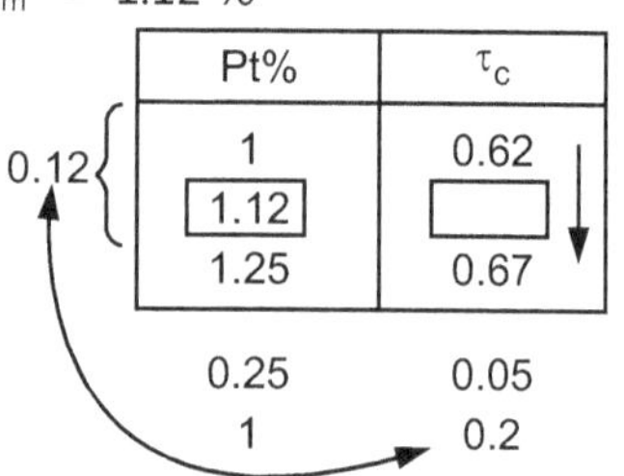

$0.12\Big\{$

Pt%	τ_c
1	0.62
1.12	
1.25	0.67

| 0.25 | 0.05 |
| 1 | 0.2 |

By interpolation

$\quad \tau_c = 0.62 + 0.12 \times 0.2$

$\quad \boldsymbol{\tau_c = 0.64\ N/mm^2}$

3. Compare τ_v and τ_c

$\quad 2.25 > 0.64$

$\quad \tau_v > \tau_c \qquad$ shear reinforcement is required

4. Shear force resisted by shear reinforcement (V_{us})

$\quad V_{us} = V_u - V_{uc}$

$\qquad = 393.75 \times 10^3 - (0.64 \times 350 \times 500)$

$\quad \boldsymbol{V_{us} = 281.75\ kN}$

Assuming out of 281.75 kN load, 50% load is resisted by vertical stirrups (i.e. 140.88 kN) and 50% load is resisted by bent-up bars.

5. Shear force resisted by bent-up bars (V_{usb})

Assume, out of 4 bars, 2 bars are bent-up.

$\quad V_{usb} = 0.87\ f_y \cdot A_{sy} \cdot \sin \alpha$

$\quad V_{usb} = 0.87 \times 415 \times [2 \times \pi/4\,(25)^2] \cdot \sin 45$

$\quad \boldsymbol{V_{usb} = 250.64\ kN} > 140.88$ kN

6. Shear force resisted by vertical stirrups

Assume 8mm$\phi - 2$ legged vertical stirrups

(i) $\quad S_v = \dfrac{0.87\ f_y \cdot A_{sv} \cdot D}{V_{usv}}$

$\qquad = \dfrac{0.87 \times 415\,[2 \times \pi/4\,(8)^2] \times 500}{140.88 \times 10^3}$

$\quad S_v = 128.82$ mm

(ii) $S_v = \dfrac{0.87\, f_y \cdot A_{sv}}{0.4\, b}$

$\qquad = \dfrac{0.87 \times 415 \times [2 \times \pi/4\,(8)^2]}{0.4 \times 350}$

$\quad S_v = 259.62$ mm

(iii) $S_v = 0.75\, d = 0.75 \times 500$

$\quad S_v = 375$ mm

(iv) $S_v = 300$ mm

$\therefore \quad S_v = 128.82$ mm

$\therefore \quad \mathbf{S_v \cong 125\ mm}$

Provide, 8 mm ϕ – 2 legged vertical stirrups @ 125 mm c/c.

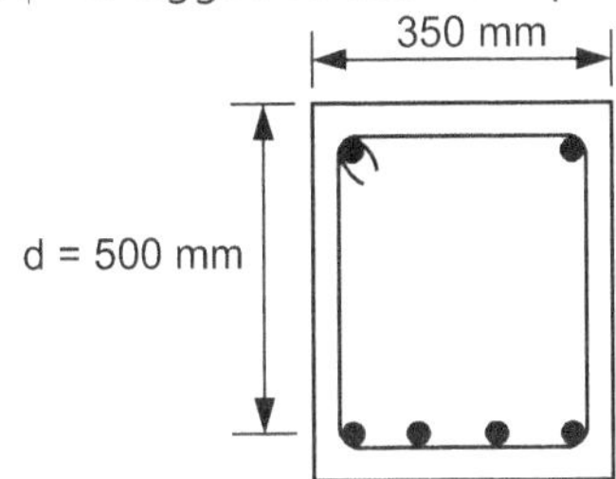

(a) C/S of beam
8 mm ϕ – 2 legged vertical stirrups
@ 125 mm C/C

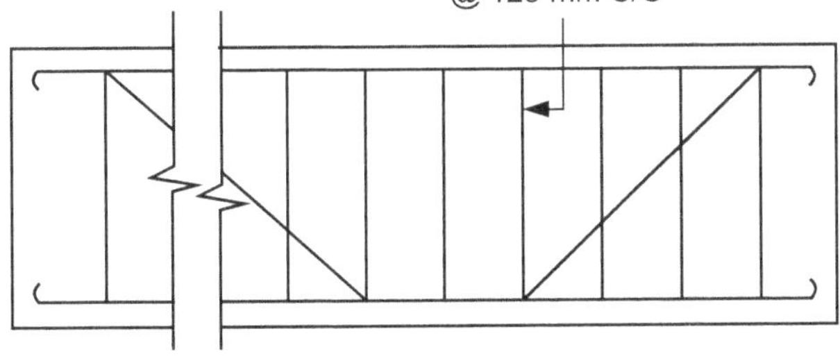

(b) Longitudinal section of beam
Fig. 9.15

Example 9.6 : *A rectangular beam section of 300 mm width and 500 mm effective depth is reinforced with 6-20 mm ϕ out of which 3 bars have been cranked at 45°. Design the shear reinforcement if the beam is subjected to maximum load of 100 kN and maximum concentrated load of 300 kN at mid-span of simply supported span of 5m. Use M20 mix and Fe 415 steel bars.*

Solution :

b = 300 mm, f_{ck} = 20 MPa,

$A_{st} = 6 - 20$ mm $\phi = 6 \times \pi/4\,(20)^2$

d = 500 mm, f_y = 415 MPa, A_{st} = 1884.96 mm^2

l_{eff} = 5m, w_{u_1} = 100 kN/m, w_{u_2} = 300 kN

3 bars are bent up

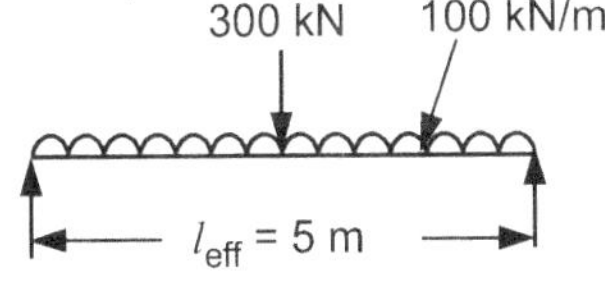

Fig. 9.16

Factored shear force

$$V_u = \frac{w_{u_1} l_e}{2} + \frac{w_{u_2}}{2}$$

$$V_u = \frac{100 \times 5}{2} + \frac{300}{2}$$

$$\mathbf{V_u = 400\ kN}$$

1. Calculation of nominal shear stress (τ_v)

$$\tau_v = \frac{V_u}{b.d} = \frac{400 \times 10^3}{300 \times 500}$$

$$\mathbf{\tau_v = 2.67\ N/mm^2}$$

Check for $\tau_{c\,max}$

$\qquad 2.67 \ <\ 2.8$

$\qquad \tau_v \ <\ \tau_{c\,max}$ $\qquad\qquad$ …safe okay.

2. Calculation of design shear strength of concrete (τ_c)

τ_c **depends upon $P_{t\,lim}$.**

Pt%	τ_c
1.25	0.67
1.26	
1.5	0.72
0.25	0.05
1.0	0.2

$$P_{t\,lim} = \frac{A_{st}}{b.d} \times 100 = \frac{1884.96}{300 \times 500} \times 100$$

$$P_{t\,lim} = 1.25\%$$

By interpolation

$\qquad \tau_c = 0.67 + 0.01 \times 0.2$

$\qquad \mathbf{\tau_c = 0.67\ N/mm^2}$

3. Compare τ_v and τ_c

$\qquad 2.67 \ >\ 0.67$

$\qquad \tau_v \ >\ \tau_c$ $\quad$ …shear reinforcement is required.

4. Shear force resisted by shear reinforcement (V_{us})

$\qquad V_{us} = V_u - V_{uc}$

$\qquad\quad = 400 \times 10^3 - (0.67 \times 300 \times 500)$

$\qquad V_{us} = 400 \times 10^3 - 100.5 \times 10^3$

$\qquad \mathbf{V_{us} = 299.5\ kN}$

Assuming, out of 299.5 kN force, 50% load is resisted by bent up bars (i.e. 149.75 kN) and 50% load is resisted by vertical stirrups (149.75 kN).

5. Shear force resisted by bent-up bars (V_{usb})

$\qquad V_{usb} = 0.87\, f_y \cdot A_{sv} \cdot \sin \alpha$

$\qquad V_{usb} = 0.87 \times 415 \times \left[3 \times \dfrac{\pi}{4}\,(20)^2\right] \times \sin 45°$

$\qquad \mathbf{V_{usb} = 240.62\ kN} > 149.75\ kN \quad$ …safe okay

6. Assuming, 8mm ϕ - 2 legged vertical stirrups

7. Spacing of shear reinforcement (S_v) :

(i) $S_v = \dfrac{0.87\, f_y \cdot A_{sv} \cdot d}{V_{usv}}$

$\quad = \dfrac{0.87 \times 415 \times [2 \times \pi/4\,(8)^2] \times 500}{149.75 \times 10^3}$

$\quad S_v = 121.19 \text{ mm}$

(ii) $s_v = \dfrac{0.87\, fy\, A_{sv}}{0.4\, b}$

$\quad = \dfrac{0.87 \times 415 \times \left[2 \times \dfrac{\pi}{4} \times (8)^2\right]}{0.4 \times 300}$

$\quad s_v = 302.47 \text{ mm}$

(iii) $s_v = 0.75\, d = 0.75 \times 500$

$\quad s_v = 375 \text{ mm}$

(iv) $s_v = 300 \text{ mm}$

$\therefore \quad s_v = 121.19 \text{ mm}$

$\therefore \quad \mathbf{s_v \cong 120 \text{ mm}}$

$\therefore$ **Provide, 8 mm ϕ – 2 legged vertical stirrups @ 120 mm c/c.**

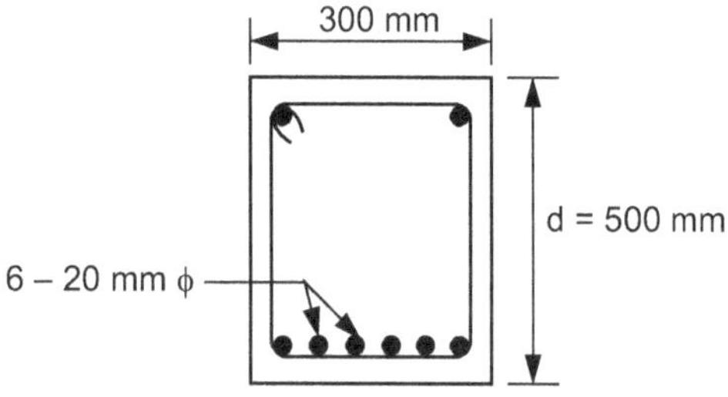

(a) C/S of beam

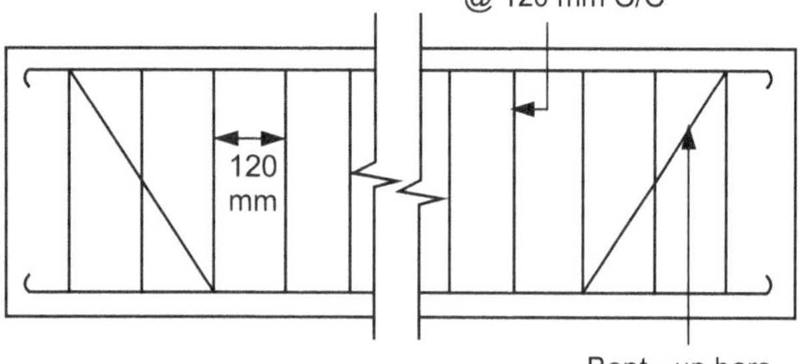

(b) Longitudinal section of beam

Fig. 9.17

Example 9.7 : *A reinforced concrete beam of rectangular section 380 mm wide is reinforced with 4-20 mm ϕ at an effective depth of 580 mm, out of which 2 bars are bent-up near the support section where a factored shear force of 350 kN is acting. Use M25 grade of concrete and Fe 415 grade of steel. Design a suitable shear reinforcement at support section.*

Solution : b = 380 mm, d = 580 mm, f_{ck} = 25 MPa, f_y = 415 MPa, A_{st} = 4 – 20 mmϕ = $4\,\dfrac{\pi}{4}\,(20)^2$, A_{st} = 1256.64 mm^2, factored S.F. = V_u = 350 kN.

1. Calculation of nominal shear stress (τ_v) :

$$\tau_\gamma = \frac{V_u}{b \cdot d} = \frac{350 \times 10^3}{380 \times 580}$$

$$\tau_\gamma = 1.59 \text{ N/mm}^2$$

Check for τ_{cmax}

$$1.59 < 3.1$$

2. Calculation of design shear strength of concrete (τ_c) :

τ_c depends upon $P_{t\,lim}$.

$$P_{t\,lim} = \frac{A_{st}}{b \cdot d} \times 100 = \frac{1256.64}{380 \times 580} \times 100$$

$$P_{t\,lim} = 0.57 \%$$

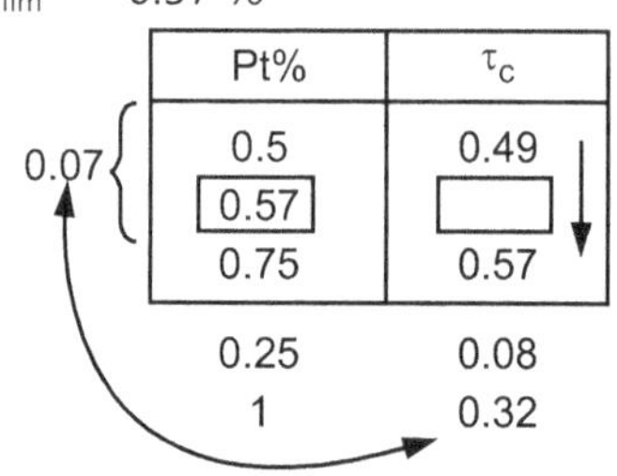

Pt%	τ_c
0.5	0.49
0.57	
0.75	0.57
0.25	0.08
1	0.32

by interpolation

$$\tau_c = 0.49 + 0.07 \times 0.32$$

$$\tau_c = 0.51 \text{ N/mm}^2$$

3. Compare τ_v and τ_c

$$1.59 > 0.51$$

$\quad \tau_v > \tau_c \quad$ ... shear reinforcement is required

4. Shear force resisted by shear reinforcement (v_{us}) :

$$v_{us} = v_u - v_{uc} = 350 \times 10^3 - (0.51 \times 380 \times 580)$$

$$v_{us} = 350 \times 10^3 - 112.40 \times 10^3$$

$$\mathbf{v_{us} = 237.6 \text{ kN}}$$

Assuming, out of 237.6 kN force, 50% load is resisted by bent-up bars (i.e. 118.80 kN) and 50% resisted by vertical stirrups

5. Shear force resisted by bent-up bars :

$$v_{usb} = 0.87\, f_y\, A_{sy} \cdot \sin \alpha$$

$$= 0.87 \times 415 \times \left[2 \times \frac{\pi}{4}\,(20)^2\right] \cdot \sin 45$$

$$v_{usb} = 160.41 \text{ kN}$$

6. Assume, 8 mm ϕ – 2 legged vertical stirrups

7. Spacing and shear reinforcement (S_v) :

(i) $S_v = \dfrac{0.87 \, f_y \, A_{sv} \, d}{V_{us}}$

$= \dfrac{0.87 \times 415 \times \left[2 \times \dfrac{\pi}{4}(8)^2 \right] \times 580}{118.80 \times 10^3}$

$S_v = 177.20$ mm

(ii) $S_v = \dfrac{0.87 \, f_y \, A_{sv}}{0.4b} = \dfrac{0.87 \times 415 \times \left[2 \times \dfrac{\pi}{4}(8)^2 \right]}{0.4 \times 380}$

$S_v = 238.79$ mm

(iii) $S_v = 0.75d = 0.75 \times 580 = 435$ mm

∴ $S_v = 177.20$ mm

∴ $S_v \cong 175$ mm

∴   **Provide, 8 mm φ – 2 legged vertical stirrups @ 175 mm c/c.**

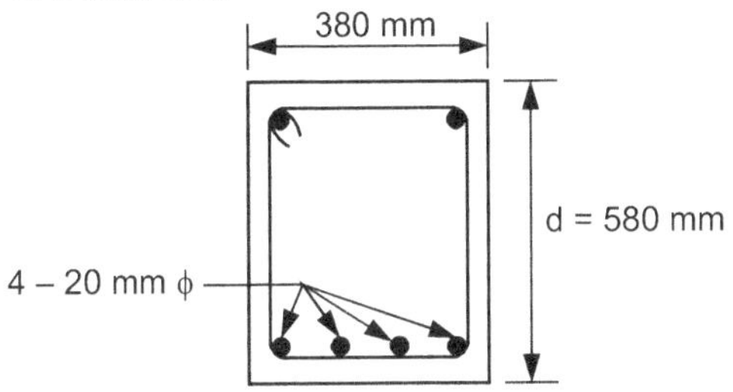

(a) C/S of beam

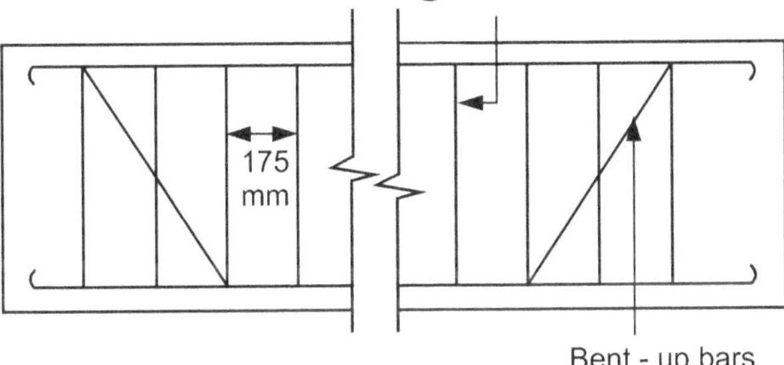

(b) Longitudinal section of beam

Fig. 9.18

Example 9.8 : *Design the shear reinforcement for a cantilever beam carrying a udl of 26 kN/m without D.L. having a span of 1.2 m and 230 mm × 650 mm in size, tensile reinforcement of 4-22 mm φ, compressive reinforcement 2-12 mm φ. Use M20 & Fe 415.*

Solution : Given : b = 230 mm, D = 650 mm, f_{ck} = 20 Mpa, f_y = 415 MPa,

$A_{st} = 4 - 22$ mm φ $= 4 \times \dfrac{\pi}{4}(22)^2$,

A_{st} = 1520.53 mm^2, w' = 26 kN/m,

Assume cover – C = 50 mm

∴   d = 600 mm

l_{eff} = 1.2 m

∴   safe udl = w = 26 + (0.23 × 0.65 × 1) × 25

= 29.74 kN/m

Factored udl = w_u = 29.74 × 1.5 kN/m

∴   **w_u = 44.61 kN/m**

Factored shear force (v_u) = $w_u \cdot l_{eff}$

v_u = 44.61 × 1.2

v_u = 53.53 kN

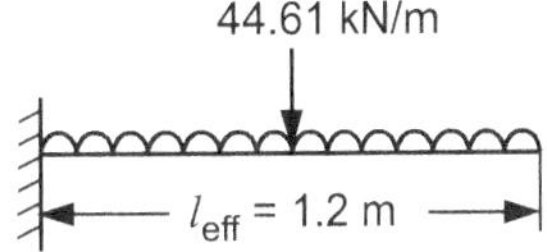

Fig. 9.19

1. Calculation of nominal shear stress (τ_v) :

$\tau_v = \dfrac{v_u}{b \cdot d} = \dfrac{53.53 \times 10^3}{230 \times 600}$

$\tau_v = 0.39$ N/mm^2

Check for $\tau_{c\,max}$

0.39 < 2.8

$\tau_v < \tau_{cmax}$ ... Safe okay

2. Calculation of design shear strength of concrete (τ_c) :

τ_c depends upon $P_{t\,lim}$

$P_{t\,lim} = \dfrac{A_{st}}{b \cdot d} \times 100 = \dfrac{1520.53}{230 \times 600} \times 100$

$P_{t\,lim} = 1.10$ %

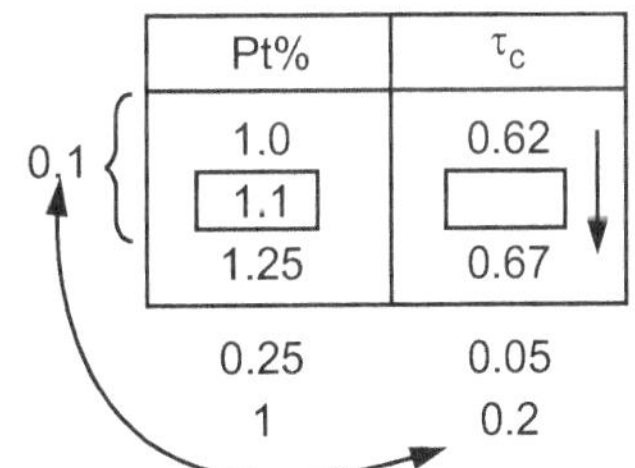

Pt%	τ_c
1.0	0.62
1.1	
1.25	0.67
0.25	0.05
1	0.2

by interpolation

$\tau_c = 0.62 + 0.1 \times 0.2$

$\tau_c = 0.64$ N/mm^2

3. Compare τ_v and τ_c

0.39 < 0.64

$\tau_v < \tau_c$ ... Shear reinforcement is not required

But, provide minimum shear reinforcement.

4. Assume, 8 mm φ – 2 legged vertical stirrups

5. Spacing of shear reinforcement (s_v) :

(i) $s_v = \dfrac{0.87\, f_y\, A_{sv}}{0.4\, b} = \dfrac{0.87 \times 415 \times \left[2 \times \dfrac{\pi}{4}\,(8)^2\right]}{0.4 \times 230}$

$s_v = 394.53$ mm

(ii) $s_v = 0.75\, d = 0.75 \times 600$

$s_v = 450$ mm

(iii) $s_v = 300$ mm

∴ $\mathbf{s_v = 300}$ **mm**

∴ **Provide 8 mm ϕ – 2 legged vertical stirrups @ 300 mm c/c.**

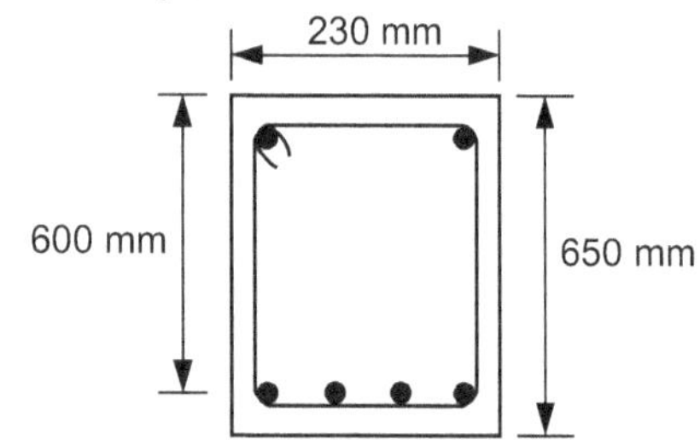

(a) C/S of beam

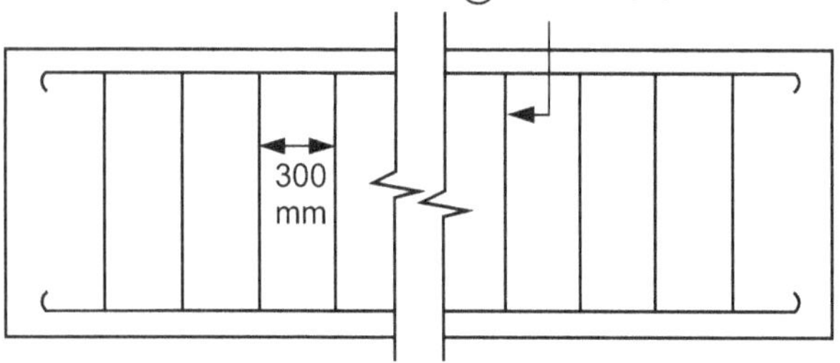

(b) Longitudinal section of beam

Fig. 9.20

Example 9.9: *Design a T-beam for the floor of hall 5.5 m × 15 m for the following data :*

L.L on slab = 3.5 kN/m²

Thickness of slab = 125 mm

Design the section for shear also. Use concrete M20 and steel Fe415 grade.

Solution : Given : Hall size = 5.5 m × 15 m

L.L on slab = 3.5 kN/m²

$F_{ck} = 20$ MPa

$f_y = 415$ MPa

Depth of flange = D_f = 125 mm

Assume, c/c distance between beams = 3 m

Support width = 230 mm

∴ $L_0 = 5.5 + 0.23 = 5.73$ m

(I) Calculation of dimensions :

1. Calculation of overall depth (D) :

$$D = \frac{L_0}{12} = \frac{5.73 \times 10^3}{12}$$

$D = 477.50$ mm

$D \cong 500$ mm

Assume, cover c = 40 mm

∴ d = 460 mm

2. Calculation of D_f

$D_f = 125$ mm ... Given

3. Calculation of b_w

Assume, $b_w = 230$ mm

4. Calculation of b_f

$$b_f = \frac{L_0}{6} + b_w + 6\, D_f$$

$$= \frac{5730}{6} + 230 + 6 \times 125$$

$b_f = 1935$ mm

(II) Calculation of loading :

L.L. = $3.5 \times 3 = 10.5$ kN/m

F.F. = $3 \times 1 = 3$ kN/m

self weight of slab = $(3 \times 0.125 \times 1) \times 25 = 9.38$ kN/m

Self weight of rib = $[(0.5 - 0.125) \times 0.23 \times 1] \times 25$

$= 2.16$ kN/m

∴ Total udl = w = 25.04 kN/m ... Safe udl

∴ Factored udl = $w_u = w \times r_f = 25.04 \times 1.5$

$w_u = 37.56$ kN/m

(III) Calculation of moment of resistance (M_u) :

$$M_u = \frac{w_u \cdot l_e^2}{8} = \frac{37.56 \times 5.73^2}{8}$$

$M_u = 154.15$ kN-m

(IV) Calculation of actual depth of neutral axis :

(a) Assuming N.A. lies within flange, (i.e. $x_u < D_f$)

$M_u = C_u \times Z$

$154.15 \times 10^6 = 0.36\, f_{ck} \cdot x_u \cdot b_f \cdot (d - 0.42\, x_u)$

$154.15 \times 10^6 = 0.36 \times 20 \times x_u \times 1935\,(460 - 0.42\, x_u)$

$x_u = 24.60$ mm $< D_f = 125$ mm ... Safe okay

(V) Calculation of area of steel (A_{st}) :

$$A_{st} = \frac{0.5\, f_{ck}}{f_y}\left[1 - \sqrt{1 - \frac{4.6\, M_u}{f_{ck} \cdot b_f \cdot d^2}}\right] \times b \cdot d$$

$$A_{st} = \frac{0.5 \times 20}{415}\left[1 - \sqrt{1 - \frac{4.6 \times 154.15 \times 10^6}{20 \times 1935 \times 460^2}}\right]$$
$$\times 1935 \times 460$$

$A_{st} = 949.64$ mm²

$$\text{no. of bars} = \frac{A_{st}}{a_{st}} = \frac{949.64}{\frac{\pi}{4}(18)^2}$$

$$= 3.73$$

∴ Provide 4-18 mmφ as a tensile reinforcement

(VI) Design shear reinforcement

$$\text{Factored shear force } (V_u) = \frac{w_u \cdot l_e}{2} = \frac{37.56 \times 5.73}{2}$$

$$V_u = 107.61 \text{ kN}$$

1. Calculation of nominal shear stress (τ_v) :

$$\tau_v = \frac{V_u}{b \cdot d} = \frac{107.61 \times 10^3}{230 \times 460}$$

$$\tau_v = 1.02 \text{ N/mm}^2$$

Check for τ_{cmax}

$$1.02 < 2.8$$

$$\tau_c < \tau_{cmax} \qquad \qquad \dots \text{Safe okay}$$

2. Calculation of design shear strength of concrete (τ_c) :

τ_c depends upon $P_{t\,lim}$,

$$P_{t\,lim} = \frac{A_{st}}{b \cdot d} \times 100 = \frac{\left[4 \times \frac{\pi}{4}(18)^2\right]}{230 \times 460} \times 100$$

$$P_{t\,lim} = 0.96 \%$$

By interpolation

$$\tau_c = 0.56 + 0.21 \times 0.24$$

$$\tau_c = 0.61 \text{ N/mm}^2$$

	Pt%	τ_c
0.21 {	0.75	0.56
	0.96	
	1.0	0.62
	0.25	0.06
	1	0.24

3. Compare τ_v and τ_c

$$1.02 > 0.61$$

$$\tau_v > \tau_c \quad \dots \text{shear reinforcement is required.}$$

4. Shear force resisted by shear reinforcement (v_{vs}):

$$V_{us} = V_u - V_{uc}$$

$$= 107.61 \times 10^3 - (0.61 \times 230 \times 460)$$

$$V_{us} = 107.61 \times 10^3 - 64.54 \times 10^3$$

$$\mathbf{V_{us} = 43.07 \text{ kN}}$$

5. Assume, 8 mm φ – 2 legged vertical stirrups

6. Spacing of shear reinforcement (s_v) :

(i)
$$s_v = \frac{0.87 \, f_y \cdot A_{sv} \cdot d}{V_{us}}$$

$$= \frac{0.87 \times 415 \times \left[2 \times \frac{\pi}{4}(8)^2\right] \cdot 460}{43.07 \times 10^3}$$

$$s_v = 407.67 \text{ mm}$$

(ii)
$$s_v = \frac{0.87 \, f_y \, A_{sv}}{0.4 \, b} = \frac{0.87 \times 415 \times \left[2 \times \frac{\pi}{4}(8)^2\right]}{0.4 \times 230}$$

$$s_v = 394.53 \text{ mm}$$

(iii)
$$s_v = 0.75 \, d = 0.75 \times 460$$

$$s_v = 345 \text{ mm}$$

(iv)
$$s_v = 300 \text{ mm}$$

∴ $$\mathbf{s_v = 300 \text{ mm}}$$

Provide, 8 mm φ – 2 legged vertical stirrups @ 300 mm c/c.

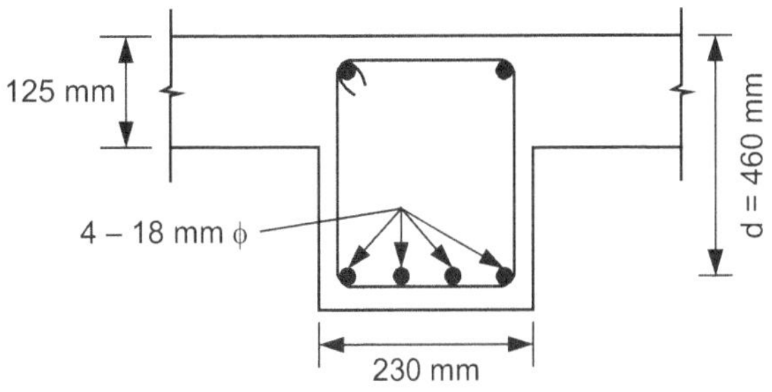

(a) C/S of beam

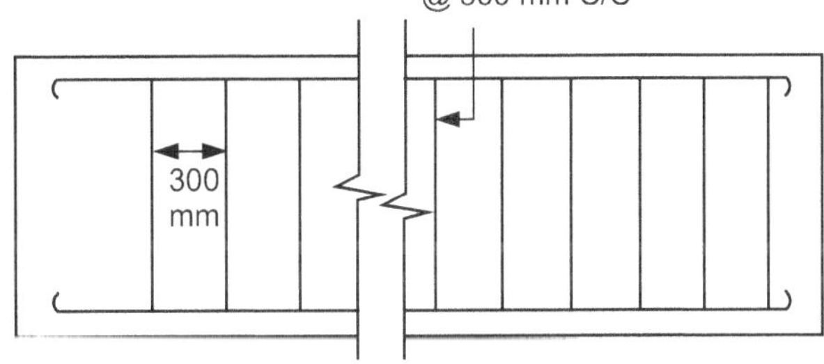

(b) Longitudinal section of beam

Fig. 9.21

Example 9.10 : *A reinforced concrete beam 230 mm wide and 460 mm effective depth is subjected to a shear force of 60 kN at support. The tensile reinforcement at the support is 0.5 per cent. Design the shear reinforcement.*

The mild steel reinforcement and the concrete having characteristic strength of 20 N/mm^2 is used.

Solution :

Given : b = 230 mm, d = 460 mm, p_t = 0.5, f_{ck} = 20 N/mm^2,

$$f_y = 250 \text{ N/mm}^2, \ V = 65 \text{ kN}$$

Required : Design of shear reinforcement.

Now, design shear force,

$$V_{uD} = 1.5 \times 60 = 90 \text{ kN}$$

From Table 9.2,

$$T_{u, max} = 2.8 \text{ N/mm}^2 \text{ (for M-20)}$$

Maximum allowable shear stress,

$$V_{uc, max} = 2.8 \times \frac{230 \times 460}{1000} = 296.24 \text{ kN} > V_{uD}$$

From Table 9.1, for $p_t = 0.5\%$ and M 20 concrete,

$$T_{uc} = 0.48 \text{ N/mm}^2$$

Shear resisted by concrete,

$$V_{uc} = T_{uc} \cdot b \cdot d$$

$$= 0.48 \times \frac{230 \times 460}{1000} = 50.7 \text{ kN}$$

Shear resisted by minimum stirrups,

$$V_{us, min} = 0.4 \, b \cdot d$$

$$= 0.40 \times 230 \times \frac{460}{1000} = 42.3 \text{ kN}$$

Shear resisted by concrete and minimum stirrups,

$$= (50.7 + 42.3) \text{ kN}$$

$$= 93.0 \text{ kN} > V_{uD} \text{ (90 kN)}$$

Therefore, minimum shear reinforcement is provided.

Select 6 mm ϕ 2-legged stirrup,

$$A_{sv} = 2 \times 28 = 56 \text{ mm}^2$$

For minimum shear reinforcement,

$$\frac{A_{sv}}{bS_v} \geq \frac{0.4}{0.87 \, f_y}$$

$$\frac{56}{230 \, S_v} \geq \frac{0.4}{0.87 \times 250}$$

$$S_v \leq \frac{56 \times 0.87 \times 250}{230 \times 0.4}$$

$$S_v \leq 132.4 \text{ mm}$$

The spacing shall not exceed

(a) $0.75 \times d = 0.75 \times 460 = 345$ mm

(b) 300 mm

(c) 132.4 mm, as calculated above.

∴ Provide 6 mm ϕ 2-legged stirrups at 130 mm c/c.

Example 9.11 : *If the shear force of the above section is increased to 90 kN, check the shear stress and find the spacing of 8 mm diameter stirrups at the support. Use HYSD bars.*

Solution :

The design shear force,

$$V_{uD} = 1.5 \times 90 = 135 \text{ kN}$$

Maximum allowable shear stress,

$$V_{uc, (max)} = 296.24 \text{ kN} > V_{uD} \text{ (= 135 kN)}$$

Shear resistance of concrete with minimum stirrups,

$$= (50.7 + 42.3) \text{ kN}$$

$$= 93 \text{ kN} < V_{uD} \text{ (= 135 kN)}$$

Therefore, design shear reinforcement is required.

Now, shear to be resisted by stirrups,

$$V_{usv} = V_{uD} - V_{uc}$$

$$= 135 - 50.7 = 84.3 \text{ kN}$$

Using 8 mm ϕ 2-legged stirrups,

$$A_{sv} = 100 \text{ mm}^2$$

$$S_v = \frac{0.87 \, f_y \cdot A_{sv} \cdot d}{V_{usv}}$$

$$= \frac{0.87 \times 415 \times 100 \times 460}{84.3 \times 10^3}$$

$$= 197 \text{ mm}$$

The spacing shall not exceed

(a) $0.75 \, d = 0.75 \times 460 = 345$ mm

(b) 300 mm

(c) 197 mm

Provide 8 mm ϕ 2-legged stirrups at 190 mm c/c.

Example 9.12 : *A rectangular beam with 300 mm width and effective depth 450 mm is reinforced on tension side with 3 numbers of 20 mm ϕ bars of grade 415. The beam is subjected to a shear of 100 kN. Design the shear reinforcement, if bending 1 bar of 20 mm ϕ at an angle of 45°. Use M 20 grade of concrete and 415 grade of steel.*

Solution :

Given : b = 300 mm, d = 450 mm, V = 100 kN,

A_{st} = 3 – 20 mm ϕ.

Now, design shear force,

$$V_{uD} = 1.5 \times 100 = 150 \text{ kN}$$

Maximum allowable shear,

$$V_{uc, max} = 2.8 \times \frac{300 \times 450}{1000}$$

$$= 378 \text{ kN} > V_{uD} \ (= 150 \text{ kN})$$

One bar of 20 mm ϕ is bent up to resist shear.

∴   Area of tension steel available,

$$A_{st} = 2 \times \left[\frac{\pi}{4} \times (20)^2 \right] = 628 \text{ mm}^2$$

$$p_t = \frac{100 \ A_{st}}{bd} = \frac{100 \times 628}{300 \times 450} = 0.465\%$$

From Table 9.1,

$$\tau_{uc} = 0.463 \text{ N/mm}^2$$

∴   Shear resisted by concrete,

$$V_{uc} = \tau_{uc} \cdot b \cdot d$$

$$= 0.463 \times \frac{300 \times 450}{1000} = 62.5 \text{ kN}$$

Shear to be carried by stirrups,

$$V_{us} = V_{uD} - V_{uc}$$

$$= 150 - 62.5 = 87.5 \text{ kN}$$

Shear resisted by one bar bent at 45°,

$$V_{usb} = 0.87 \ f_y \cdot A_{sv} \times \sin \alpha$$

$$= 0.87 \times 415 \times (314) \times \sin 45° = 80.2 \text{ kN}$$

According to code,

$$V_{usb} \leq \frac{V_{us}}{2}$$

$$\leq \frac{87.5}{2} = 43.75 \text{ kN}$$

Therefore, the useful contribution to bent bar is 43.75 kN only.

Shear to be resisted by vertical stirrups,

$$V_{usv} = V_{us} - 43.75 = 87.5 - 43.75 = 43.75 \text{ kN}$$

Now, shear resisted by minimum stirrups

$$= 0.4 \ bd = 0.4 \times \frac{300 \times 450}{1000}$$

$$= 54 \text{ kN} > V_{usv} = (= 43.75 \text{ kN})$$

Therefore, minimum shear reinforcements are required.

Provide 8 mm ϕ 2-legged stirrups.

$$A_{sv} = 2 \times \left[\frac{\pi}{4} \times (8)^2 \right] = 100 \text{ mm}^2$$

For minimum shear reinforcement,

$$\frac{A_{sv}}{bS_v} \geq \frac{0.4}{0.87 \ f_y}$$

$$\frac{100}{300 \times S_v} \geq \frac{0.4}{0.87 \times 415}$$

$$S_v \leq \frac{100 \times 0.87 \times 415}{0.4 \times 300}$$

$$S_v \leq 300.8 \text{ mm}$$

The spacing shall not exceed

(a) $0.75 \times d = 337.5$ mm.

(b) 300 mm.

(c) 300.8 mm, as calculated above.

Therefore, provide 8 mm ϕ 2-legged stirrups at 275 mm c/c.

Bend one bar of 20 mm ϕ at a distance of between d (= 450 mm) to 2d (= 900 mm) from face of support.

EXERCISE

1. How shear reinforcement improves the strength of beam ?

2. What is nominal shear and maximum shear stress ?

3. Why minimum shear reinforcement is provided in beam ?

4. What is critical section for shear ?

5. What is shear reinforcement ?

6. What are the different forms of shear reinforcement?

7. Briefly explain the functions of shear reinforcement.

8. What is diagonal tension?

9. A R.C. beam 230 mm wide and 450 mm effective depth is reinforced with 3-16 mmϕ of grade Fe-415 on the tension side. If the beam is subjected to factored shear of 85 kN. Design the shear reinforcement for only 2 bars on tension side are available and 3rd bar is curtailed. Assume concrete of grade M-20.

10. A simply supported beam of width 300 mm and effective depth as 600 mm, if it is subjected to safe shear force of 95 kN and provide tension steel as 1.5 %, design shear reinforcement.

11. Design the design reinforcement for a cantilever beam carrying an udl of 26 kN/m without dead load having a span of 1.2 m and 230 mm × 650 mm in size. Tensile reinforcement 4-22 mm ϕ and compressive reinforcement as 2-12 mm ϕ. Use M-20 and Fe-415.

12. A R.C. beam of 200 mm × 550 mm effective depth carries a UDL of 20 kN/m (including self weight) over a simply supported span of 6 m. If 1% ten site steel is provided determine the nominal shear stress and shear strength of concrete.

13. A P.C. slab 120 mm thick, simply supported over a span of 5 m, carries UDL of 10 kN/m, (inclusive of self weight). The slab is reinforced with 8 mm bars at 110 mm etc. concrete used is M 20 and steel Fe. 500. Find design shear stress and nominal shear stress. Take load factor of 1.5 and minimum cover of 20 mm.

◈ ◈ ◈

CHAPTER - 10

LIMIT STATE OF COLLAPSE (BOND)

10.1 INTRODUCTION

- One of the most important assumptions in the behaviour of R.C. structures is that there is proper bond between concrete and reinforcing bars. When steel bars are embedded in concrete, the concrete (after setting) adheres to the surface of the bars and thus resist any force that tends to push or pull this rod. Thus, the term 'Bond' describes the means by which the relative movement between concrete and steel is prevented and the intensity of adhesive force is called bond stress.

- Bond stress is defined as longitudinal shear stress acting on the surface between steel and concrete. Bond between steel and concrete is due to combined effect of adhesive resistance, frictional resistance and mechanical resistance (for deformed bars). The adhesive resistance is provided by 'chemical gum' produced by concrete during setting. The bond due to friction is provided by gripping of bars due to shrinkage. The friction gives considerable bond resistance.

- With increasing force in bar, the adhesion is loose first than the friction between concrete and steel. The mechanical resistance is provided by deformed bars only (not by the plain bars). The deformed bars have lugs or corrugations and give higher bond resistance by providing an interlock between steel and concrete. In deformed bar, adhesion and friction become minor elements, and the bond strength is primarily dependent on bearing of concrete against the lugs or corrugations.

10.2 TYPES OF BOND

The bond stress in reinforced concrete members arises due to two distinct situations.

1. The change in the bar force along its length due to variation in bending moment in this length. This type of bond force is called **Flexural Bond Stress**.

2. From the anchorage of bar in case of tension or compression. This type of bond force is known as **Anchorage Bond Stress**.

10.2.1 Flexural Bond Stress

- As discussed above, the flexural bond arises along the length of the bar when change in bar force occurs due to variation in bending moment.

- Consider the segment between two sections ab and cd, spaced dx apart of a R.C. beam, as shown in Fig. 10.1.

- Let M be the bending moment at section ab and M + dM be the bending moment at section cd. Let T and T + dT be the tensile forces developed in steel reinforcement at ab and cd respectively.

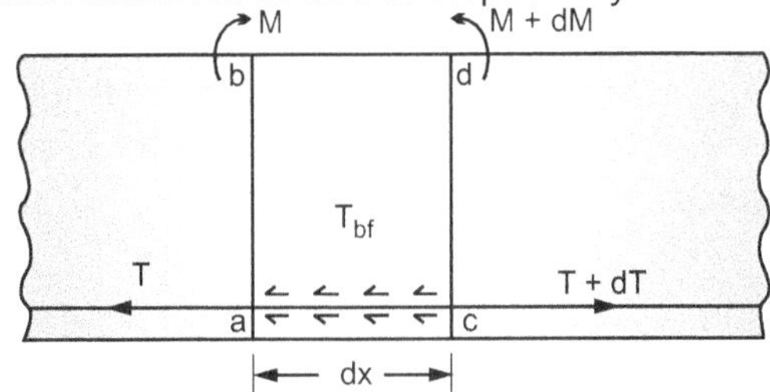

Fig. 10.1

$$\text{Now,} \qquad M = T \cdot jd$$
$$\text{and} \quad (M + dM) = (T + dT) \cdot jd$$
$$\text{Hence,} \qquad dM = dT \cdot jd$$
$$\text{or,} \qquad dT = \frac{dM}{jd} = \text{change in bar force.}$$

If τ_{bf} is the bond stress acting along the surface of bar, then for equilibrium,

Bond force acting along periphery of bar = Change in bar force dT.

$$\text{or} \quad \tau_{bf} [dx \textstyle\sum O] = dT$$

where, $\sum O$ = sum of perimeters of all steel bars resisting tension.

$$\text{or,} \quad \tau_{bf} (dx \textstyle\sum O) = \frac{dM}{jd}$$

$$\text{or,} \qquad \tau_{bf} = \frac{dM}{dx} \cdot \frac{1}{jd \sum O}$$

$$\text{But} \qquad \frac{dM}{dx} = \text{Shear force} = V$$

$$\text{Hence,} \qquad \boxed{\tau_{bf} = \frac{V}{jd \sum O}} \qquad \ldots (10.1)$$

- The stress τ_{bf} at a particular section is called local bond stress. Thus, the flexural bond stress is directly proportional to the shear force and inversely proportional to the sum of perimeters of the bars at the section.

10.2.2 Anchorage Bond Stress and Development Length

- Anchorage bond stress arises when a bar is carrying certain force. In anchorage bond, it is necessary to transfer this force in the bar to the surrounding concrete over a certain length. The length of bar 'L_d' is required to transfer the force in the bar.

- Fig. 10.2 shows a steel bar embedded in concrete and subjected to a tensile force T. Due to this force, there will be a tendency of the bar to slip out and this tendency is resisted by the bond stress developed over the perimeter of the bar, along its length of embedment. This required length 'L_d' is called **Anchorage Length** in case of axial tension (or compression) and **Development Length** in case of flexural tension.

- The development length is an embedded length of the bar required to develop the design strength of reinforcement at the critical section.

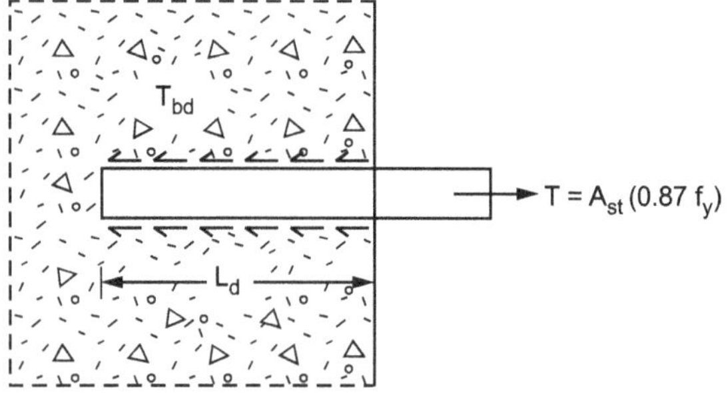

Fig. 10.2

If ϕ is the nominal diameter of a bar, then

Tension, $T = (0.87\,f_y)\,A_{st}$

$$T = (0.87\,f_y)\left(\frac{\pi}{4}\,\phi^2\right)$$

- This force must be transferred from steel to concrete through bond acting over the perimeter of the bar along its length of embedment L_d.

If τ_{bd} is the average bond stress, then

$$\text{Force} = \tau_{bd} \cdot (\pi\phi) \cdot L_d$$

For equilibrium,

$$0.87\,f_y\left(\frac{\pi}{4}\,\phi^2\right) = \tau_{bd} \cdot (\pi\phi)\,L_d$$

$$\therefore \quad \boxed{L_d = \frac{0.87\,f_y}{4\,\tau_{bd}}\,\phi} \quad\quad \dots (10.2)$$

- The value of design bond stress for plain bars in tension prescribed by IS code are reproduced in Table 10.1.

Table 10.1 : Permissible Bond Stress in Tension

Grade of Concrete	M 20	M 25	M 30	M 35	M 40 and above
Design Bond Stress τ_{bd}, (N/mm^2)	1.2	1.4	1.5	1.7	1.9

Notes

- For deformed bars, these values shall be increased by 60 percent.
- For bars in compression, the above values may be increased by 25 percent.

10.3 DEVELOPMENT LENGTH FOR FLEXURAL BOND

- From equation (10.1), it is clear that flexural bond stress is maximum at the section where shear force is large or where the sum of the perimeters of bars is minimum. Therefore, the check for flexural bond becomes necessary at the section where shear force is maximum and where bending moment is zero.

- The locations of such sections are
  (a) Simple supports.
  (b) Points of contraflexure.

- The code IS : 456 (clause 26.2.3.3) stipulates that at simple supports and at point of inflexion, positive moment tension reinforcement shall be limited to a diameter such that

$$L_d \ \le \ \frac{M_1}{V} + L_O$$

where, L_d = development length

M_1 = moment of resistance of the section assuming all reinforcement at the section to be stressed 0.87 f_y

V = shear force at the section due to design loads

L_O = sum of anchorage beyond the centre of the support and the equivalent anchorage value of any hook or mechanical anchorage at simple support, and at a point of inflexion, L_O is limited to the effective depth of the members or 12 ϕ, whichever is greater; and

ϕ = diameter of bar.

Notes

- At simple support, for the computation of L_O, the support width should be known. Fig. 10.3 shows a beam with end support, in which x' is the side cover and x_O is the distance of the beginning of the hook from the centre line of the support.

Now, L_O = sum of the anchorage beyond the centre of the support and the equivalent anchorage value

- The dark portion shows the hook which has an anchorage value of 16 ϕ. The distance of the beginning of the hook from its apex of the semi-circle can be taken to be equal to 3ϕ.

Let, l_S be the length of support.

$\therefore \qquad L_O = x_O + 16\ \phi$

where, $\quad X_O = \dfrac{l_S}{2} - x' - 3\phi$

$\therefore \qquad L_O = \left(\dfrac{l_S}{2} - x' - 3\phi\right) + 16\phi$

$$\therefore \qquad \boxed{L_O = \dfrac{l_S}{2} - x' + 13\phi}$$

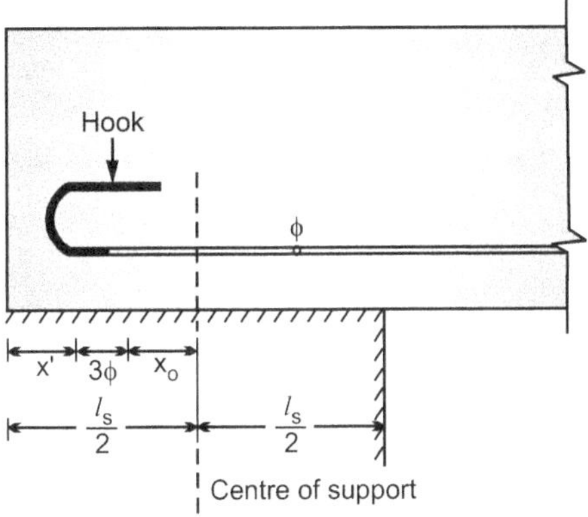

Fig. 10.3

If no hook is provided, as in case of deformed bars,

$$L_O = \dfrac{l_S}{2} - x'$$

- The code further recommends that the value of $\dfrac{M_1}{V}$ may be increased by 30 per cent when the ends of the reinforcement are confined by a compressive reaction. This condition of 'confinement' of reinforcing bars may not be available at all the types of simple supports. In simple supports of beams resting on walls or columns, the reaction induces compressive stress due to which bar gets confined, as shown in Fig. 10.4.

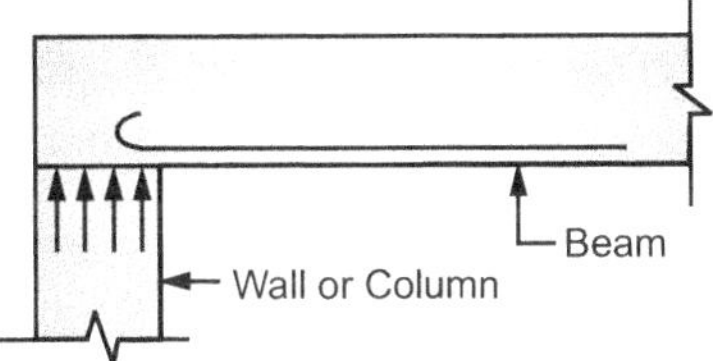

Thus, at simple supports, where the compressive reaction confines the ends of reinforcing bars, we have

$$L_d \leq 1.3\ \dfrac{M_1}{V} + L_O \qquad \qquad \dots (10.3)$$

Following Fig. 10.5 shows the code requirements for curtailment of tension reinforcements.

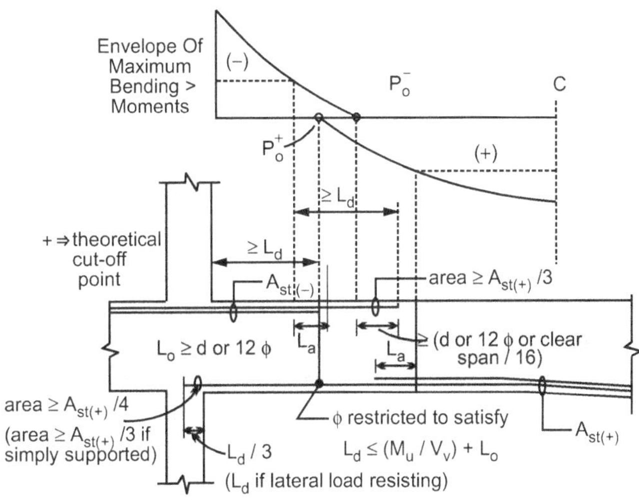

Fig. 10.5 : Code requirements for curtailment of tension reinforcement

10.4 BENDS AND HOOKS

- When the required straight development length (or anchorage length) cannot be provided due to limited space at the end of beam, in that case, standard hooks or bends are provided to find the full development length. The common types of anchorage provided are U-type hook and L-type bend. The bends and hooks shall conform to IS : 2502.

- The anchorage value of bend shall be taken as 4 times the diameter of the bar for each 45° bend subject to a maximum of 16 times the diameter of bars. The anchorage value of a standard U-type hook shall be equal to 16 times the diameter of the bar. Thus, the anchorage value of the standard hook and standard bend shall be taken equal to 16 ϕ and 8 ϕ respectively. Following Fig. 10.6 shows the dimensions of the hook and bend.

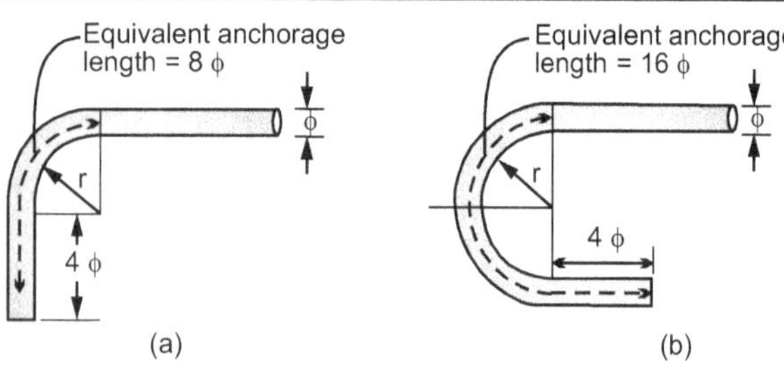

(a) Standard 90° bend   (b) Standard U-type hook

Fig. 10.6 : Anchorage lengths of standard bends and hooks

- The minimum (internal) turning radius 'r' specified for a hook is 2 φ for plain mild steel and 4 φ for cold worked deformed bars.

1. **Anchoring Bars in Tension (Clause 26.2.2.1a)**

   Deformed bars may be used without end anchorages, provided development length required is satisfied. Normally hooks should be provided for plain bars in tension.

2. **Anchoring Bars in Compression (Clause 26.2.2.2)**

   The anchorage length of straight bar in compression shall be equal to the development lengths of bars in compression. The projected length of hooks, bends and straight lengths beyond bends, if provided for a bar in compression, shall only be considered for development length.

3. **Anchoring Shear Reinforcement (Clause 26.2.2.4)**

(a) Inclined Bars : The development length shall be as for bars in tension, this length shall be measured as under :

   (i) In tension zone, from the end of the sloping or inclined portion of the bar, and

   (ii) In compression zone, from the mid depth of the beam.

(b) Stirrups : Not withstanding any of the provisions of this standard, in case of secondary reinforcement, such as stirrups and transverse ties, complete development lengths and anchorage shall be deemed to have been provided when the bar is bent through an angle of atleast 90° round a bar of atleast its own diameter and is continued beyond the end of the curve for a length of atleast eight diameters or when the bar is bent through an angle of 135° and is continued beyond the end of the curve for a length of atleast six bar diameters or when the bar is bent through an angle of 180° and is continued beyond the end of the curve for a length of atleast four bar diameters. (Shown in Fig. 10.7).

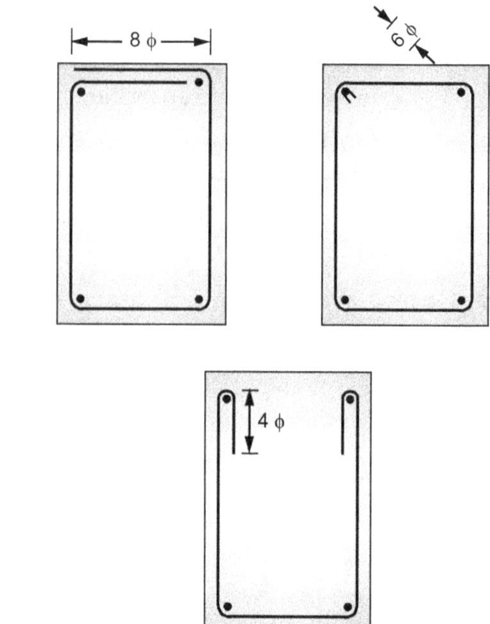

Fig. 10.7 : Anchoring vertical stirrups

10.5 REINFORCEMENT SPLICING

- The reinforcing bars are normally manufactured in specified lengths. If in any R.C. structure, the length of the bar required is more than the available length, splicing of bars is done. In such a case, some minimum lap length is required so that the force can be transferred from one bar to the other bar.

- The requirements of reinforcement splices are set out in clause 26.2.5 of IS : 456. Where splices are provided in the reinforcing bars, they shall as far as possible be, away from the sections of maximum stress and be staggered.

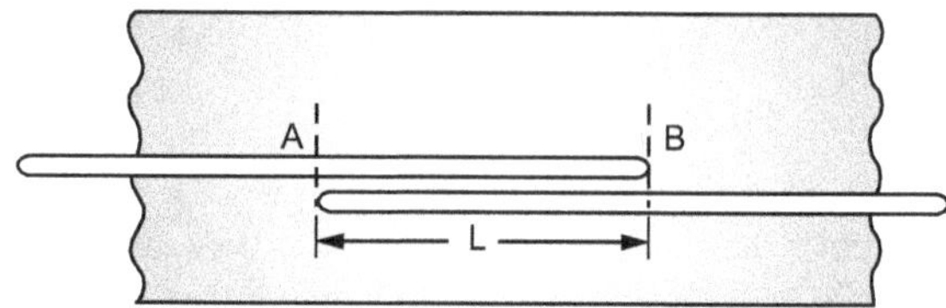

Fig. 10.8

- It is recommended that splices in flexural members should not be at sections where the bending moment is more than 50% of the moment of resistance, and not more than half the bars shall be spliced at a section.

- Where more than one-half of the bars are spliced at a section or where splices are made at points of maximum stress, special precautions shall be taken such as increasing the length of lap and/or using spirals or closely spaced stirrups around the length of the splice.

- The splices may be lap splice, welded splice or end bearing splice.

(a) Lap Splices : These are covered in clause 26.2.5.1 of IS : 456 and are explained below and is shown in Fig. 10.9.

- Lap splices shall not be used for bars larger than 36 mm. For larger diameters, bars may be welded. In case where welding is not practicable, lapping of bars larger than 36 mm may be permitted, in which case additional spirals should be provided around the lapped bars.

- Lap splices shall be considered staggered, if the centre to centre distance of the splices is not less than 1.3 times the lap length as described in para (3) below.

- Lap length including the anchorage value of hooks in flexural tension shall be L_d, the development length, or 30 ϕ whichever is greater and for direct tension $2L_d$ or 30 ϕ whichever is greater. The straight length of the lap shall not be less than 15 ϕ or 200 mm.

The above requirements are illustrated in Fig. 10.9.

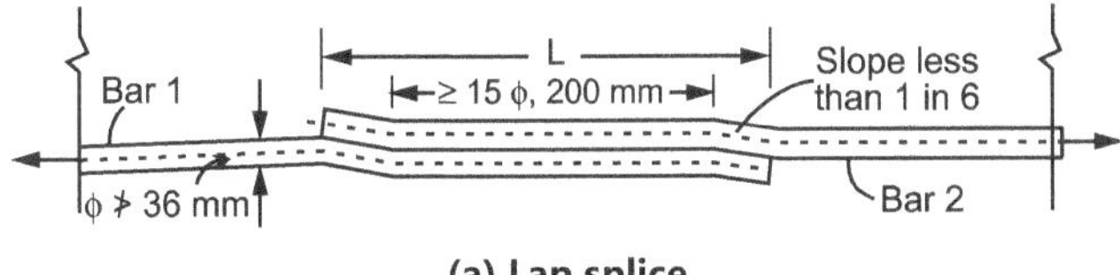

(a) Lap splice

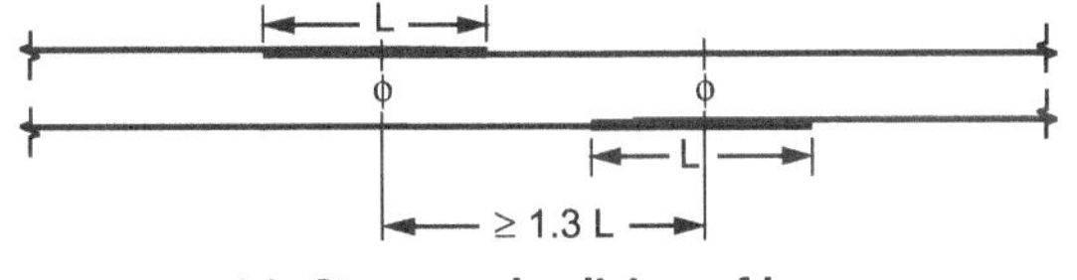

(b) Use of spirals in lap splices for large diameter bars

(c) Staggered splicing of bars

Fig. 10.9

- The lap length in compression shall be equal to the development length in compression but not less than 24 ϕ.

- When bars of two different diameters are to be spliced, the lap length shall be calculated on the basis of diameter of the smaller bar.

- In case of bundled bars, lapped splices of bounded bars shall be made by splicing one bar at a time. Such individual splices within a bundle shall be staggered.

- **Welded Splices :** The following values may be used where the strength of the weld has been proved by tests to be atleast as great as that of the parent bar.

- **Splices in Compression :** For welded splices and mechanical connection, 100% of the designed strength of joined bars.

- **Splices in Tension :**
  - ➢ 80% of the design strength of welded bars (100% if welding is strictly supervised and if at any cross-section of the member not more than 20% of the tensile reinforcement in welded).
  - ➢ 100% of design strength of mechanical connection.

- **End-Bearing Splices :** End-bearing splices shall be used only for bars in compression. The ends of the bars shall be square cut and concentric bearing ensured by suitable devices.

10.6 CURTAILMENT OF REINFORCEMENT

- When the tension bars are not required to resist the moment, they can be curtailed. The distance 'x' measured from the centre of span, at which N_x, number of bars (out of that N_c bars) can be curtailed, is given by

$$x = \frac{L}{2} \sqrt{\frac{N_x}{N_c}}$$

- This is the theoretical cut off point (TPC). But for curtailment, reinforcement shall extend beyond the point at which it is no longer required to resist the flexure for a distance equal to the effective depth of the member or 12 times the diameter, whichever is greater, except at simple support or end of cantilever. These requirements are explained in Fig. 10.10.

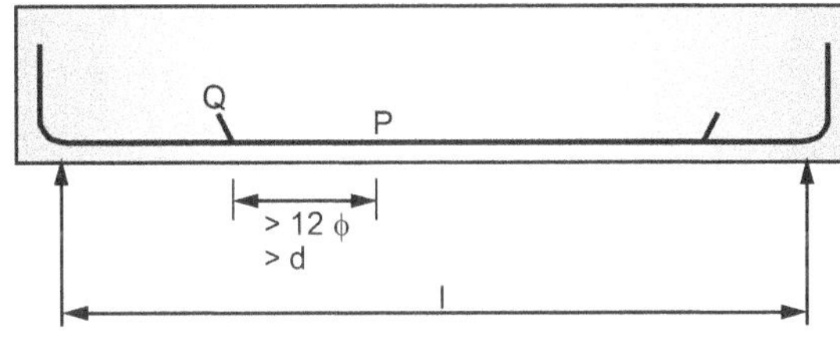

P - Theoretical cut - off point
Q - Actual cut - off point

Fig. 10.10 : Requirements for bar curtailment

10.7 SIMPLIFIED RULES FOR CURTAILMENT OF STEEL (SP : 34-1987)

- The simplified rules for curtailment of steel given below are applicable to continuous beams and slabs with approximately equal span (not differing more than 15% of the longest) and subjected to predominantly u.d.*l.*

1. **Simply Supported Beam :** At least 50% bars shall extend into the support providing bond length of $\frac{L_d}{3}$ from the face of support and remaining 50% can be curtailed at a distance of $\frac{L}{12}$ from the face of supports.

2. **Continuous Beam :** The typical details of continuous beam are shown in Fig. 10.11.

- All tension bar at top of support should extend a distance of 0.15 L or L_d, whichever is greater, from the face of support. At least 60% of these bars should extend 0.25 L and minimum 20% should continue through the span.
- At intermediate support, minimum 30% bar should continue throughout the span.
- Reinforcement has been shown in Fig. 10.11, may be in same layer and applicable for beams of approximately equal span and subjected to mainly u.d.*l.* for other cases, the actual analysis shall be made.

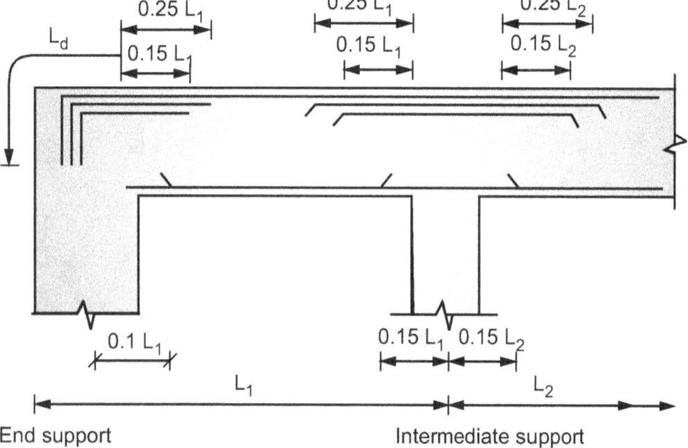

End support　　　　　　　Intermediate support

Fig. 10.11 : Curtailment details for continuous beam

(c) Cantilever Beam :

- In the case of cantilever beam, all top tension bar should extend a distance of $\dfrac{L}{2}$ or L_d, whichever is greater from the face of column. Atleast 50% bar should extend to the end of cantilever. Minimum 2 bars or 25% of tension steel should be provided on compression side.

Note : The curtailment of bar effects loss of shear strength of member at the point of cut-off. To make up the shear strength, code gives the following direction (clause 26.2.3.2). It states "flexural reinforcement shall not be terminated in a tension zone, unless any one of the following condition is satisfied".

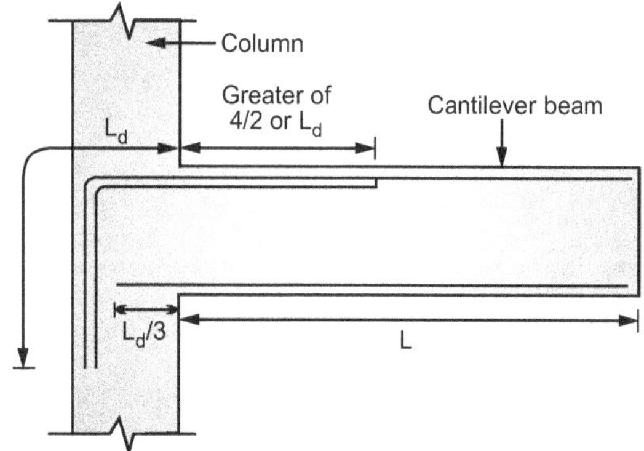

Fig. 10.12 : Curtailment details for cantilever beam

(i) The shear at the cut-off point does not exceed two-thirds that permitted, including the shear strength of web reinforcement provided.

Shear at cut-off,

$$V_u \leq \frac{2}{3} \text{ (shear strength of web + shear strength of concrete).}$$

(ii) Stirrup area in excess of that required for shear and torsion is provided along each terminated bar over a distance from the cut-off point equal to three-fourth the effective depth of the member. The excess stirrup area shall not be less than $\dfrac{0.4\ bS}{0.87\ f_y}$.

i.e. Excess stirrups area,

$$A_{sv} \geq \frac{0.4\ bS}{0.87\ f_y}$$

i.e.
$$S = \frac{0.87\ f_y \cdot A_{sv}}{0.4\ b}$$

but the resultant spacing,

$$S \leq \frac{d}{8\ \beta_b}$$

where, β_b is the ratio of $\dfrac{\text{area of cut-off bar}}{\text{total area of bar}}$.

(iii) For 36 mm and smaller bars, the continuing bars provide double the area required for flexure at the cut-off point and the shear does not exceed three-fourth that permitted.

Important Points :

1. Development length $(L_d) = \dfrac{0.87\ f_y\ \phi}{4 \times \tau_{bd}}$

2. Ancharge value of bend shall be taken as 4 times diameter of bar subjected to max. of 16 times diameter of bar.

3. Lap splice / Lap length / Over lap :

(a) Far flexural tension = L_d or 30 ϕ　　…whichever is more

(b) For direct tension = 2Ld or 30ϕ　　…whichever is more

(c) For compression = Ld or 24 ϕ　　…whichever is more

SOLVED EXAMPLES

Example 10.1 : *Determine the anchorage value 45° bend and 90° bend for 20 mm diameter of bar.*

Solution : Given :

$$\phi = 20 \text{ mm},$$

(i) Anchorage value for 45° bend

$$= 4\phi = 4 \times 20$$

$$= \textbf{80 mm}$$

(ii) Ancharage value for 90° bend

$$= 8\phi = 8 \times 20$$

$$= \textbf{160 mm}$$

Example 10.2 : *Find the development length for 12 mm diameter bar in tension. Take M.S. reinforcement ; M-20 concrete.*

Solution : Given, $\phi = 12$ mm, $f_y = 250$ N/mm^2, $f_{ck} = 20$ N/mm^2 bars in tension.

$\therefore$ Development length $= L_d = \dfrac{0.87\, f_y \cdot \phi}{4 \times \tau_{bd}}$

$\therefore \qquad L_d = \dfrac{0.87 \times 250 \times 12}{4 \times 1.2}$

$\qquad\qquad L_d = 543.75$ mm

$\therefore$ Provide $\mathbf{L_d = 550}$ **mm**

Example 10.3 : *Calculate the development length for 16 mm diameter bar in compression. Take F_e- 415, M-25 grades material.*

Solution : Given, $\phi = 16$ mm; $f_y = 415$ N/mm^2, $f_{ck} = 25$ N/mm^2, Assume : bars in tension.

$\therefore$ Development length $(L_d) = \dfrac{0.87\, f_y \cdot \phi}{4 \times \tau_{bd}}$

$\qquad\qquad L_d = \dfrac{0.87 \times 415 \times 16}{4 \times 1.4 \times 1.6}$

$\therefore \qquad\qquad L_d = 644.73$ mm

$\therefore$ **Provide, $\mathbf{L_d \approx 650}$ mm**

Example 10.4 : *Find the lap length for bars of 18 mm diameter in tension as well as compression. Take M-20, Fe-500.*

Solution : Given : $\phi = 18$ mm, $f_{ck} = 20$ N/mm^2,

$\qquad f_y = 500$ N/mm^2

(I) Lap length for bars in tension :

(a) Far flexural tension :

$\qquad$ lap length $= L_d$ or 30ϕ ...whichever is more

(i) $\qquad L_d = \dfrac{0.87\, f_y \cdot \phi}{4 \times \tau_{bd}}$

$\qquad\qquad = \dfrac{0.87 \times 500 \times 18}{4 \times 1.2 \times 1.6}$

$\qquad\qquad = \textbf{1019.53 mm}$

(ii) $\qquad 30\phi = 30 \times 18$

$\qquad\qquad = \textbf{540 mm}$

$\therefore$ Provide, lap length $= 1019.53$

$$\approx \textbf{1020 mm}$$

(b) For direct tension

(i) $\qquad 2L_d = \dfrac{2 \times 0.87 \times 500 \times 18}{4 \times 1.2 \times 1.6}$

$\qquad\qquad = \textbf{2039.06 mm}$

(ii) $\qquad 30\,\phi = 30 \times 18$

$\qquad\qquad = \textbf{540 mm}$

$\therefore$ Provide, lap length $= 2039.06$

$$\approx \textbf{2040.00 mm}$$

(II) Lap length for bars in compression

(i) $\quad L_d = \dfrac{0.87 \times 500 \times 18}{4 \times (1.2 \times 1,.6 \times 1.25)} = 815.625$ mm $\Bigg\}$ more

(ii) $\qquad 24\,\phi = 24 \times 18 = 432$ mm

$\therefore$ **Provide lap length $= 815.625 \approx 820$ mm**

Example 10.5 : *Find the lap splice of 16 mm diameter bar in compression. Take Fe-250 and M-15 concrete.*

Solution : Given $\phi = 16$ mm, $f_y = 250$ N/mm^2, $f_{ck} = 15$ N/mm^2, bars in compression

$\qquad$ Lap splice $= l_d$ or $24\,\phi$ ...whichever is more

(i) $\qquad L_d = \dfrac{0.87\, f_y \cdot \phi}{4 \times \tau_{bd}}$

$\qquad\qquad = \dfrac{0.87 \times 250 \times 16}{4 \times (1.00) \times 1.25}$

$\qquad\qquad (\because$ for M-15, Take $\tau_{bd} = 1.00$ N/mm^2)

$\therefore \qquad\qquad L_d = 696.00$ mm

$\therefore$ **Provide, $\mathbf{L_d \approx 700.00}$ mm**

EXERCISE

1. Define bond, state and explain types of bond.

2. State the factors affecting bond strength.

3. Define curtailment of bars.

4. What do you understand by development length of bar ?

5. Write about local bond and anchorge length.

6. Distinguish between flexural band and development bond.

7. Why bond stress is more in compression bars than in tension bars ?

8. What is lap splice ? Why splicing is done ? What are IS code provisions for providing lap splice ?

9. How curtailment of bars are done in case of simply supported and continuous beam.

10. Calculate the anchorage value of 18 mm diameter bar, for 90° and 180° bend.

11. Calculate the development length for 12 mm diameter bar in tension. Take Fe-415.

12. Calculate the development length for 16 mm diameter in tension and compression. Take Fe-250.

13. Find the lap length of bars of 12 mm diameter in flexural tension. Take M.S. reinforcement.

14. Determine the lap splice for 20 mm diameter bar of Fe-500 in tension and compression.

15. Why bends or hook is provided to the reinforcement.

MODEL QUESTION PAPERS

Paper - I

Time : 3 Hours   **Max. Marks : 60**

Instructions to the candidates :

   (1) Each question carries 12 Marks.

   (2) Attempt any five questions from the following.

   (3) Illustrate your answers with neat sketches, diagram etc. wherever necessary.

   (4) If some part or parameter is noticed to be missing, you may appropriately assume it and should mention it clearly.

   (5) Use of non-programmable calculator is allowed.

   (6) Use of IS456 : 2000 is allowed.

1. **(a)** What is reinforced concrete? Why concrete is to be reinforced? **[4]**

    (b) State the assumptions in WSM. **[4]**

    (c) State and explain permissible stresses in steel and concrete. **[4]**

2. **(a)** Define SRRS and DRRS, why doubly reinforced section is designed? **[4]**

    (b) Draw the neat sketch of strain diagram and stress block diagram for singly reinforced rectangular section. **[2]**

    (c) Design the slab for a room of size 3m × 4m carries a superimposed load of 4.00 kN/m^2. The corners of slab are held down over a support 200 mm. Take M-20 and Fe-415. **[6]**

3. Design a dog legged staircase for a building in which the vertical distance between floors in 2.9m. The space available for staircase is 2.5 × 2m. Take L.L. as 3.2 kN/m^2. Use M-20 and Fe-415. **[12]**

4. **(a)** State Assumptions in limit state method. **[4]**

    (b) Define : 1. Characteristics strength **[4]**

                 2. Characteristics load

                 3. Design strength

                 4. Design load

    (c) What is the purpose of providing nominal cover to the reinforcement. **[4]**

5. **(a)** Determine the values of design parameters kumax, Rumax and Ptmax for balanced section of grades of material M-15 and Fe-250. **[6]**

    (b) A simply supported beam of size 230 mm × 450 mm is reinforced with 5-16 mm φ on tension side and 3-12 mm φ on compression side. What is the maximum amount of udl the beam can carry. The effective span of beam is 5m. **[6]**

6. **(a)** An R.C. beam 350 mm wide and 500 mm effective depth is reinforced with 4-25 mm φ. It is having a udl of 75 kN/m including self wt. over a simply supported span of 7.00m. Design the shear reinforcement using vertical stirrups and bent-up bars. **[10]**

    (b) Define, curtailment of bars. **[2]**

Paper - II

Time : 3 Hours   **Max. Marks : 60**

Instructions to the candidates :

   (1) Each question carries 12 Marks.

   (2) Attempt any five questions from the following.

   (3) Illustrate your answers with neat sketches, diagram etc. wherever necessary.

   (4) If some part or parameter is noticed to be missing, you may appropriately assume it and should mention it clearly.

   (5) Use of non-programmable calculator is allowed.

   (6) Use of IS456 : 2000 is allowed.

---

1. **(a)** State the advantages and disadvantages of R.C.C. **[6]**

   (b) Differentiate between different design philosophies **[6]**

2. **(a)** Why over reinforced section not permitted in R.C.C. design. **[2]**

   (b) Explain the need of torsion reinforcement is slab. **[2]**

   (c) Using stress strain diagram for a balanced singly reinforced section as per WSM, derive the design constants (k = Neutral axis constant , j = lever arm factor, R = moment resisting factor) Use M-20 maximum and Fe-415 grade steel. **[8]**

3. **(a)** Design a column is to be designed to carry an axial load of 750 kN. Permissible stresses in concrete and steel are 5 MPa and 130 MPa **[8]**

   (b) Explain types of staircase. **[4]**

4. **(a)** State how LSM is advantageous than WSM and ULM. **[4]**

   (b) Define partial safety factor, also state the values of partial safety factor for concrete and steel. **[4]**

   (c) While designing a structure which section you will prefer and why? **[4]**

5. **(a)** Determine the ultimate moment of resistance of beam as well as corresponding stresses and strain in steel and concrete as shown in fig. Take M-25 maximum and Fe-415 steel. **[8]**

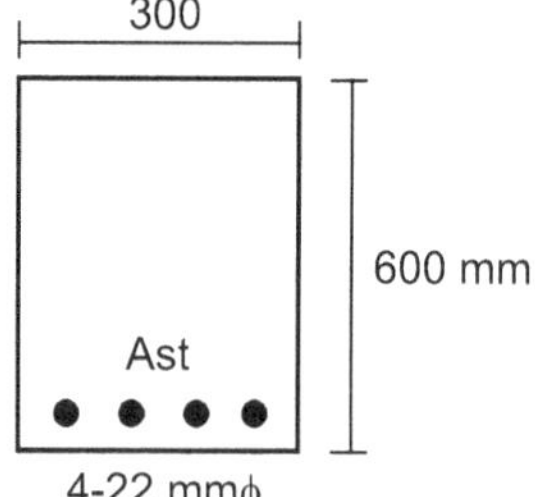

Fig.

   (b) A T-beam of flange width 1250 mm. flange thickness 100 mm, rib thickness 250 mm effective depth 500 mm and reinforcement 4-20 mm φ and 2-12 mm φ on tension side. Calculate the ultimate moment of resistance. Use M-25, Fe-415 grades material. **[4]**

6. **(a)** Briefly explain the functions of shear reinforcement. **[3]**

   (b) Calculate the development length of 16 mm diameter in compression. Take Fe-415 steel. **[3]**

   (c) A beam 230 mm × 450 mm effective subjected to a factored S.F. of 50 kN. Design the shear reinforcement without bent up bears. The beam is reinforced with 3-16 mmφ. Take M-15 and Fe-415. **[6]**

Paper - III

Time : 3 Hours   **Max. Marks : 60**

Instructions to the candidates :

(1) Each question carries 12 Marks.

(2) Attempt any five questions from the following.

(3) Illustrate your answers with neat sketches, diagram etc. wherever necessary.

(4) If some part or parameter is noticed to be missing, you may appropriately assume it and should mention it clearly.

(5) Use of non-programmable calculator is allowed.

(6) Use of IS456 : 2000 is allowed.

1. **(a)** Explain different design philosophies.   **[6]**

    (b) Define modular ratio with reference to WSM.   **[3]**

    (c) Enlist types of loads with there is code provision and different load combination.   **[3]**

2. **(a)** Differentiate between one way slab and two way slab.   **[3]**

    (b) Explain the concept of under reinforced, balanced and over reinforced section.   **[3]**

    (c) A rectangular reinforced beam is 350 mm × 700 mm effective. The beam has to resist a bending moment of 300 kN.m. find the area of reinforcement Take M-20, σst = 190 N/mm^2 , m = 13.33, effective cover = 50 mm **[6]**

3. Design an isolated footing of uniform depth of a R.C. column carrying an axial load of 675 KN and having size 500 × 550 mm. The SBC of soil is 125 kN/m^2. Take M-20 and Fe-500 grades material.   **[12]**

4. **(a)** Differentiate between under-reinforced, balanced and over reinforced section.   **[6]**

    (b) Define LSM. explain types of limit state method.   **[6]**

5. A hall of building has an internal dimension 4m × 16 m a floor slab of 120 mm thick is cast monolithically with beam, the slab is subjected to L.L. of 4 kN/m^2. The beams one spaced at 4m c/c. The beam of 250 mm thick. Design the intermediate beam and also design the beam for shear reinforcement.   **[12]**

6. **(a)** Explain the types of bond.   **[4]**

    (b) A R.C. beam 230 mm wide and 400 mm effective depth is reinforced with 4-12 mm ϕ of grade Fe-415 on tension side. If the beam is subjected to safe shear force of 60 kN. Design the shear reinforcement. Assume M-20 and Fe-415 steel.   **[8]**

Notes